CIVIL ENGINEERING MATERIALS

Other Engineering titles from Macmillan

A Guide to Soil Mechanics
Malcolm Bolton

An Introduction to Engineering Fluid Mechanics, Second Edition
J. A. Fox

Cost and Financial Control for Construction Firms
B. Cooke and W. B. Jepson

Highway Design and Construction
R. J. Salter

Highway Traffic Analysis and Design, Revised Edition
R. J. Salter

Reinforced Concrete Design
W. H. Mosley and J. H. Bungey

Strength of Materials, Third Edition
G. H. Ryder

Structural Theory and Analysis
J. D. Todd

Surveying for Engineers
J. Uren and W. F. Price

Civil Engineering Materials

Edited by
N. JACKSON

Second Edition

First edition 1976
Reprinted 1977, 1978
Second edition 1980

Published by
THE MACMILLAN PRESS LTD
London and Basingstoke
Associated companies in Delhi Dublin Hong Kong Johannesburg Lagos Melbourne New York Singapore and Tokyo

Text set in Press Roman by Reproduction Drawings Ltd.

Printed in Hong Kong

British Library Cataloguing in Publication Data

Civil engineering materials. – 2nd ed.
1. Building materials
I. Jackson, Neil
624′.18 TA403

ISBN 0–333–28959–5
ISBN 0–333–28960–9 Pbk

Contents

Preface

The importance of an understanding of the materials used in civil engineering is widely recognised, but only in comparatively recent years has much emphasis been placed on the teaching of material properties at undergraduate level. This introductory textbook on materials satisfies a need for a single book covering the principal materials used in civil engineering.

The aim has been to provide the student with an authoritative text on the more traditional materials, which will also serve as a valuable source of reference in his subsequent career. Soils are included along with metals, timber, concrete and bituminous materials in recognition of their importance as constructional materials, the fundamental properties of soils being covered in greater depth than usual. Extensive references to all relevant British Standards are made throughout the book.

The treatment of material properties here is suitable for students studying for a degree or equivalent qualification in civil engineering, building science, architecture and other related disciplines. The particular point in a course at which the study of civil engineering materials is introduced depends on the course structure of the individual educational institution but would generally be during the first two years of a three- or four-year course. Similarly the extent of further formal study of civil engineering materials depends on the emphasis and structure of the course within a particular educational institution. However, it is not envisaged that further study of the basic material properties of metals, timber and concrete will be required, although the application of these materials within the general context of analysis and design might well continue throughout the remainder of a course. Further study of soils might normally be expected to continue, for civil engineering students, within the context of soil mechanics, and further study of bituminous materials only where highway materials are studied in later years.

Throughout the book, the underlying theme is an emphasis on the factors affecting engineering decisions. It is hoped that this will promote an awareness in the reader of the importance of material behaviour in both design and construction if the final product, whether structural or purely decorative, is adequately to fulfil the purpose for which it is provided.

The courtesy extended by the undernamed, and others named in the text, in permitting the reproduction of the material indicated, is gratefully acknowledged. Material from the following publications is reproduced by permission of the British Standards Institution, 2 Park Street, London, W1A 2BS: CP 110 : 1972 The structural use of concrete; CP 2001 : 1957 Site investigations; CP 2004 : 1972 Foundations; BS 812 : 1975 Methods for sampling and testing of mineral aggregates, sands and fillers; BS 1377 : 1975 Methods of tests for soils for civil engineering purposes; BS 1470 : 1972 Wrought aluminium and aluminium alloys for general engineering purposes – plate, sheet and strip; BS 1490 : 1970 Aluminium and aluminium alloy ingots and castings; BS 1881 : 1970 Methods of testing concrete; and BS 4360 : 1979 Weldable steels. Table 22.3 is reproduced by permission of the Director, Transport and Road Research Laboratory and table 25.1 by permission of the American Society of Civil Engineers.

This second edition incorporates changes associated with revisions of many of the British Standards referred to in the first edition.

N. Jackson

List of Contributors

D. H. J. Armour, B.Sc.
Senior Lecturer, Department of Civil Engineering, Paisley College of Technology
R. K. Dhir, B.Sc., Ph.D., C.Eng., M.I.M.M., F.G.S.
Senior Lecturer, Department of Civil Engineering, University of Dundee
T. Gibson, M.Sc., F.R.I.B.A., F.R.I.C.S., F.R.I.A.S., F.F.B.
Senior Lecturer, Department of Land Economics, Paisley College of Technology
R. J. Hey, B.Sc., C. Eng., F.I.M.
Head of the Department of Metallurgy and Materials, Teesside Polytechnic
N. Jackson, B.Sc., Ph.D., C.Eng., M.I.C.E., M.I.Struct.E. (*editor*)
formerly Senior Lecturer, Department of Civil Engineering, University of Dundee
D. G. McKinlay, B.Sc., Ph.D., C.Eng., F.I.C.E., F.A.S.C.E., F.G.S.
Professor, Department of Civil Engineering, University of Strathclyde
A. V. Pagan, B.Sc., C. Eng., M.I.M.
Senior Lecturer, Department of Metallurgy and Materials, Teesside Polytechnic

I METALS

Introduction

The applications of metals in civil engineering are many and varied, ranging from their use as main structural materials to their use for fastenings and bearing materials. As main structural materials cast iron and wrought iron have been superseded by rolled-steel sections and limited use has been made of wrought aluminium alloys. Steel is also of major importance for its use in reinforced and prestressed concrete. On a smaller scale, metals are extensively used for fastenings, such as nails and screws, for bearing surfaces in the expansion joints of bridges and for decorative facings.

The properties of metals which make them unique among constructional materials are high tensile strength, the ability to be formed into plate, sections and wire, and the weldability or ease of welding of those metals commonly used for constructional purposes. Other properties typical of metals are electrical conductivity, high thermal conductivity and metallic lustre, which are of importance in some circumstances. Perhaps the greatest disadvantage of the common metals, and particularly steels, is the need to protect them from corrosion by moist conditions and the atmosphere, although weathering steels have been developed which require no protection from atmospheric corrosion.

When in service, metals frequently have to resist not only high tensile or compressive forces and corrosion, but also conditions of shock loading, low temperatures, constantly varying forces or a combination of several of these effects. Pure metals are relatively soft and weak, and do not meet these rigorous demands except for applications where the properties of high electrical conductivity or corrosion resistance are required. Normally one or more alloying elements are added to increase strength or to modify the properties in some other way. Metals and alloys are crystalline in character and in order to appreciate their properties and behaviour in service it is necessary to study their structure on an atomic scale, the types of crystal formed, the microstructure (the structure as observed under the microscope), the coarser macrostructure, and the ways in which these may be affected by heat treatment, stress, deformation and the environment.

1

Structure of Metals

1.1 Atomic Bonding

The atoms of the chemical elements consist of a central nucleus carrying a positive electrical charge, surrounded by a number of negatively charged electrons which tend to move in layers or shells. The first shell is complete with only two electrons but with more electrons present eight are required to completely fill an outer shell. When atoms combine to form molecules or crystals this may happen in a variety of ways but it is associated with the formation of stable units with complete outer shells of electrons. This may occur by atoms donating, gaining or sharing electrons according to the number present in that atom and in the atom with which it is combining. In general, atoms with fewer valency electrons, that is, electrons in the outer shell, donate them whereas atoms with greater numbers of valency electrons gain or share electrons, although the number of valency electrons that may be donated increases as the total number of electrons in the atom increases. Atoms which donate electrons are usually metallic in character whereas atoms which gain or share electrons are nonmetallic. In chemical compounds the bonds between atoms are ionic and covalent whereas in metals the metallic bond predominates.

In ionic bonding the valency electrons are transferred from the outer shell of one type of atom to the outer shell of the atom of the other element, thus leaving both atoms with complete outer shells. A typical example of this type of bond occurs in sodium chloride (common salt) and is shown diagrammatically in figure 1.1. The single valency electron of the sodium atom is donated to the

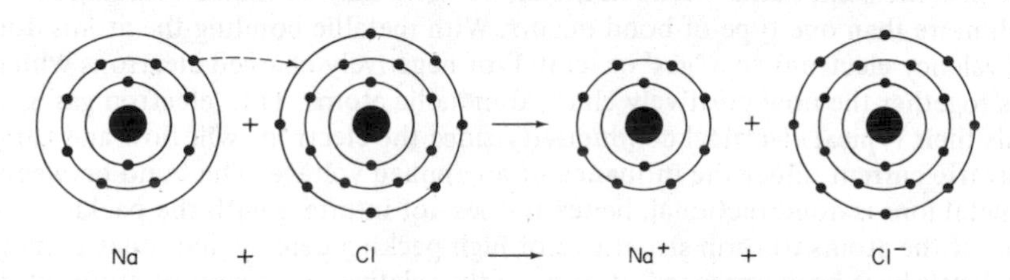

Figure 1.1 *Bonding of sodium and chlorine by transfer of an electron from the sodium to the chlorine atom to form a positively charged sodium ion and a negatively charged chlorine atom*

chlorine atom, bringing the number of atoms in the outer shell of the chlorine atom up to eight, which is a complete outer shell. The transfer of atoms in this way leaves the donating atom positively charged and the receiving atom negatively charged, and it is the electrostatic attraction between these charged atoms or ions that forms the bond. This type of bond permits close packing together of the atoms to form a large molecule or crystal in which bonding is continuous throughout the crystal. Ionic bonds are strong but any displacement of the charged atoms relative to each other would bring like-charged atoms together. The force necessary for this displacement normally equals the breaking strength of the bond between the atoms, hence fracture occurs without deformation and the ionically bonded compounds are hard and brittle.

Covalent bonding is dependent on the sharing of electrons between atoms in order to achieve stable units with a complete outer shell associated with each atom. This may be illustrated by considering the bonds between the carbon atoms in diamond. The carbon atom has four valency electrons and in diamond each of these is shared with a neighbouring atom while an electron from each of the neighbouring atoms is also shared. This is shown diagrammatically in figure 1.2.

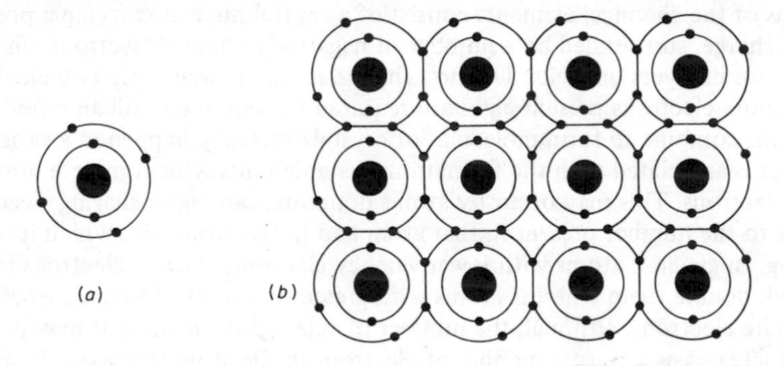

Figure 1.2 *(a) Structure of C atom and (b) bonding of carbon atoms by sharing of electrons*

The covalent bonds are of high strength but they limit the number of atoms which may closely approach one another and are directional in character. This prevents relative movement of atoms without fracture and makes covalently bonded compounds hard and brittle.

In metals the metallic bond predominates, although other types of bond may occur in some compounds between metals or there may be mixed bonding in which more than one type of bond occurs. With metallic bonding the atoms donate their valency electrons to a 'gas' or 'cloud' of negatively charged electrons which holds together the now positively charged metallic atoms. This 'electron gas' gives metals their typical electrical conductivity since the electrons will flow and carry an electric current under the influence of an applied voltage. The bond between the metal ions is nondirectional, hence it does not interfere with the packing together of the atoms to form structures of high packing density and, once certain stress levels have been exceeded, it permits the relative movement of atoms without rupture occurring. The strength, ductility and toughness of metals are due to the properties of the metallic bond and are discussed more fully in chapter 2.

1.2 Crystalline Structure of Metals

Solid metals consist of an aggregate of crystals or grains in which the atoms are arranged in a regular three-dimensional geometrical pattern called a space lattice. These are usually described by reference to a unit cell which, if repeated a large number of times in any direction, will generate the space lattice. Although the concept of the unit cell is necessary to describe the different types of space lattice it is often more convenient to consider the crystal as built up from successive planes of atoms. In metallic crystals a high density of packing occurs in the planes of atoms and in the stacking of these planes to form the crystal, although ideal close packing is not always achieved.

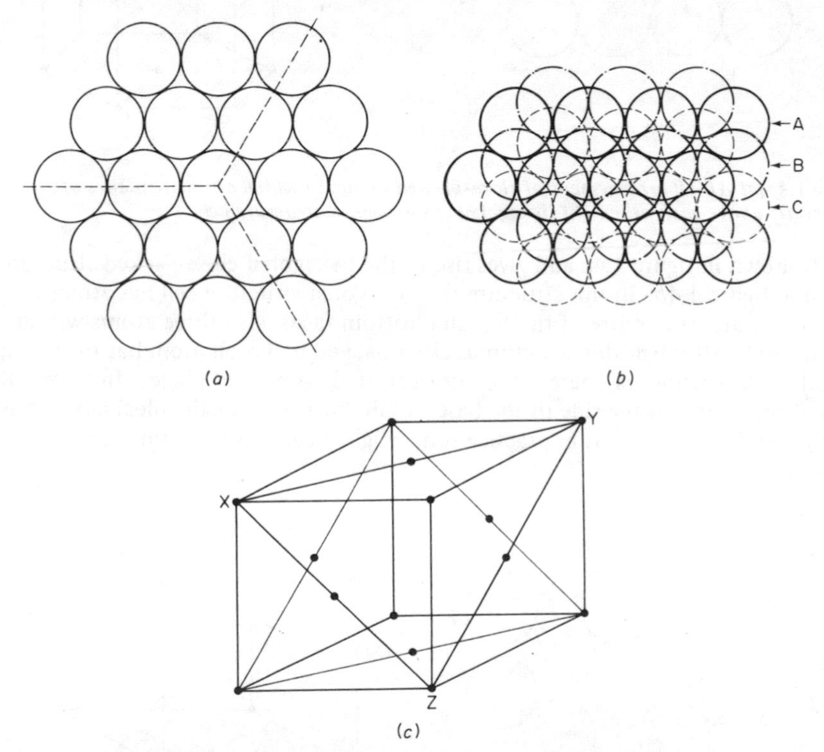

Figure 1.3 *(a) The close-packed arrangement of atoms in a plane. (b) The stacking of close-packed planes in an ABCABC arrangement. (c) The face-centred cubic unit cell in which the plane XYZ is close packed*

Figure 1.3*a* shows the ideal close-packed arrangement of spheres or atoms in a plane with the atoms lying on three sets of close-packed lines which are identical to each other. When planes of this type are stacked on top of each other in an ABCABCABC, etc., arrangement, as shown in figure 1.3*b*, a close-packed lattice results which may be described as a face-centred cubic lattice. In this lattice the unit cell shown in figure 1.3*c* has atoms at each corner of the cube and in the centre of each face. The close-packed planes from which it is built cut the diagonals of any three adjacent faces and this gives four sets of close-packed planes.

One plane of this type is indicated as the plane XYZ in figure 1.3*c*. In this structure each atom has twelve equidistant close neighbours.

An alternative method of stacking close-packed planes places the third layer of atoms immediately above the first layer to give an ABABAB, etc., arrangement.

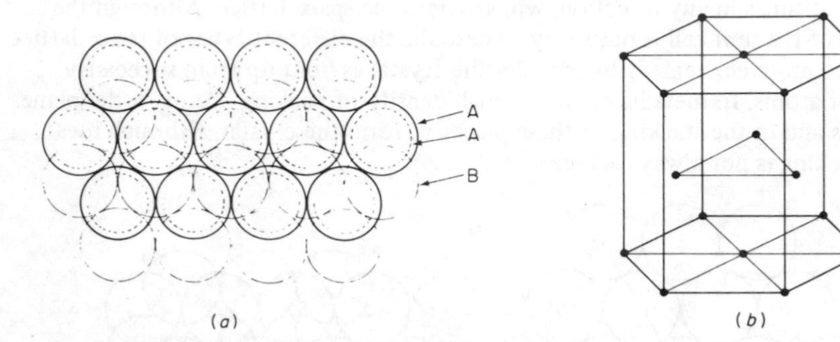

Figure 1.4 *(a) ABAB arrangement of close-packed planes in which alternate planes of atoms are directly above each other. (b) Hexagonal close-packed structure cell*

This is shown in figure 1.4*a* and gives rise to the hexagonal close-packed structure shown in figure 1.4*b*. In this structure the hexagonal structure cell has atoms at each corner and the centre of the top and bottom faces, also three atoms within the unit cell. Although this structure is close-packed and each atom has twelve equidistant close neighbours, there is only one set of close-packed planes. In theory the ratio of the height to the side of the base should be 1.633 for the ideal hexagonal close-packed structure, but in practice some slight variation from this ratio is observed.

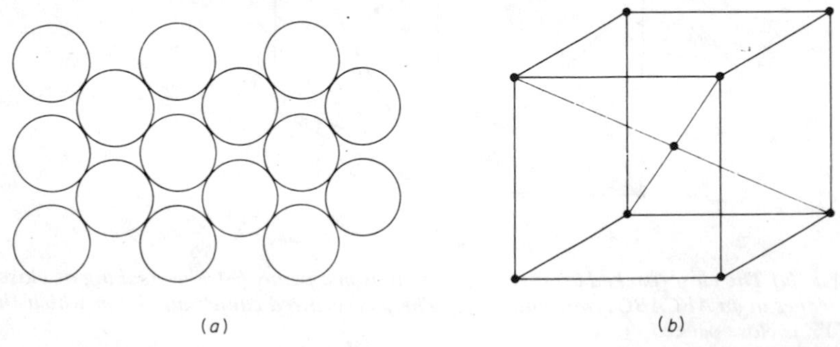

Figure 1.5 *(a) Diagonal planes from body-centred cubic structure showing less than ideal close packing of atoms. (b) Body-centred cubic unit cell*

With planes of slightly less dense packing, as shown in figure 1.5*a*, the planes of atoms will stack together more closely than the close-packed planes. This results in the body-centred cubic structure shown in figure 1.5*b*. It has a high atomic packing density, although not an ideally close-packed structure like the face-centred cubic and hexagonal close-packed structures previously described. The body-centred cubic unit cell has an atom at each corner and one in the centre of the cube. Each atom in this structure has only eight equidistant close neighbours but there are six more atoms only slightly further away.

The three lattice structures described are the principal lattices occurring in the common metals. More complex lattice structures with a lower density of packing occur in some intermetallic compounds and in some of the metals of higher valency, such as arsenic or antimony. The type of lattice structure of a metal affects its deformation characteristics; these relationships are discussed in chapter 2, but first it is necessary to consider the arrangements of crystals in metals.

1.3 Grain Structure of a Pure Metal

In studying the arrangement of crystals or grains in a pure metal it is convenient to consider the development of the structure during solidification from the liquid state. In the liquid the atoms are in an almost completely disordered condition and any ordered arrangement will contain only a small number of atoms and exist for a very short period of time. On cooling to the freezing point ordered arrangements become stàble and act as nuclei on to which further atoms join to form growing grains. Growth of these grains continues until they interfere with one another, restricting further growth except by solidification of any liquid remaining between the solid crystals. This governs the external shape of the grains which are identical in internal structure but, having grown from different nuclei, their lattices are at different angles. Where the grains meet a boundary is formed which is a zone only a few atoms thick where the lattice structure is disordered as it changes from one grain orientation to another identical grain at a different orientation.

When a polished surface of a pure metal is etched by subjecting it to attack by a weak acid the grain structure becomes apparent and may be observed under the microscope or, in the case of very large grains, by the naked eye. The atoms in the narrow disordered bands at the grain boundaries are less rigidly held in place than the atoms within the grains and the attack by the etching reagent tends to be concentrated in these regions, forming shallow grooves. These reflect light at different angles from that reflected by the grains and show the grain structure.

1.4 Solid Solutions and Compounds

The crystallisation of alloys during solidification is generally similar but more complex than the crystallisation of a pure metal. In liquid alloys the constituent metals are usually, but not always, completely in solution in each other whereas in solid alloys they may be insoluble, dissolved in each other to form solid solutions, or they may combine to form compounds. The type of solid formed is governed by factors such as the chemical nature of the metals and the relative sizes of the atoms. Different crystalline forms appearing in the structure are referred to as phases and appear to be homogeneous, although they may contain more than one type of atom.

Solid solutions may be described as substitutional or interstitial solid solutions. In substitutional solid solutions the atoms are within 14 per cent of the same diameter and atoms of the solute metal replace atoms of the solvent metal in the crystal lattice. While usually occurring over a fairly limited range of composition, under favourable conditions and when both metals crystallise in the same lattice form, complete solubility may occur. For interstitial solid solutions the solute atoms must be about 41 per cent less in diameter than the solvent atoms so that they will enter the interstices or gaps between the atoms of the parent lattice. The extent of solubility in interstitial solid solutions is much more restricted and is limited to only a few per cent.

Compounds between metals may follow normal laws of chemical valency when one atom contains more valency electrons than the other, or their formation may be governed by laws relating to the ratio of valency electrons to atoms in the compound. When metals combine to form compounds with elements such as carbon, these are interstitial compounds where their formation is governed by considerations of the small size of the carbon atom in comparison with the size of the metal atoms. Compounds are usually hard and brittle, but appreciable ductility can occur in some of the simpler compounds formed under the laws relating to the ratio of valency electrons to atoms in compounds.

1.5 Equilibrium Diagrams

In an alloy system between two metals, the relationship between the phases present, composition and temperature may conveniently be represented by an equilibrium or phase diagram. Figure 1.6 shows the type of diagram that represents

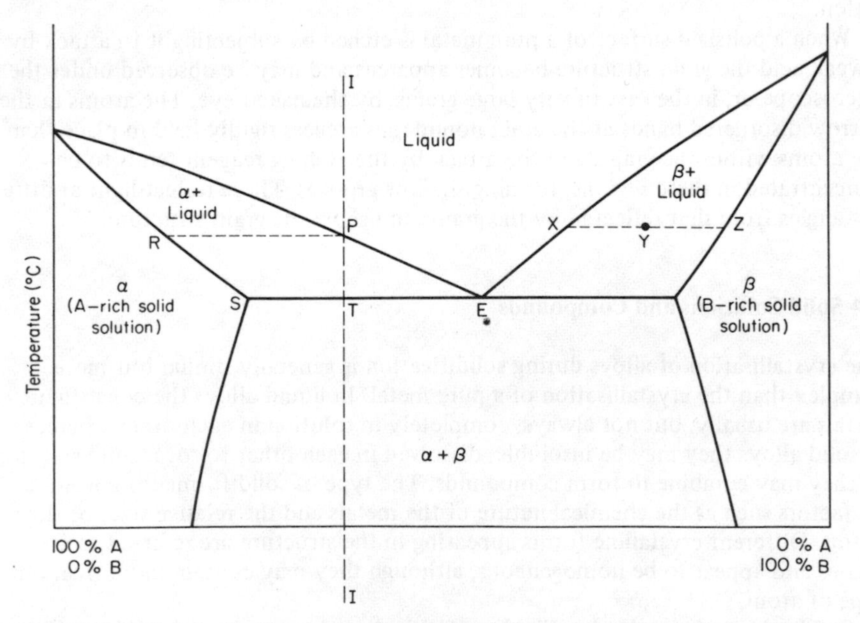

Figure 1.6 *Equilibrium diagram for two hypothetical metals A and B which are completely soluble in the liquid state but only partially soluble in the solid state*

two hypothetical metals A and B which are completely soluble when liquid and partially soluble when solid. The diagram shows the phase or phases present for any temperature and composition. In the two-phase regions for a given temperature the compositions of the phases present are given by the points where a horizontal or constant-temperature line cuts the boundaries of the two-phase region and the amounts of the phases present are in inverse proportion to the lengths of this line from the point representing the alloy composition to the boundaries of the region. For example, point Y on the diagram in figure 1.6 represents an alloy containing a mixture of B-rich solid solution of composition Z and liquid solution of composition X in the ratio Z:X = XY:YZ. Equilibrium diagrams also provide information regarding the final distribution of phases in an alloy. On slow cooling of the alloy shown by the dotted line II on the diagram, no change occurs until the point P is reached. At this point A-rich solid solution starts to separate, leaving the liquid enriched with metal B. This continues until temperature T is reached, the solid solution changing composition along the line RS while the liquid composition follows the line PE. On cooling through point T the solid metal remains unchanged while the remaining liquid of composition E solidifies to a fine mixture of the two solid solutions. Below point T some B-rich solid solution is precipitated from the A-rich solid solution and may be seen at the boundaries of this constituent. The final structure therefore consists of grains of A-rich solid solution embedded in a fine mixture of the two solid solutions with some precipitated B-rich solid solution. The reaction at E when the remaining liquid solidifies is called a eutectic reaction and the temperature remains constant during this reaction owing to the evolution of latent heat. With rapid cooling the changes are similar but, owing to insufficient time for diffusion, the A-rich solid solution contains a lower proportion of metal B and the precipitation of B-rich solid solution may be suppressed.

Many more complicated forms of phase diagram arise in alloy systems involving compounds and polymorphic changes when a metal shows different types of crystal lattice over different temperature ranges. The rules and treatment discussed with reference to the diagram in figure 1.6 may be applied in all cases. Diagrams of alloy systems containing three constituent metals can only be represented by solid models and when four or more metals are present it is impossible to represent them diagrammatically. In these cases it is usual to consider the effect of each alloying element separately while assuming the others remain present in fixed proportions. Further reference to appropriate equilibrium diagrams will occur in the discussion of strengthening mechanisms in chapter 3.

1.6 Microstructure and Macrostructure

The microstructure of metals and alloys, that is, the fine structure observed under the microscope, plays an important part in determining the mechanical properties of the material; the microstructures of some commercial alloys are discussed in chapter 2. In general, strength depends on the grain size of the phases present, their nature and distribution. Fine-grained structures whether in a single or two-phase alloy are stronger than coarse-grained structures. With two-phase alloys where one phase is hard and brittle relative to the other phase optimum properties are obtained when it is uniformly distributed in isolated particles. If a brittle phase occurs as a continuous network then fracture will follow this network, making the alloy as a whole brittle. The distribution of the phases depends initially on the conditions under which the alloy solidified but may frequently be varied by mechanical working and heat treatment.

The term macrostructure used with reference to metals refers to the coarser structure visible to the naked eye. With many alloys and almost all pure metals that have been allowed to solidify under conditions of slow cooling the grain structure may be so coarse as to be visible without the aid of the microscope and is readily observed as the macrostructure. In wrought metals, however, the grain size is usually small and cannot be observed in the macrostructure, although other effects of deformation can frequently be observed. During deformation any segregates or zones of different composition in the original metal tend to be elongated in the direction of working, and when a longitudinal section of the alloy is subsequently polished and heavily etched these elongated zones become apparent, giving it a fibrous appearance. Provided the banding effect is not excessive it has only a limited effect on properties but it is generally desirable that the principal tensile forces should be parallel to, rather than across, the fibre effect.

2

Deformation of Metals

The importance of metals as constructional materials is almost invariably related to their loadbearing capacity in either tension or compression and their ability to withstand limited deformation without fracture. It is usual to assess these properties by tensile tests in which the modulus of elasticity, the yield or proof stress, the tensile strength and the percentage elongation are determined. Determination of hardness or resistance to indentation may also be carried out (BS 240 : Part 1).

2.1 Tensile Properties of Materials

In the tensile test a suitably shaped specimen is extended until fracture occurs, while measurements of the load required to cause extension and the amount of extension are recorded. British Standard methods of tensile testing are given in BS 18 : Parts 1 - 4. Typical load - extension curves for mild steel and copper are shown in figure 2.1. On loading mild steel it behaves elastically, showing a straight-

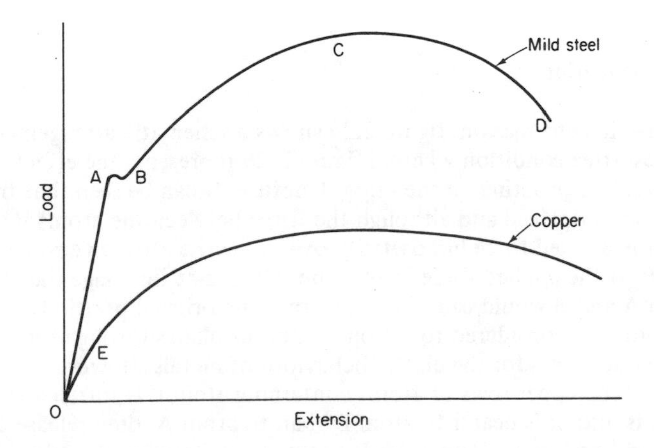

Figure 2.1 *Load - extension curves for mild steel and copper*

line relationship between load and extension from the origin O until the yield point at A is approached. The modulus of elasticity is defined as the ratio of load per unit area (stress) to the extension per unit length (strain) in this linear elastic range. At A sudden extension or discontinuous yielding occurs, followed by permanent or plastic deformation which is uniform along the length of the specimen from point B on the curve until the point of maximum load or tensile strength is reached at C. From C to D plastic deformation continues but localised reduction in the diameter of the specimen, or necking, occurs, causing a reduction in the load required to cause extension although the true stress on the reduced cross-section is increasing. At D fracture occurs. The load - extension curve for copper is similar except that there is no marked yield point at E, the transition from elastic to plastic deformation occurring gradually. In the absence of a readily determined yield point it is usual to determine the proof stress, that is, the stress required to cause a given permanent extension, usually 0.1 or 0.2 per cent. In the tensile test elastic deformation occurs in the initial stages by the temporary elastic displacement of the atoms in the crystal lattice, but with greater applied stress plastic deformation occurs owing to the permanent displacement of atoms. The stress necessary to cause atomic movement increases as the extent of deformation increases until a stage is reached when the stress exceeds that required to separate the atoms and fracture occurs.

Ductility

The ductility of a material is its ability to deform permanently under tensile forces prior to fracture. This may be measured in terms of the percentage elongation at fracture which for comparative tests must be measured on tensile test specimens of standard dimensions. Brittle materials exhibit comparatively little elongation at fracture (less than 2 per cent for ordinary grey cast iron) whereas ductile materials show considerable elongation (greater than 25 per cent for some steels).

2.2 Elastic Deformation

Considering elastic deformation, figure 2.2*a* shows a schematic arrangement of atoms in a stress-free condition whereas figure 2.2*b* represents the effect of a stress within the elastic range acting on the same structure. It can be seen that the upper rows of atoms are displaced and although the force between the atoms B and C is unaffected the applied force has partially overcome the attractive force between atoms A and B. If the applied force were to be released at this stage the attractive force between A and B would cause it to return to its original position. If this type of mechanism is considered to act on numerous planes throughout the crystal lattice it accounts for the elastic behaviour of metals. However, if the stress applied to the upper rows of atoms containing atom B is sufficiently great to move atom B until it is nearer to atom D than to atom A, then release of the applied stress will no longer allow atom B to return to its original position and plastic deformation will have occurred.

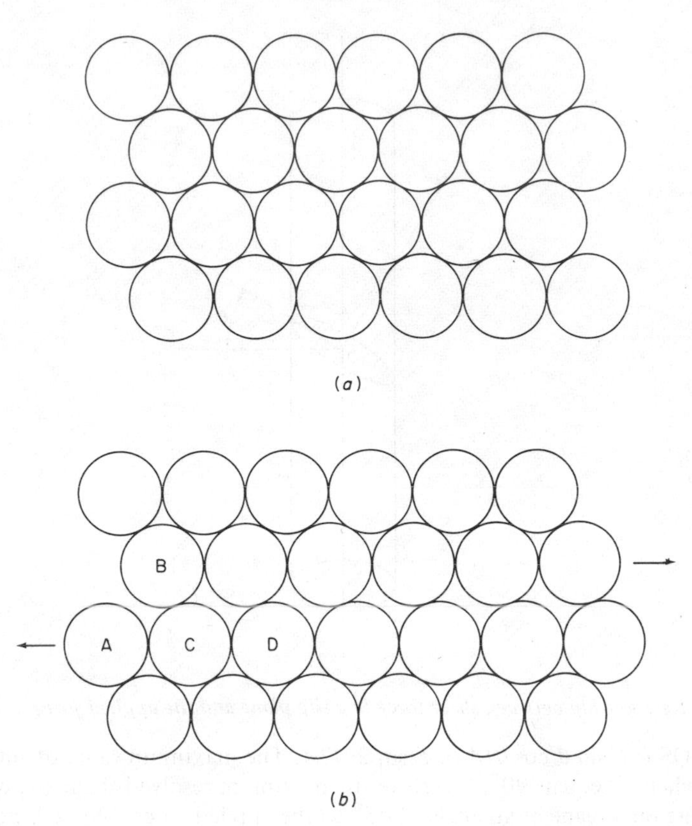

Figure 2.2 *Effect of elastic strain on schematic crystal lattice showing (a) unstrained lattice and (b) strained lattice*

2.3 Plastic Deformation

Plastic deformation is usually the result of planes of atoms slipping over each other by the action of shear stresses on certain types of plane known as slip planes. Slip does not occur on all the planes in a parallel group of slip planes but on relatively widely spaced planes. This gives rise to slip bands or steps on the surface when a previously polished sample is deformed.

Resolved Shear Stress

In the representation of a cylindrical crystal of cross-sectional area A shown in figure 2.3 a tensile force F is acting on the shaded plane. The force F may be resolved into forces in the direction ON normal to the plane and in the direction OS which is the direction of greatest slope within the plane. If the plane is at an angle θ with the cross-section of the crystal then the area of the plane equals $A/\cos\theta$, the force in the direction OS equals $F\sin\theta$ and the shear stress in the

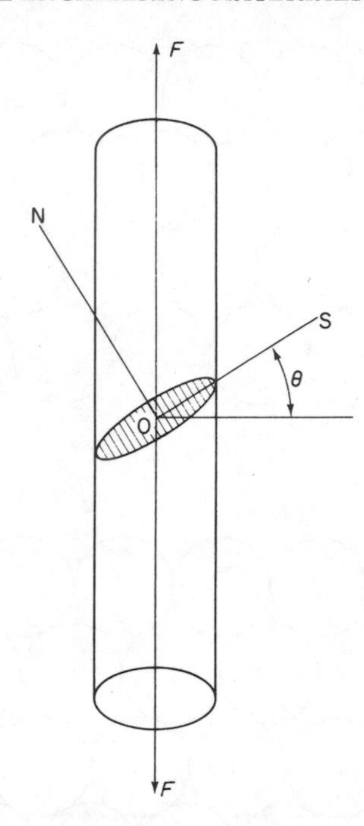

Figure 2.3 *Relationship between shear force in a slip plane and the applied force*

direction OS is $F \sin \theta \cos \theta / A$ or $F \sin 2\theta / 2A$. The maximum value of $\sin 2\theta$ equals 1 when 2θ equals 90°, therefore the maximum resolved shear stress will be $F/2A$ acting on a plane at an angle of 45° to the applied force. Slip will not necessarily take place exactly in the direction of the maximum resolved shear stress but it will occur in the suitable slip plane and direction most closely approaching the 45° angle. Note Lüders Lines

Mechanism of Slip

The stresses required to cause plastic deformation by simultaneous slipping of all the atoms in one plane relative to those in the next plane may be calculated, and are found to be from 10 to 10 000 times greater than the forces necessary to cause deformation in practice. This apparent anomaly is explained by the presence of lattice defects known as dislocations which permit slip to occur by the successive movement rather than the simultaneous movement of atoms. When an edge dislocation occurs in a slip plane there is an extra part-plane of atoms present on one side of the slip plane and for a screw dislocation there is sideways displacement of a part of each plane of atoms. In figure 2.4, which represents successive stages in the movement of dislocations, the atoms in the planes on either side of the slip plane are shown. It can be seen that the arrangement of the atoms in the final stage is identical to the original arrangement, except for the displacement of one layer of atoms relative to the other. The slip direction is perpendicular to the length of an edge dislocation and parallel to the length of a screw dislocation. In

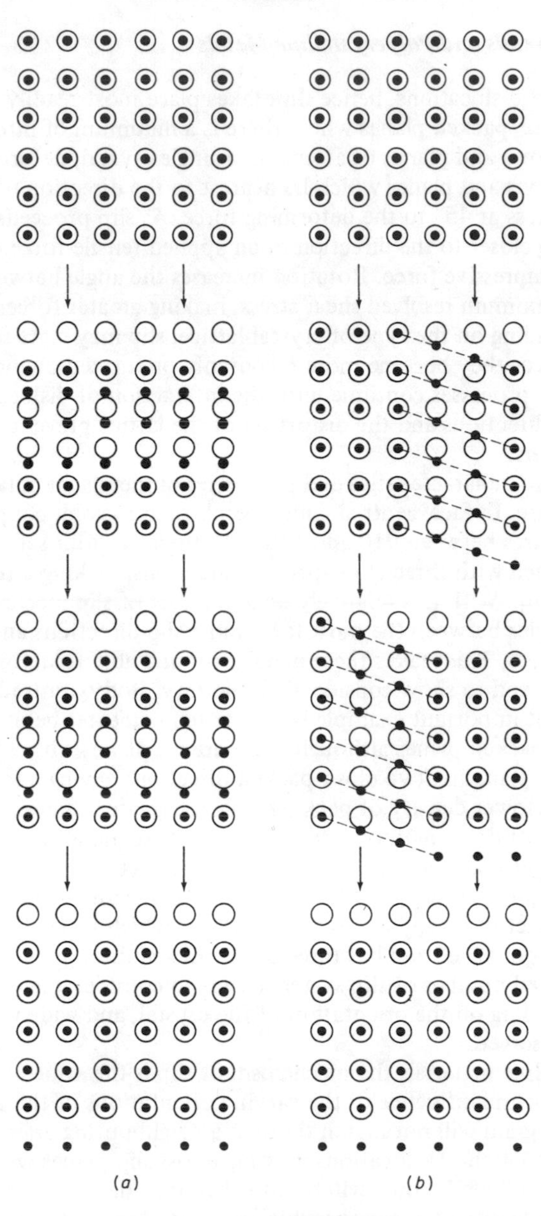

Figure 2.4 *Diagrammatic representation of slip by the movement of (a) edge dislocation and (b) screw dislocation. Atoms on either side of the slip plane are represented as* ○ *and* ● *respectively*

practice, slip normally occurs by the movement of mixed dislocations combining both edge and screw dislocation movements. Dislocations are always present in metal crystals as a result of irregularities during the growth of the crystal and further dislocations are produced during deformation.

Slip in Single Crystals and Polycrystalline Metals

The movement of dislocations, hence slip, takes place most readily in close-packed directions on close-packed planes where there is a minimum of interlocking between adjacent rows and planes of atoms. In a single crystal, deformation starts in those slip directions and planes which lie nearest to the direction of the maximum resolved shear stress at 45° to the deforming force. As slip proceeds the slip planes rotate, becoming closer to the direction of an applied tensile force or perpendicular to an applied compressive force. Rotation increases the angle between the slip plane and the maximum resolved shear stress, making greater forces necessary to cause slip. Depending on the type of crystal lattice, slip may start to occur on additional planes as they become more favourably orientated owing to rotation of the lattice. These processes continue until the interaction of dislocations moving in different slip directions and the distortion of the lattice prevents further slip, and fracture occurs.

The deformation characteristics and properties of metals are related to the type of crystal structure. In face-centred cubic metals such as gold, copper, nickel, and iron at temperatures between 910 and 1400 °C, there are four sets of ideally close-packed planes, each with three close-packed directions, making a total of twelve systems of easy slip. With this relatively large number of slip mechanisms there can only be small angles between the most favourable slip directions and the maximum resolved shear stress. This makes these metals soft and able to deform under relatively low forces, and to show considerable ductility. Body-centred cubic metals, of which the most important example is iron at temperatures below 910 °C, have no ideally close-packed planes although there are six planes with a high density of packing and each plane has two close-packed directions, giving a total of twelve slip systems. The lower density of packing calls for greater forces to initiate slip, but a high degree of deformation is still possible. These metals are therefore harder and stronger than the face-centred cubic metals but are of similar ductility. In the hexagonal close-packed structure which occurs in zinc and magnesium there is only one set of ideally close-packed planes which is parallel to the basal plane of the hexagonal cell and has three close-packed directions. In single crystals this limited number of slip systems leads to considerable directionality of properties, depending on the orientation of the crystal, and wide variations in properties are observed.

In polycrystalline materials the mechanism of slip is the same as that in single crystals except for the influence of the varying orientations of the grains. The slip directions in one grain will not match those in a neighbouring grain and slip must start afresh in each grain. Dislocations moving across slip planes will be halted at grain boundaries and while some will be absorbed into the grain boundaries some will remain, causing distorted regions which will interfere with further slip. In general this makes the stresses necessary to cause deformation greater in polycrystalline materials than in single crystals.

With hexagonal close-packed metals the number of slip systems is limited and additional mechanisms of shear are required to allow grains to conform to the shape of their neighbours. This is achieved by slip on less closely packed planes or by mechanical twinning. Mechanical twinning may be regarded as a special case of slip taking place on adjacent planes where the distance moved on each plane is related to that on the previous plane. It forms a band within the grain which has a related although different orientation from that of the grain. Owing to the necessity for these alternative modes of deformation, polycrystalline hexagonal metals require greater stress to initiate deformation than is normally required for single crystals.

3

Strengthening Mechanisms in Metals

In engineering design the modulus of elasticity and yield or proof stress are of paramount importance but some measure of ductility is essential if brittle failure is to be avoided. Whereas the modulus of elasticity is an inherent property of a metal capable of little alteration, the strength may be increased by mechanical working, alloying, heat treatment or a combination of these processes. Strengthening occurs by distortion of the slip planes or the introduction of hard additional phases which inhibit the onset of plastic deformation. Depending on the nature of the strengthening mechanism this may be accompanied by a loss of ductility in some instances but in other cases increased ductility may result.

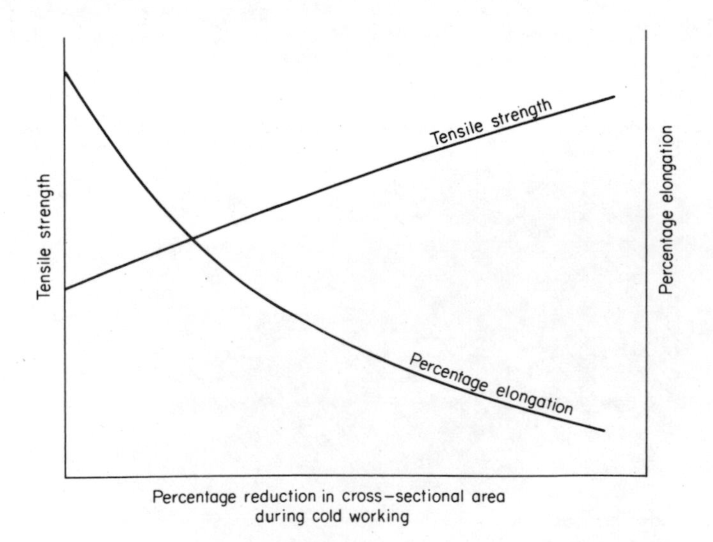

Figure 3.1 *The changes in mechanical properties occurring during cold working of a typical metal*

3.1 Work Hardening and Annealing

In the discussion on the mechanism of plastic deformation (chapter 2) it was seen that in a polycrystalline material the forces required to cause plastic deformation increase progressively as deformation occurs. Metals and alloys which are deformed when cold therefore become harder and stronger but the capacity for further deformation, measured by the ductility, is decreased. The changes in properties occurring during cold working are shown in figure 3.1. This method of hardening is usually referred to as work hardening and the metal is said to be cold worked. Work hardening is used extensively to increase the strength of pure metals, single-phase alloys and a limited number of two-phase alloys. The methods most frequently used are the cold rolling of plate and sheet, and the drawing of wire, rod and tubes. When a pure metal that has been cold worked is subsequently annealed or heated to a temperature where the increased energy of the atoms gives them some slight mobility the effects of cold work may be partially or completely removed. As the annealing temperature is increased the three stages of recovery, recrystallisation and grain growth are usually observed. Recovery occurs at the lower temperatures and the metal softens to a limited extent, becoming more ductile, although there is no visible change in the elongated grains shown in the microstructure of the cold-worked material. It is believed that recovery is due to the rearrangement of dislocations in the crystal lattice whereby some of them annihilate one another, reducing the interference with slip due to distortion in the lattice, and some polygonisation occurs where grains break down into subgrains of similar orientation as a result of the amalgamation of dislocations.

The second stage observed on reheating the cold-worked metal to higher temperatures is recrystallisation. This occurs when the atoms in the still distorted cold-worked crystals have gained sufficient thermal energy and mobility for them to rearrange into a less distorted lattice similar to the original structure. This process starts from the polygonised regions formed during recovery of the most heavily distorted regions of the cold-worked lattice and the growth of these subgrains by the movement of further atoms to join on to them. When recrystallisation is complete there will be a fine-grained structure very similar to the structure prior to cold working. During recrystallisation there is a large decrease in strength and hardness with a considerable increase in ductility. The temperature at which recrystallisation occurs is usually referred to as the recrystallisation temperature. It varies slightly, decreasing with a greater degree of prior cold work and the longer the time, at constant temperature, allowed for recrystallisation. In general, recrystallisation temperatures are related to the melting points of metals and increase as the melting point increases. Lead, for example, has a melting point of 327 °C and recrystallises below room temperature whereas pure iron, with a melting point of 1537 °C, recrystallises at about 450 °C. On further heating to still higher temperatures above the recrystallisation temperature, migration of grain boundaries occurs, to give grains of increasing size with a further decrease in strength and little change in ductility. The changes in properties occurring during annealing of cold-worked material are shown graphically in figure 3.2.

In single-phase alloys the effects of cold work and annealing are essentially the same as in the pure metal previously discussed, but in two-phase alloys the initial distribution and relative hardnesses of the phases may affect the final micro-

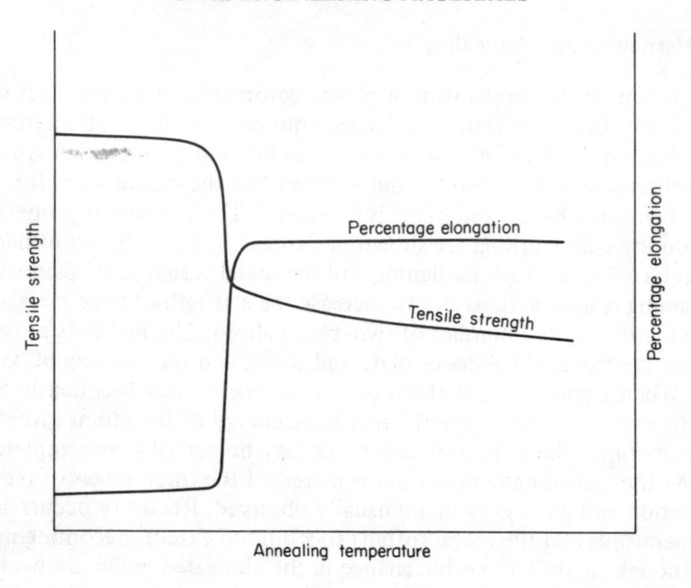

Figure 3.2 *The changes in mechanical properties occurring during annealing of a typical cold-worked metal*

structure. During cold working of a two-phase alloy the phases tend to be elongated in the direction of working and although recrystallisation occurs within each elongated phase during annealing the elongated arrangement persists in the final structure. The degree of elongation and the exact distribution of the two phases depends on their proportions and relative hardnesses, and may show a uniformly elongated or banded structure, or the second phase may show as stringers of isolated particles. Banded structures tend to show directional properties and are frequently stronger and more ductile in the longitudinal direction.

Cold-working processes are not restricted to their application for increasing strength: they are often used in conjunction with annealing to control grain size and achieve optimum properties, although here the strengthening effect is limited. Their most frequent application is for pure metals and single-phase alloys. They have only limited application for two-phase alloys, which are by nature harder, and in some cases may be grain-refined by heat treatment alone.

Hot Working

When a metal is deformed above its recrystallisation temperature deformation and recrystallisation occur simultaneously and the process is referred to as a hot-working process. The resultant grain structure is the same as that produced by cold working and subsequent annealing and depends on the working temperature during the final stages of working. For metals and alloys that are to be used in the hot-worked condition without further working or heat treatment, careful control of the hot-finishing temperature is essential, although more critical for some alloys than others. The working forces required for hot working are much smaller than those required for cold working below the recrystallisation temperature, because the metal is softer at higher temperatures and work hardening does not take place. Hot-working processes are used whenever possible in the shaping of metals because of the economies resulting from the reduced working forces. Rolling, forging and extrusion are common hot-working processes while cold working is reserved for processes such as wire and rod drawing, spinning, deep drawing and pressing which can only be carried out cold, and for rolling when especially good surface finish and dimensional accuracy are required or where cold work is desirable to increase strength.

3.2 Strengthening by Alloying

When alloying elements are added to a metal they may be present in the solid metal in solution or as an insoluble second metal or compound; or, more usually, a limited amount of solid solubility occurs in conjunction with a second phase. In both substitutional and interstitial solid solutions the presence of atoms in solution causes distortion of the slip planes, making greater stresses necessary to promote the movement of dislocations, and the metal is more resistant to deformation, or harder.

If the effect of alloying is to introduce into the alloy a second phase that is of the same order of hardness as the base metal then the properties of the alloy are only varied to a relatively small extent, and any changes are largely those associated with the effects of reduced grain size. The alternative and far more frequently occurring second phase is a compound of considerably greater hardness than the base metal of the alloy. The extent to which a second phase of this type is present, and its distribution, determine the effect on properties. In general, distribution in fine particles throughout the structure has the effect of reducing the effective grain size and resisting the change in shape of the grains of the base metal during deformation, making greater stresses necessary for deformation. If on the other hand the second phase is present in coarser form its effects are correspondingly reduced; however, when the second phase is present in very coarse form or as a continuous network, the brittleness of the second phase may dominate the properties of the alloy.

The properties of commercial alloys and their selection are discussed in chapter 5 but it is convenient to discuss here the effects of alloying and heat treatment with reference to commercially important alloy systems.

Solid-solution Strengthening

The aluminium-rich alloys containing magnesium form a series of commercially important alloys where solid-solution strengthening occurs although in many cases this is coupled with further strengthening by cold working. The aluminium-rich part of the aluminium - magnesium equilibrium diagram shown in figure 3.3 indicates that the solubility of magnesium is about 15 per cent at 450 °C, falling to approximately 2 per cent at 20 °C under conditions of equilibrium cooling. In

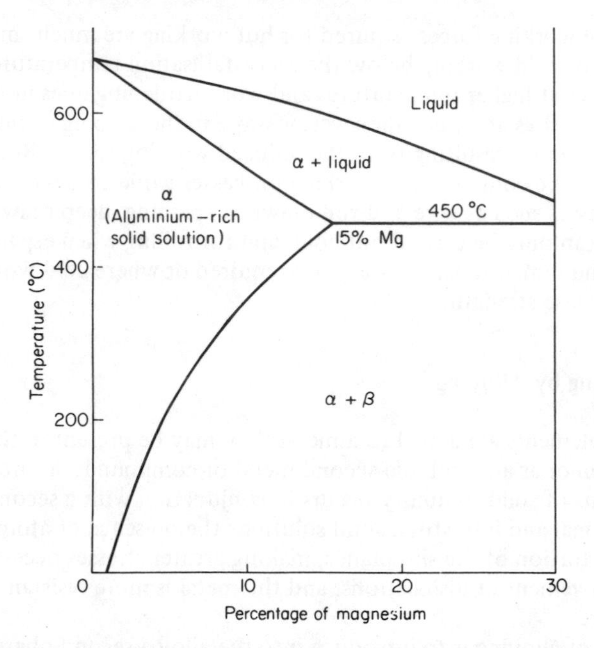

Figure 3.3 *Equilibrium diagram for alloys of aluminium containing up to 30 per cent magnesium*

commercial practice, however, the separation of the second phase from the aluminium-rich solid solution does not occur in alloys with less than about 5 per cent magnesium if prolonged heating in the temperature range 80 to 260 °C is avoided. A range of alloys based on this system is in use, with mechanical properties ranging from 85 MN m^{-2} tensile strength, 30 per cent elongation in commercial purity aluminium to 240 MN m^{-2} tensile strength, 18 per cent elongation in the annealed 3.5 per cent magnesium alloy and 300 MN m^{-2} tensile strength, 8 per cent elongation in the cold-worked alloy.

Soft Eutectic Alloys

Although not of great interest to the civil engineer, the alloys of lead and tin including the soft solders are a typical example of two-phase alloys where both phases are of similar strength. It can be seen from the equilibrium diagram shown in figure 3.4 that there is only very limited solid solubility and that the alloys consist of either the lead-rich solid solution with varying amounts of eutectic or the tin-rich solid solution with varying amounts of eutectic. The alloys therefore practically all comprise mixtures of solid solution although a proportion of this is very finely mixed as the eutectic structure. The tensile stress varies from less than 16 MN m^{-2} in the pure metals to 52 MN m^{-2} in the alloy of eutectic composition. This increase can be attributed partly to the effects of solid solution hardening but is also due to the fine structure of the eutectic mixture.

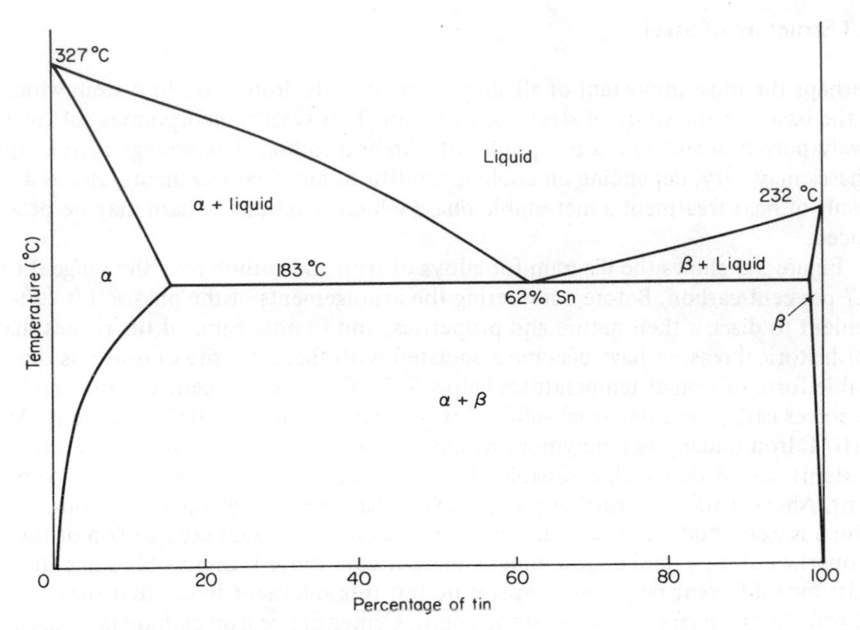

Figure 3.4 *Equilibrium diagram for alloys of lead and tin*

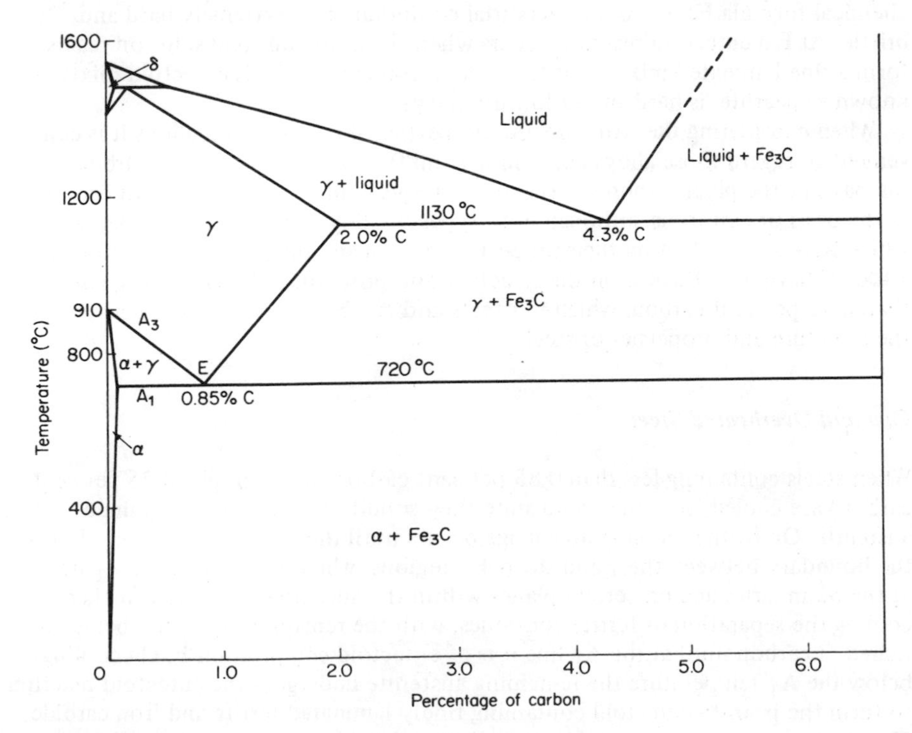

Figure 3.5 *Equilibrium diagram for alloys of iron and carbon*

3.3 Structure of Steel

Perhaps the most important of all alloy systems is the iron - carbon system which is the basis for the study of steel and cast iron. This system incorporates soft relatively pure iron and a hard compound of iron and carbon. The arrangement of the phases may vary, depending on cooling conditions and heat treatment, and as a result of heat treatment a metastable phase which is extremely hard may be produced.

Figure 3.5 shows the diagram for alloys of iron and carbon over the range 0 to 6.7 per cent carbon. Before considering the arrangements of the phases it is convenient to discuss their nature and properties, and to note some of the names that for historical reasons have become associated with them. Ferrite or α-iron is the stable form of iron at temperatures below 910 °C. It is body-centred cubic and dissolves carbon in interstitial solid solution to a maximum of 0.025 per cent. At 910 °C iron undergoes a polymorphic change to a face-centred cubic structure austenite or γ-iron which is capable of dissolving carbon to a maximum of 2.0 per cent. Above 1400 °C a further polymorphic change occurs giving rise to δ-iron which is again body-centred cubic similar to α-iron. An exact comparison of the properties of α-, γ- and δ-iron under standard conditions is impossible since they exist over different ranges of temperature but it is sufficient to say that they are all soft ductile phases of moderate strength. Cementite or iron carbide is an interstitial compound of iron and carbon containing 6.7 per cent carbon with the chemical formula Fe_3C. As an interstitial compound it is extremely hard and brittle. At E a eutectoid reaction occurs when the austenite solid solution reacts to form a fine laminated mixture of ferrite and iron carbide. This eutectoid mixture, known as pearlite, is hard and of low ductility.

When considering the structure and properties of iron - carbon alloys it is convenient to regard those alloys containing from 0 to about 1.3 per cent carbon as the basis of the plain carbon steels and the alloys with 2.5 to 4.5 per cent carbon as the basis of cast irons, although both types of alloy contain other elements which have some effect on their properties. δ-iron and the peritectic reaction at 1400 °C have no influence on the structure and properties of those steels, less than 0.55 per cent carbon, which it affects and can be ignored in the discussion of the structure and properties of steel.

Cast and Overheated Steel

When steels containing less than 0.85 per cent carbon (for example, 0.25 per cent carbon) are cooled from the liquid state they solidify to form coarse-grained austenite. On further cooling no change occurs until the A_3 line is reached, that is, the boundary between the γ and the $\alpha + \gamma$ regions, when ferrite starts to separate at the boundaries and on certain planes within the austenite grains. On further cooling the separation of ferrite continues, with the remaining austenite being enriched in carbon until at the A_1 line it is of eutectoid composition E. On cooling below the A_1 temperature the remaining austenite undergoes the eutectoid reaction to form the pearlite eutectoid containing finely laminated ferrite and iron carbide. The final structure of cast steel is therefore a relatively coarse structure of ferrite and pearlite and does not show the optimum properties since these are dominated by the coarse structure of the weak ferrite. A similar coarse structure, referred to as an overheated structure, arises if during working operations or heat treatment a steel is held for a long period at high temperatures in the γ-region, permitting grain growth to occur.

Heat Treatment of Steel

Since changes occur in the solid state it is possible to modify the structure of steel by heat treatment. Annealing and normalising are processes used to refine the structure of steel. Annealing of steel to refine the structure can be applied to material that has not been cold worked and this should not be confused with annealing as used to remove the effects of cold work. In the annealing process the steel is heated to a temperature just greater than the A_3 temperature and held at that temperature to achieve uniformity of composition and temperature prior to slow cooling, usually in the furnace. On heating no change occurs until the A_1 line is reached, when the pearlite changes to austenite. This change occurs from many nuclei within each area of pearlite, hence the austenite formed has a very fine grain size. As heating continues from A_1 to A_3 the ferrite present changes progressively to austenite, the change again occurring from many nuclei, resulting in fine-grained austenite. Cooling from A_3 to A_1 permits the separation of ferrite, which moves to the austenite grain boundaries because of the small grain size, and at the A_1 temperature the remaining austenite changes to pearlite. This much-refined structure shows both increased strength and ductility compared with the cast structure.

Normalising is a process similar to annealing, the essential difference being that in normalising the steel is removed from the furnace and allowed to cool in still air. The changes occurring are the same as during annealing but less time at high temperature and the faster cooling rate give slightly finer grain structures and finer laminations in the pearlite. The finer structures result in slightly improved properties compared with annealing. Normalising is a cheaper process than annealing since it occupies the furnace for less time but it can only be used for fairly uniform small sections where air cooling is unlikely to cause distortion due to differential cooling and contraction. Annealing and normalising may be applied to steels with other types of structure prior to the treatment but the changes form a similar pattern, and result in the same type of final structure.

Structural sections, plate, etc., in mild steel show very good properties of strength combined with ductility in the normalised condition. Heat treatment is costly, however, and for many purposes the expense of normalising can be avoided provided the finishing temperature during the hot rolling of steel is appropriate. A finishing temperature for hot rolling which is only slightly above the A_3 temperature gives a very fine austenite grain size and, on air cooling, a structure very similar to that obtained by normalising. Higher temperatures give coarser and less desirable structures while lower temperatures can result in a banded structure of ferrite and austenite being produced owing to rolling in the $\alpha + \gamma$ region, and a final structure on cooling showing bands of ferrite and pearlite. While such a structure shows strength and ductility in the direction of rolling, its properties in the transverse direction are impaired.

The structures of slow- or air-cooled steels containing up to 0.85 per cent carbon are similar. All consist of ferrite and pearlite, the proportion of pearlite increasing from 0 to 100 per cent as the carbon content increases from 0 to 0.85 per cent. In fine structures such as those obtained by annealing and normalising the properties follow closely the mean values that could be expected according to the varying proportions of the soft ductile ferrite and the hard pearlite of low ductility.

Steels with more than 0.85 per cent carbon are not of great significance in civil engineering, being mainly of use for cutting tools or where high hardness is required. The structure of these steels is of interest, however, as an example of the effect of the distribution of the second phase on properties. When cooled slowly or air cooled from the γ-region they pass first through the $\gamma + Fe_3C$ region, where iron carbide separates to the boundaries of the austenite grains, and finally at the A_1 line the remaining austenite changes to pearlite. The final structure is therefore of pearlite surrounded by a network of iron carbide which, being very hard and extremely brittle, dominates the properties, making the steel extremely brittle. This may be avoided if the steel is hot worked in the γ-region and hot working is continued as it cools into the $\gamma + Fe_3C$ region, followed by cooling from the $\gamma + Fe_3C$ region. This treatment breaks up the iron carbide as it separates, forming a structure of dispersed globular iron carbide in austenite which on cooling gives a structure of globular iron carbide in pearlite. To further increase the hardness of these steels they are usually reheated into the $\alpha + Fe_3C$ region, followed by quenching and tempering.

Quenching and tempering

When small sections of steel are water-quenched from the γ-region the cooling rate is too great to allow the separation of ferrite and formation of pearlite by the nucleation and growth processes. The face-centred cubic austenite is unstable, however, and the change to a body-centred cubic structure similar to ferrite cannot be prevented. This takes place by a shearing mechanism where planes of atoms move to give a body-centred tetragonal structure which is almost cubic but slightly elongated in one direction and retains the carbon in solution. This structure, known as martensite, is extremely hard and brittle owing to the distortion produced in the lattice by the carbon retained in supersaturated solution. On tempering or reheating at temperatures below the A_1 temperature the retained carbon starts to be precipitated as iron carbide in globular form at temperatures above about 200 °C, the size of the globules increasing with increasing temperature. Tempering at the lower temperatures causes only slight loss in hardness and greatly increases toughness, and tempering at the higher temperatures may, if a suitable temperature is selected, give strength and ductility superior to those in a normalised steel. Quenching and tempering are principally applied to the higher carbon steels where high hardness is required or to alloy steels to achieve high strength.

In plain carbon steels the cooling rate required to produce hardening may be achieved only in small sections that are water-quenched, and there is a considerable risk of distortion and cracking. Alloying elements such as nickel, chromium, manganese and molybdenum when present in steels reduce the cooling rate necessary for the formation of martensite and therefore increase the hardenability or depth to which hardening may occur. Steels containing small percentages of these elements may be used in the oil-quenched and tempered condition to give high strength which is due to the effects of the dispersed carbides combined with the solid-solution hardening effects of the alloys in solution in the ferrite.

3.4 Cast Iron

The structure of the iron - carbon alloys containing approximately 2.5 to 4.5 per cent carbon, usually referred to as cast irons, depends on the cooling rate and the presence of other elements, particularly silicon. Cast irons are usually classified as white or grey cast irons according to the appearance of the fractured surface. With white irons, which are rapidly cooled and contain less than about 1.0 per cent silicon, solidification and cooling occur according to the iron- iron-carbide diagram and the final structure is of pearlite embedded in iron carbide. The high proportion of iron carbide and its continuous nature throughout the structure make white irons extremely hard and brittle. White irons have limited use where very high hardness but low shock resistance is required and as an intermediate step in the production of malleable cast iron.

Grey cast irons are produced by slow cooling and have a silicon content of up to 2.5 per cent. In the solidification and earlier stages of cooling they behave as alloys of iron and graphite since iron carbide is unstable, but at lower temperatures iron carbide becomes stable and the iron - iron-carbide diagram is followed. This results in a high temperature structure of graphite and austenite which on cooling gives a final structure of graphite surrounded by pearlite and ferrite. Although the ferrite/pearlite structure of the matrix is similar to steel, and could be expected to have similar properties, it is broken up by graphite which occurs in flake form and forms a path of easy fracture under conditions of tensile stress. Common cast irons show compressive strengths comparable to steel but have only about one-quarter the strength of steel in tension owing to the weakening effect of the graphite.

Cast irons of higher tensile strength can be produced by reducing the carbon content, by special treatments just prior to casting which reduce the quantity and size of the graphite flakes and by alloying to strengthen the matrix structure. Further improvements can be obtained by treatment with alloys of cerium or magnesium or both just prior to casting, when the graphite takes on a spheroidal form and has practically no deleterious effect on properties. When cast, spheroidal graphite cast irons have a matrix of pearlite, but with prolonged annealing this may be largely converted to graphite and ferrite with some reduction in strength but an increase in ductility. Malleable cast irons which are produced by annealing white irons for long periods also show some ductility and toughness by avoiding the weakening effects of flake graphite. The annealing procedure and composition of iron used govern the final structure, which contains compact rosettes of graphite in a matrix of ferrite or pearlite and ferrite. This structure is formed partly by the breakdown of iron carbide to form iron and graphite rosettes and partly by the oxidation of carbon diffusing to the surface of the casting.

3.5 Age or Precipitation Hardening

The phenomenon of age hardening, also known as precipitation hardening, occurs in a number of alloy systems and is a source of increased strength coupled with some loss in ductility. The aluminium-based alloys containing about 4 per cent copper are typical of age-hardening alloys and form the basis of one important group of high-strength aluminium alloys. The aluminium–copper equilibrium diagram for alloys containing 0 to 20 per cent copper is shown in figure 3.6. It can be seen that the solubility of copper in

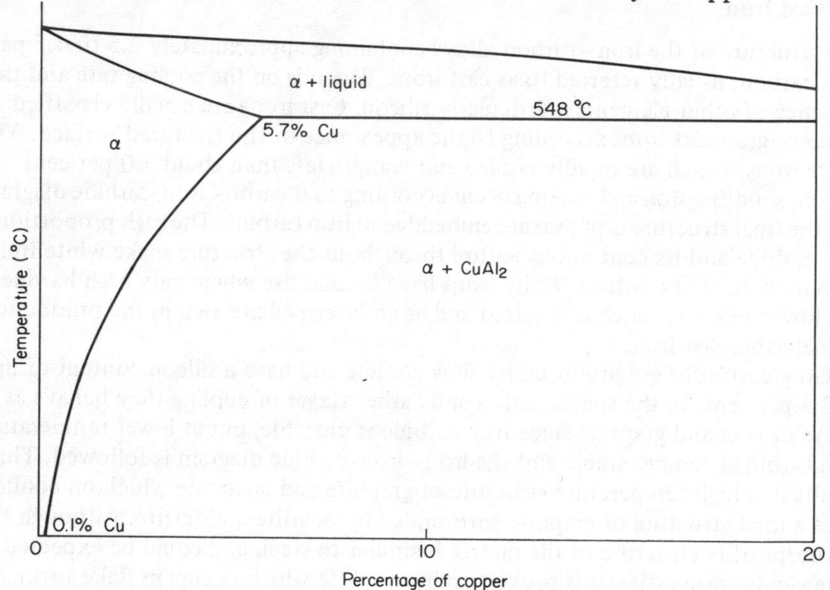

Figure 3.6 *Equilibrium diagram for alloys of aluminium containing up to 20 per cent copper*

aluminium is a maximum of 5.7 per cent at 548 °C and falls to about 0.1 per cent at room temperature. When slowly cooled, alloys in this range show the aluminium-rich solid solution and the second phase of the $CuAl_2$ compound. In order to harden the alloy, solution treatment is carried out by heating for a considerable period in the solid solution region at about 500 °C, followed by quenching then ageing at room temperature or at temperatures up to about 200 °C. During solution treatment a uniform substitutional solid solution is formed with the copper atoms evenly distributed throughout the lattice, and this structure is retained as a super-saturated solid solution on quenching.

On ageing, a number of changes occur in this solid solution, eventually leading to the formation of a $CuAl_2$ precipitate that is visible in the microstructure. Maximum hardening and strengthening effects are observed during the changes, prior to the formation of a visible precipitate which is accompanied by a loss in hardness. The first change during ageing is the formation of plates of copper atoms associated with certain planes in the crystal lattice and in the second stage these develop into zones containing copper and aluminium atoms. In both these stages the plates or zones remain part of the parent solid-solution lattice, causing considerable distortion of the lattice, hence hardening of the alloy. A further stage is the formation of a separate precipitate of $CuAl_2$ which occurs as the zones increase in size to such an extent that they are able to separate as very small particles. Precipitation relieves the strain in the solid-solution lattice and is accompanied by a fall in hardness. The ageing time depends on the temperature of the ageing treatment and the purity of the alloy. At room temperature maximum hardness is reached after about six days, but at 160°C hardening occurs in about ten hours and longer periods lead to overageing with loss of hardness owing to precipitation.

The mechanism of age hardening follows a similar pattern in other alloy systems and in some commercial alloys there may be a tendency for more than one compound to precipitate. The times and temperatures required for heat treatment vary according to the alloy system, and room temperature ageing occurs in only a few cases. Cold working of the solution-treated alloy prior to ageing is sometimes used to give a further increase in the final strength and hardness or, in some cases, it triggers off room-temperature ageing in alloys that would not otherwise age except at slightly elevated temperatures.

4

Behaviour in Service

In service, metals and alloys are required to withstand not only simple tensile and compressive stresses but also the effects of complex stress systems, shock loading, varying loads, prolonged loading at high temperatures, low temperatures, corrosion or the combined effects of a number of these conditions. Simple tensile and compressive stresses may be calculated and many of the other conditions affecting behaviour in service may be simulated in various forms of test. These tests can never reproduce all the factors affecting service behaviour and in order to apply the test results to the greatest advantage in the selection of materials an understanding of the possible modes of failure is necessary.

4.1 Corrosion and Surface Protection

The corrosion of metals may be broadly classified as dry corrosion, usually oxidation occurring at elevated temperatures, and wet corrosion which occurs in the presence of an electrolyte. In the dry oxidation of metals other than the noble metals such as gold and platinum, the nature of the first-formed oxide film governs the rate of further oxidation. When the oxide film is nonporous further oxidation can only occur by migration of ions through the oxide. If the rate of movement of ions is low then the rate of oxidation is low and decreases as the oxide layer increases in thickness, leading to thin protective oxide layers. Alternatively if the first-formed oxide layer is porous, or nonporous but increases rapidly in thickness, this is likely to lead to thick oxide layers which crack and flake off, offering little resistance to further oxidation. The adherence of oxide layers is affected by changes in temperature, which introduce thermal stresses owing to differences in thermal expansion of metal and oxide.

Few metals offer appreciable resistance to oxidation over a wide range of temperature but chromium and aluminium are notable exceptions. Chromium oxidises readily in the initial stages but the very thin layer of chromic oxide produced has an extremely slow growth rate and confers a high degree of protection from further oxidation. The affinity of chromium for oxygen is so great that in alloys containing more than about 12 per cent chromium preferential oxidation of chromium occurs, leading to an oxide layer rich in chromic oxide and a high degree of protection. This effect may best be illustrated by considering the stainless steels, which fall into three main types, martensitic, ferritic and austenitic. The martensitic stainless steels contain about 13 per cent chromium with 0.3 per cent carbon and can be hardened by quenching. The ferritic stainless steels contain 12 to over 20 per cent chromium with low carbon, and the austenitic steels contain varying amounts of chromium and nickel, although a typical composition is 18 per cent chromium and 8 per cent nickel. These steels are resistant to oxidation and, provided conditions are oxidising, they have high resistance to wet corrosion owing to the formation of a protective chromic oxide layer. The oxidation and corrosion resistance increases with increasing chromium content and in the presence of nickel. The ferritic steels are subject to very rapid grain growth at high temperatures and this causes difficulties in welding, but appropriate grades of the austenitic steels are readily welded and have generally better mechanical properties.

Aluminium, like chromium, oxidises readily in the initial stages, producing a very thin protective layer of aluminium oxide. Increased protection from further oxidation and atmospheric corrosion is obtained when the oxide layer is further thickened to a carefully controlled extent by anodising, which is a process of electrolytic oxidation. The anodised surface may be dyed when necessary for decorative purposes.

Wet Corrosion

Wet corrosion is a result of electrochemical action when different potentials are developed by electrically connected metal parts in contact with a solution containing free ions. If a single homogeneous metal is immersed in an electrolyte metal ions tend to pass into solution, leaving the metal with an excess of free electrons or a negative charge. This may be represented as the reversible reaction $M \rightleftharpoons M^+ + e$. In the absence of an applied potential or connection with a second metal - electrolyte system, equilibrium is established when the build-up of free electrons in the metal produces a potential relative to the solution such that further free electrons cannot be accepted. This potential, normally referred to as the electrode potential, is dependent on the particular metal, its condition and the nature of the solution, its concentration and temperature. Electrode potentials cannot be measured directly since this would involve a second metal - electrolyte connection but comparative values may be measured against a standard electrode - electrolyte system. A list of electrode potentials for some common metals is given in table 4.1.

When a metal component comprising two parts which are electrically connected but generate different electrode potentials is placed in an electrolyte, current flows through the system causing attack on the more anodic metal, that is, the metal with the more negative electrode potential. The metal with the more positive electrode potential or cathodic metal remains unattacked and the reaction at this metal may be deposition of metal, liberation of hydrogen or formation of OH^- hydroxyl ions, depending on the nature of the metal, the solution and the presence of dissolved oxygen in the solution. In some instances an adherent insoluble corrosion product is formed which minimises further corrosion, for

TABLE 4.1
Normal electrode potentials of some common metals measured relative to the normal hydrogen electrode

Metal	Electrode potential (V)	Metal	Electrode potential (V)
Gold	+ 1.5	Nickel	− 0.25
Copper	+ 0.34	Iron	− 0.44
Hydrogen	0	Zinc	− 0.75
Lead	− 0.13	Aluminium	− 1.66
Tin	− 0.14	Magnesium	− 2.05

example, the chromic oxide layer formed on stainless steels when conditions in the solution are oxidising. More usually, however, the corrosion products are soluble and uninhibited corrosion proceeds.

The formation of corrosion cells of this type may occur in a number of ways. In addition to dissimilar metals it may be caused by different phases in an alloy, varying amounts of cold work which give the metal a more negative electrode potential, varying concentrations of solution, varying oxygen content of solution or possibly by the effects of oxide scale on metals. Corrosion may also occur without the presence of different electrode potentials if there is an applied electrical current due to the pick-up of stray electrical currents from electrical conductors and equipment or the incidence of induced electrical currents. The effects of corrosion are always harmful but particularly so when concentrated in local areas, when pitting occurs, or in crevices by the concentration of a weak electrolyte by evaporation.

Of particular interest in civil engineering is the rusting of iron and steel. Corrosion cells are readily formed between different areas on the steel surface owing to local variations in the steel or oxygen content of the water in contact with the steel. At the anodic areas where the iron has the more negative electrode potential ferrous ions pass into solution while at the cathodic areas of more positive electrode potential hydroxyl ions are formed. Both these reaction products are soluble and diffuse through the solution until they meet precipitating ferrous hydroxide. This usually oxidises to form a hydrated ferric oxide, which is the familiar rust deposit. As deposition of this rust occurs at neither the anodic nor the cathodic areas it does not prevent further corrosion and may increase the corrosion rate owing to its tendency to hold moisture.

Corrosion Prevention

While metals may be protected from corrosion it is important that conditions stimulating corrosion should be avoided whenever possible. Consideration should be given to the avoidance of problems due to crevice corrosion, poor drainage of rainwater, use of dissimilar metals without insulation from each other, the possibility of stray electric currents, etc.

Protection from corrosion may be achieved by inactive or active surface coatings or by means of a protective electrical potential. Whereas oils and greases may give temporary protection by insulating the metal from the corrosive environment more lasting protection is given by paints and a variety of surface coatings. On drying, paints form a semi-elastic resin coating which contains a pigment which serves to colour and strengthen the semi-elastic coating. In addition, paints for use on ferrous metals may contain active ingredients such as red lead or zinc chromate which tend to make the metal surface passive owing to the formation of a fine protective oxide layer. Alternatively, metal powders such as zinc or aluminium may be included since they are anodic to iron and confer a measure of electrochemical protection. Surface coatings that are inactive and serve only to insulate the metal from the corrosive environment are bituminous paints, enamel, plastic coatings and plastic wrapping-materials.

Protective surface coatings may be developed by chemical methods although these are usually of limited durability and require further protection by painting. An example of this type of protection is the treatment of steel with a solution containing phosphoric acid, zinc phosphate and manganese phosphate which produces a thin protective phosphate coating on the steel and forms a good basis for painting. A different form of chemical treatment commonly used is the addition of inhibitors to the corrosive medium. In the process of corrosion these compounds react to form insoluble salts at either the anode or the cathode, thus stifling further corrosion. This method of corrosion control is limited to closed systems where the necessary concentration of inhibitor can be maintained.

Metallic coatings are frequently used either for decorative purposes or the prevention of corrosion. When applied to metals such as brass which have inherently good corrosion resistance metallic coatings are mainly used for decorative purposes but with ferrous metals the use of metallic coatings may confer considerable protection against corrosion. The effectiveness of a metallic coating on steel depends on the relative values of the electrode potentials of the metal and iron, and its resistance to corrosion. Zinc has good resistance to atmospheric corrosion owing to the formation of an insoluble basic carbonate on its surface; being anodic to iron, zinc gives electrochemical protection even when the coating is incomplete. It is widely used for galvanised products where it is applied by hot dipping and electroplating but it may also be applied by metal spraying. Copper, nickel and chromium are all cathodic to iron and when coatings are incomplete they tend to increase the rate of corrosion of the iron. Chromium deposited by electroplating is porous and can only be used after previous deposits of copper and nickel. Under these circumstances protection is achieved by insulation of the iron from the atmosphere but any break in the plating will lead to an increased corrosion rate. Tin is normally cathodic to iron and increases the corrosion rate except when used for tinned foods where, in the presence of weak organic acids, it behaves anodically owing to the formation of complex ions.

When a protective electrical potential is used to prevent corrosion it is referred to as cathodic protection. If a metal electrode which behaves anodically to the structure is connected, then corrosion attacks this sacrificial anode while the structure is protected. Anodes of zinc and magnesium can be used in this way to protect iron and steel. Cathodic protection may also be achieved by an applied electrical potential making the structure cathodic while the anode is of carbon or stainless steel. Unfortunately, this method is limited in usefulness owing to the difficulties of maintaining the required electrical conditions on all parts of complicated shapes.

4.2 Brittle and Ductile Fracture

The fracture of metals occurs at stresses well below the theoretical fracture stresses which are shown by calculation to be necessary to pull adjacent layers of atoms apart. To account for this difference it is usual to consider the effects of stress on a small microcrack or void that is present in the metal or has been formed by a pile-up of dislocations. With an applied stress, the effect of the crack or void is to cause a stress concentration at the ends of the crack. If the concentrated stress causes the crack to grow by the successive breaking of the bonds between atoms at the tip of the crack then brittle failure results. In these circumstances the concentrated stress need only be sufficient to exceed the theoretical fracture stress at one point at a time. Alternatively the stress required to cause deformation at the tip of the crack may be less than that to cause crack growth. In this case deformation occurs until work hardening raises the stress required for further deformation above that for crack growth when fracture occurs, that is, ductile fracture.

In metals any increase in the yield stress required to cause plastic deformation is associated with a reduction in ductility. For metals that are ductile at room temperature the ductility decreases with decreasing temperature, with a corresponding increase in the yield stress. For face-centred cubic metals the increase in yield stress is not sufficient to cause brittle fracture even at very low temperatures but the yield stress of body-centred cubic metals increases to a much greater extent and may exceed the fracture stress. For mild steel this occurs at about −150 °C when it becomes brittle under tensile conditions.

During impact testing a notched specimen is broken by a weighted pendulum and the energy absorbed by the specimen is measured. Using a notched specimen a complex three-dimensional stress system is developed in the specimen at the moment of fracture. With stresses acting in three directions the shear stresses which normally lead to plastic deformation tend to cancel each other, leading to

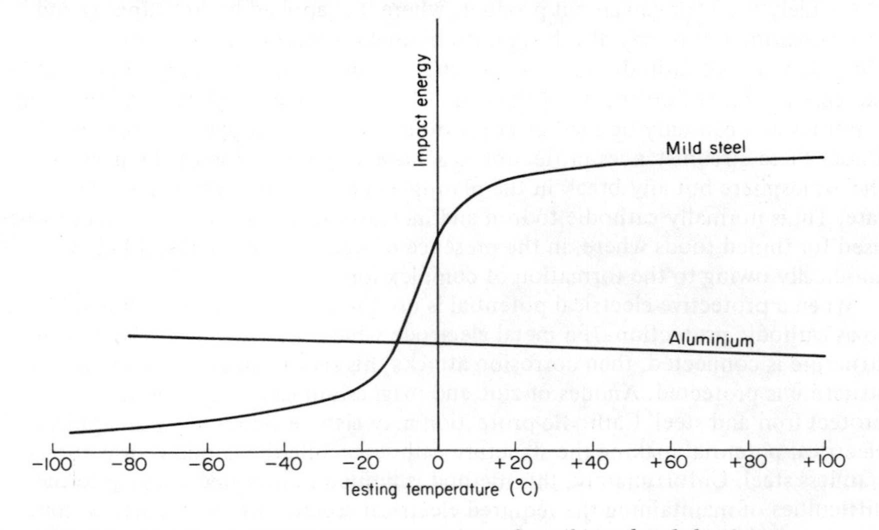

Figure 4.1 *Impact-energy–testing-temperature curves for mild steel and aluminium*

an increase in the yield stress and a reduction in ductility. This effect, coupled with the very high strain rate under impact conditions, makes the impact test results indicative of the behaviour of the material under some of the most severe stress conditions likely to be encountered. A high-energy absorption during an impact test on a notched specimen is associated with a ductile fracture. Materials showing this are commonly referred to as showing 'notch ductility'. While mild steels are ductile at temperatures above about −150 °C in the tensile test they show brittle behaviour at temperatures up to about 0 °C in the impact test. Standard impact tests including the Charpy V-notch and Izod tests are described in BS 131.

The typical impact-strength - testing-temperature curves shown in figure 4.1 indicate that transition from ductile to brittle fracture occurs for mild steel at about 0°C while there is no transition to brittle fracture in the case of aluminium. The factors affecting the brittle/ductile transition temperature in mild steel are discussed in chapter 5 and are related to composition and grain size.

4.3 Creep

At elevated temperatures, or at room temperature for some low melting point metals, prolonged loading at stresses below the normal yield point may lead to a process of plastic flow or creep. A typical strain-time curve for extension by creep under constant stress is shown in figure 4.2. The three stages of creep

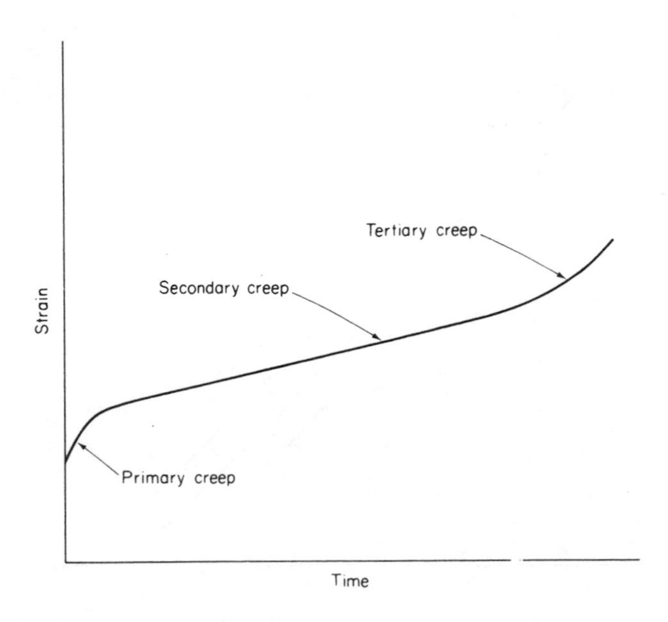

Figure 4.2 *Typical creep curve showing the strain – time relationship under conditions of constant load*

usually observed are primary creep where strain is relatively rapid, secondary creep where the strain occurs at a slower uniform rate and tertiary creep where the strain rate increases and leads to fracture. Strain - time curves may be distorted from the typical shape in alloys where structural changes, for example, precipitation hardening or tempering of steel, occur simultaneously with the creep process.

A number of mechanisms occur during creep and, while these may overlap to some extent, different mechanisms are more important during each of the three stages. In the primary stage deformation occurs owing to the normal movement of dislocations but the rate of creep is reduced as the dislocations become concentrated at barriers to their movement, that is, by work hardening. This leads to the secondary stage where further slip is only possible at a rate equal to the rate at which dislocation concentrations are reduced and rearranged, that is, the rate of work hardening equals the rate of recovery. During the third stage grain boundary sliding becomes an important mechanism and voids develop in the grain boundaries. Owing to the reduced loadbearing area the stress increases, giving an increased creep rate, and the voids eventually lead to fracture.

The rate of creep at each stage is increased by increased temperature or increased stress. Where creep conditions are likely to be encountered in service the design stress must be such that creep will not proceed beyond the secondary creep stage within the expected working life of the component and distortion will not interfere with its normal operation. Creep data obtained from a number of creep tests may be presented as a series of curves showing the stress - time relationships for different amounts of strain at constant temperature, as shown in figure 4.3.

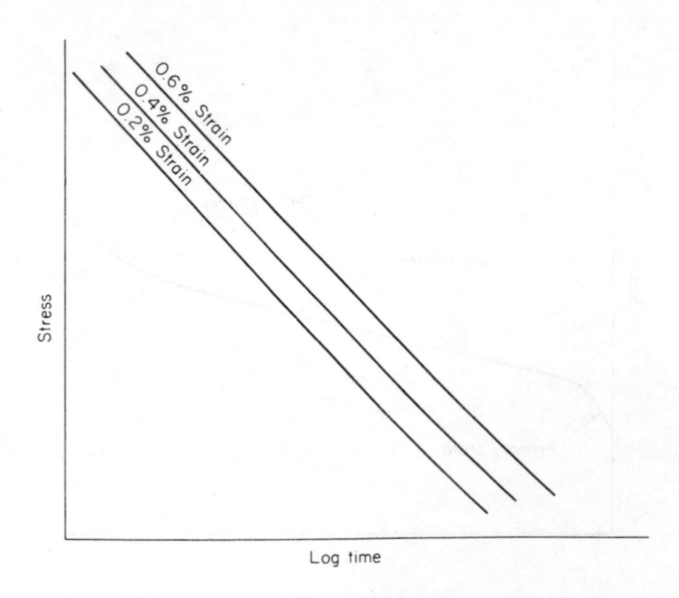

Figure 4.3 *Typical curves showing the stress - time relationship for different amounts of creep strain at constant temperature*

Similar curves may be prepared showing stress - temperature relationships for constant strain.

In creep-resistant alloys reduced rates of creep are achieved by the presence of atoms in solid solution and age hardening, which restrict the movement of dislocations. Grain boundary sliding is reduced by precipitated particles locking the grain boundaries and by coarse grain size, which reduces the extent of the grain boundaries available for sliding.

Relaxation

Relaxation occurs when a component is held for a long period in a state of elastic strain. The effect of creep mechanisms occurring is to reduce the elastic strain and stress in the component by an amount corresponding to the extent that the unstressed length of the component would be increased. This is important in bolts working at elevated temperatures and in the steel bars used in prestressed concrete.

4.4 Fatigue

Fatigue cracks leading to fracture may occur in components subjected to reversing or fluctuating stresses although the maximum applied stress is below that normally required to cause fracture. Typical stress systems likely to cause fatigue failure are shown in figure 4.4. The characteristic appearance of a fatigue failure shown in figure 4.5 comprises one or more fatigue cracks which are smooth, with ripples spreading from a focal point usually on the surface of the component, and a tensile fracture occurring when the section is so reduced as to fail under the applied stress.

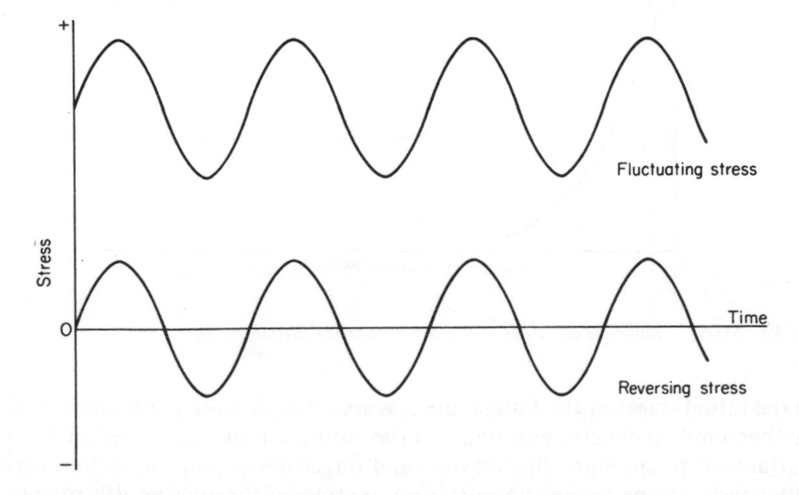

Figure 4.4 *Typical stress systems likely to cause fatigue failure*

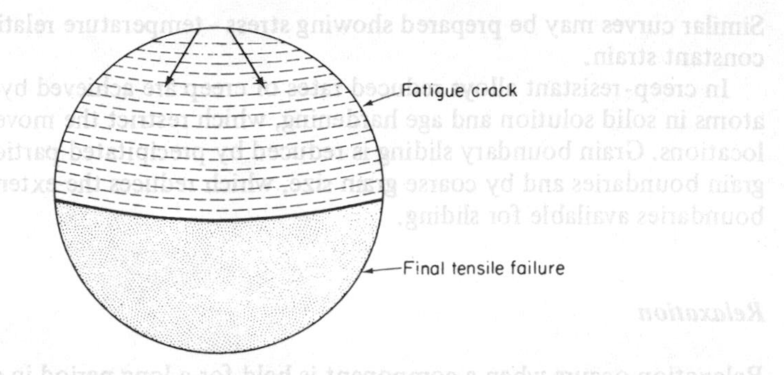

Figure 4.5 *Characteristic appearance of fatigue failure showing fatigue fracture and tensile fracture of reduced loadbearing section*

A variety of fatigue tests representing different stress systems is used to assess the relationship between stress and the likelihood of fatigue failure. The results of a series of such tests carried out with different stress levels are usually represented graphically as in figure 4.6. Steels normally show a fatigue limit below which failure is unlikely to occur but with steels at high temperatures, steels under corrosive conditions and nonferrous alloys there appears to be no lower limit below which fatigue will not occur. The effects of increased temperature and increased mean tensile loading are generally to reduce the resistance to fatigue failure.

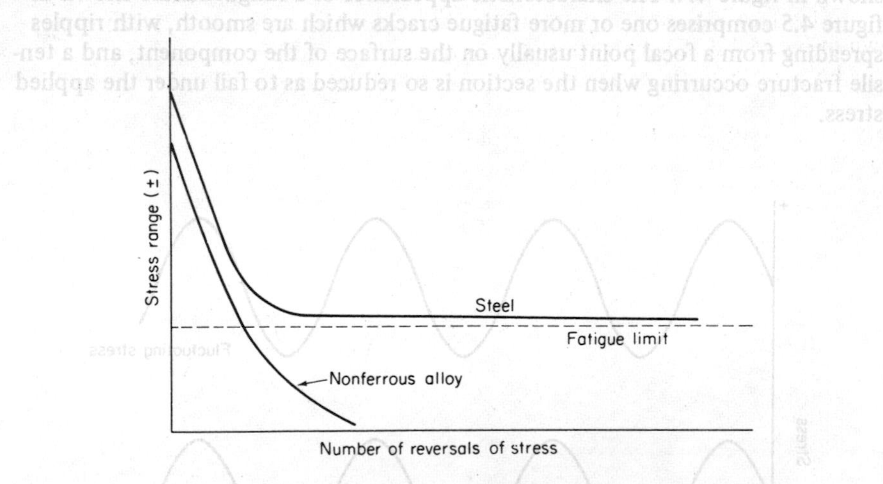

Figure 4.6 *Stress – reversals of stress curve derived from fatigue tests*

In the initial stages of the fatigue process some slight work hardening is followed by slip becoming more concentrated and the formation of coarse slip bands. At the surface of the specimen fine crevices and ridges develop in connection with the slip bands. Owing to the concentration of stress in the crevices this process is repeated and the crevice grows into a crack which continues to propagate. Since fatigue cracks originate from points of maximum stress they are frequently associated with stress raisers such as sharp changes in section, protruding weld deposits, oil holes, keyways and rough machining marks. Fatigue resistance is increased by increased tensile strength and by the presence of compressive stresses in the surface layers of a component. These can be produced by the surface hardening processes of flame hardening, induction hardening, nitriding, case-carburising and shot peening although there are restrictions on the size of component that can be treated.

Corrosion Fatigue

When fatigue conditions arise in conjunction with a corrosive environment this is frequently referred to as corrosion fatigue. Under the influence of corrosion, stress raisers are formed on the surface of the component which facilitate the onset of fatigue. This effect is marked for steel where the fatigue limit is no longer observed and fatigue may occur at very low stress levels. Corrosion fatigue is minimised by the normal precautions which may be taken to reduce the risk of fatigue failure and by the methods of protection against corrosion discussed in section 4.1.

5

The Use of Metals in Civil Engineering

Many ferrous and nonferrous metals and alloys are available to engineers. However, designers working in a particular branch of engineering are unlikely to use all of these alloys, and only those of interest to civil engineers are discussed here. For example, tool steels, alloy cast irons and nickel-base alloys are not included. Information concerning these alloys and others that have been omitted is available in the metallurgical literature.

Pure nonferrous metals have properties that make them suitable for engineering applications but unalloyed iron does not possess useful engineering properties. Despite this, iron alloys are used in larger quantities than the alloys of any other metal. This arises from the relative cheapness with which steels and cast irons can be produced with a variety of useful properties, by variations of composition, heat treatment and working.

5.1 Wrought Steels

Engineering components are produced in these steels by hot- or cold-working processes, although some machining may be involved. The component may be put directly into service or may be heat treated or surface treated depending on the grade of steel or the application.

In civil engineering the largest usage of wrought steels consists of plain carbon steels or slightly modified plain carbon steels. These are discussed by considering the requirements of typical civil engineering applications and explaining why specific steels satisfy the requirements. The discussion is conveniently based on figure 5.1, which shows that the strength and hardness of normalised plain carbon steels tend to increase with increasing carbon content whereas ductility decreases. Plain carbon steels relate directly to the graphs and the differences in properties shown by the modified steels may be explained in terms of the strengthening mechanisms discussed in chapter 3. The consideration of economic aspects is outside the scope of this discussion but cases are cited of the choice of an expensive material giving an over-all cost saving owing to the reduction of other costs.

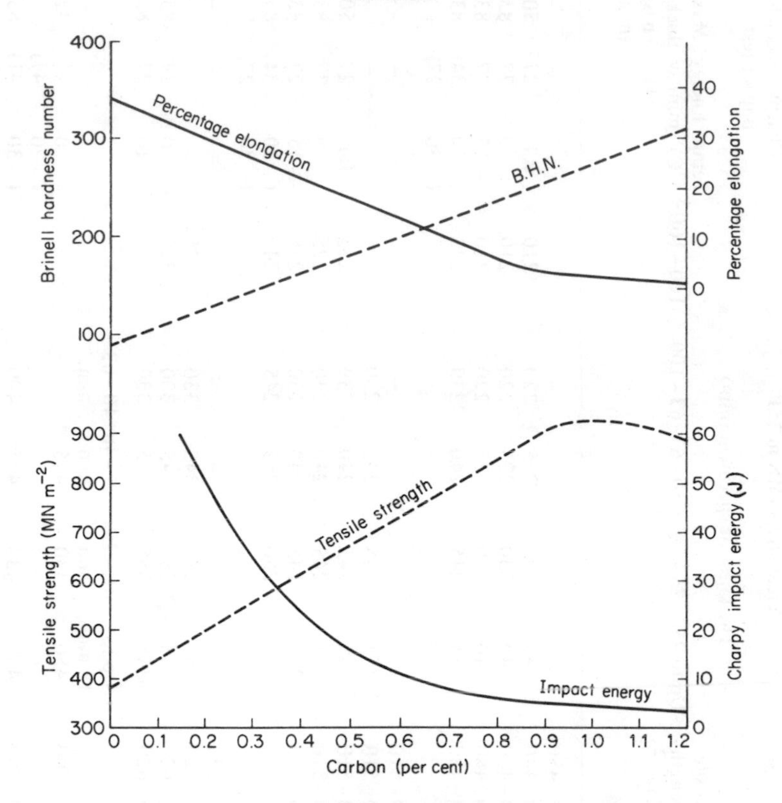

Figure 5.1 *Effect of carbon content on mechanical properties of normalised plain carbon steels*

Structural Steels

The design of steel structures is based primarily on the yield stress or proof stress of the steel but ductility, toughness (notch ductility) at normal and sub-zero temperatures, and weldability are often important properties. Weldability, which for steels deteriorates with increasing carbon content (see section 5.4) is important because welded structures give a weight saving and ease of fabrication compared with bolted or riveted structures. The relative importance of these properties varies between applications and to satisfy the variation a range of weldable structural steels is detailed in BS 4360. The specification gives four tensile ranges, which are subgraded to satisfy the various yield strength and notch ductility requirements. Table 5.1 is a summary of BS 4360 for rods and square bars.

Plain carbon steels

It is clear from the above discussion and section 5.4 that plain carbon steel has good ductility and weldability if its carbon content is low and a high yield strength

TABLE 5.1

Summary of BS 4360 for round and square bars

Grade	C max.	Si	Mn max.	V/Nb max.	S max.	P max.	Tensile strength (MN m^{-2}) : N/mm^2	Yield stress (MN m^{-2}) Thickness of specimen (mm) Up to 25	25 – 50	50 – 63	63 – 100	100 – 160[3]	Charpy V-notch impact test Temp (°C)	Energy min. av. (J)	Max. thickness (mm)
40A[1]	0.22	0.5	1.6	–	0.05	0.05	400 – 480	–	–	–	–	–	–	–	–
40B[1]	0.20	0.5	1.5	–	0.05	0.05	400 – 480	240	230	225	220	210	RT	27	50.0
40C	0.18	0.5	1.5	–	0.05	0.05	400 – 480	240	230	225	220	210	0	27	83.0
40D	0.18	0.5	1.5	–	0.05	0.05	400 – 480	240	230	225	220	210	– 15	27	83.0
40E	0.16	0.1 – 0.5	1.5	–	0.04	0.04	400 – 480	255	245	240	230	225	(– 20	34)	83.0
													(– 30	27)	
43A1[1]	0.25	0.5	1.6	–	0.05	0.05	430 – 510	–	–	–	–	–	–	–	–
43A[1]	0.25	0.5	1.6	–	0.05	0.05	430 – 510	255	245	240	230	–	–	–	–
43B[1]	0.22	0.5	1.5	–	0.05	0.05	430 – 510	255	245	240	230	225	RT	27	50.0
43C	0.18	0.5	1.5	–	0.05	0.05	430 – 510	255	245	240	230	225	0	27	83.0
43D	0.18	0.5	1.5	–	0.04	0.04	430 – 510	255	245	240	230	225	– 15	27	83.0
43E	0.16	0.1 – 0.5	1.5	–	0.04	0.04	430 – 510	270	260	255	245	240	(– 20	34)	83.0
													(– 30	27)	
50A[1]	0.23	0.5	1.6	–	0.05	0.05	490 – 620	–	–	–	–	–	–	–	–
50B[1]	0.20	/0.5	1.5	0.10	0.05	0.05	490 – 620	355	355	345	330	325	–	–	–
50C	0.20	/0.5	1.5	0.10	0.05	0.05	490 – 620	355	355	345	330	325	0	27	83.0
50D	0.18	0.5[1]	1.5	0.10	0.04	0.04	490 – 620	355	355	345	330	[3]	– 10	27	83.0
								Up to 16 mm	16 – 25 mm	25 – 40 mm	40 – 63 mm				
55C[1]	0.22	/0.5	1.6	0.20[2]	0.04	0.04	550 – 700	450	430	415	–		0	27	32.0
													(– 20	47)	
55E[1]	0.22	0.1 – 0.6	1.6	0.20[2]	0.04	0.04	550 – 700	450	430	415	400		(– 30	41)	63.0
													(– 50	27)	

[1] Cu can be included by request. [2] 0.10% if Nb. [3] Minimum yield stress for bars > 160 mm negotiable and for grade 50D over 100 mm.

ie copper for increased corrosion resistance

if its carbon content is high. Consequently the carbon content is limited to 0.25 per cent maximum in the basic structural steels to give a compromise between the opposing requirements. In table 5.1 steels 40A and 43A are of this type and steel 43A is placed in a higher tensile range than steel 40A owing to its higher carbon content. These steels also contain the elements manganese, silicon, sulphur and phosphorus which arise and are controlled in the steelmaking process. Manganese improves strength and notch ductility and between 0.5 and 1.0 per cent is normally present. Silicon improves the strength but if present in excessive amounts may cause the carbon to occur as graphite flakes which reduce the strength (see section 3.4) and so the silicon content rarely exceeds 0.6 per cent. Sulphur and phosphorus embrittle the steel and are controlled to 0.06 per cent maximum. Although still widely used these steels have the following deficiencies

(1) poor notch ductility since they become brittle at temperatures only slightly below room temperature;
(2) low yield strength;
(3) poor atmospheric corrosion resistance.

Good notch ductility ensures that catastrophic failure from points of stress concentration is less likely to occur. Materials, transport and erection costs are reduced if the yield strength is high since thinner sections can be used, provided the ductility permits the production of the thin sections. Maintenance costs are reduced by good corrosion resistance. To achieve these benefits modified mild steels have been and are continuing to be developed.

Notch ductile steels

Improvements in notch ductility are achieved with the realisation that the ductile-to-brittle impact transition temperature is lowered by a reduction in the carbon content or an increase in the manganese content, or both, of low carbon steels. The result is the 'notch ductile' mild steels with the manganese content raised from 0.5/1.0 per cent to 1.5 per cent and with a progressive reduction of the carbon content to lower the impact transition temperature. The effect is illustrated by steels 40B, 40C, 40D in the first tensile range and steels 43B, 43C, 43D in the second tensile range. In each of these series the temperature at which a Charpy V-notch value of 27 J is achieved is lowered from room temperature to −15 °C. The impact transition temperature is further lowered by a reduction of the ferrite grain size. This may be achieved by treatment of the molten steel with aluminium, which combines with nitrogen in the steel to form aluminium nitride particles. The particles associate with the ferrite grain boundaries and prevent grain growth during processing of the steel. Aluminium-treated steels 40E and 43E complete the two series discussed above and give a Charpy V-notch value of 27 J at −30 °C.

Grade 50, 55 steels

The strongest and toughest structural steels are developed by controlling the factors that significantly influence the strength and notch ductility of low carbon steels. Within this very limited carbon range

(1) an increase in the pearlite percentage gives an increase in tensile strength, a rise in the impact transition temperature and no significant effect on the yield strength;

(2) a decrease in the ferrite grain size gives an increase in both the tensile strength and the yield strength and the impact transition temperature is lowered.

Thus the desired combination of high yield strength and low impact transition temperature is best achieved by a reduction in both the pearlite percentage and the ferrite grain size. The pearlite may be reduced by a decrease in the carbon content, and adequate grain refinement results from small additions of niobium or vanadium or both. These elements also give benefits in the steelmaking process that are not obtained with an aluminium addition. During processing of the steels the niobium and vanadium occur partly as insoluble carbide and nitride particles which inhibit grain growth of the ferrite, and partly in supersaturated solid solution so that precipitation strengthening can be induced to supplement the effect of grain refinement. Steels 50B, 50C and 50D in tensile range 3 shcw that the yield strength is increased by approximately 100 MN m^{-2} to 325/355 MN m^{-2}, with a Charpy V-notch impact value of 27 J at temperatures down to -10 °C, depending on the carbon content. The best combination of properties, a yield strength of 400/450 MN m^{-2} and a Charpy V-notch impact value of 27 J down to -50 °C, is produced in steels 55C and 55E by a combination of the niobium/vanadium and aluminium treatments.

The properties in table 5.1 are for the rolled or normalised conditions which are the conditions specified in BS 4360 and those normally supplied. If requested a controlled rolled condition can be supplied which requires 30 per cent of the reduction to be given at a temperature just above A_3. At this temperature recrystallisation of austenite is very slow and a distorted austenite grain structure results which transforms to very fine ferrite and pearlite during subsequent cooling. The properties obtained are superior to those produced by conventional rolling. The treatment is applied readily to thin plates but is more difficult with thicker sections, which require a smaller total reduction, and with complex sections. owing to the difficulty of ensuring uniform reduction throughout the section. The latter can result in varying grain sizes and properties in, for example, the web and flange of a beam.

Corrosion protection

The atmospheric corrosion resistance of steel structures may be enhanced in several ways (BS 5493) and a summary of these is as follows.

(1) Protective coatings of paint, aluminium and zinc may be applied. The metal coatings are more resistant to corrosion than steel in all but the most severe environments and are more abrasion-resistant than painted coatings.

(2) Cathodic protection may be applied to parts of structures which are continuously immersed in water but this method is not effective when only

part of the structure is immersed, for example, in a tidal situation.

(3) Steel grades marked [1] in table 5.1 can be supplied with copper content of 0.2/0.5 per cent. They have improved resistance to atmospheric corrosion but still need to be protected.

(4) A number of weathering steels are covered in BS 4360: these are summarised in table 5.2. The steels contain an increased amount of phosphorus and some chromium and copper, compared with normal steels. Exposure to the weather for about two years causes the formation of an adherent protective oxide film with a pleasing purplish-copper colour instead of the normal flaky rust. To ensure uniform weathering the sections are shot-blasted at works. The cost of weathering steels is about 20 per cent greater than for normal steels but is offset by savings in weight, protective treatment and maintenance. Fire regulations, which require main structural members to be encased in concrete to delay softening, have entailed modifications to the design of high-rise buildings in weathering steels. In one building in the United States this has involved the use, outside the curtain walling, of welded hollow box sections filled with water containing antifreeze.

Selection of structural steel

Sage (1970) in a review of the metallurgy, properties and applications of high-duty low-alloy steels suggests that selection for a specific application is determined by the following factors

(1) strength level required;
(2) other mechanical properties required together with the strength;
(3) steelmaking, heat treatment and other plant available;
(4) arbitrary local conditions such as are imposed by specifications and codes of practice.

These factors, with the exception of factor (3), have been discussed previously and the significance of factor (3) is that a steelmaker can supply only the grades of steel within the capability of his plant. He may not have steelmaking facilities to produce a particular grade, the required heat-treatment capacity or a rolling mill capable of giving a controlled rolling treatment. However, he may be able to supply a suitable alternative grade.

Typical applications of structural steels are in bridges, high-rise buildings, spectator stands, galvanised electricity power-supply pylons and welded pipelines.

Buried steel pipes may suffer corrosion, depending on the nature of the ground and its effect on the access of oxygen to the pipes. With aerobic conditions the oxygen supply is plentiful because good drainage in sand or limestone ensures a changing supply of water. Thus corrosion may not be severe after a layer of rust has formed. With anaerobic conditions, for example, waterlogged clay, the oxygen supply is poor and bacteria can cause the chemical reduction of sulphates and severe corrosion produces a mixture of rust and black iron sulphide. This may be prevented by surrounding the pipeline with chalk or limestone to give aerobic conditions. Protection may still be necessary and can be provided by coatings of glass-reinforced bitumen or coal tar, cement or reinforced concrete, synthetic resins,

TABLE 5.2

Summary of BS 4360 for weathering steels for plates and flats

Grade	Chemical composition (ladle analysis) (%) C	Si	Mn	S Max	P	Cr	Cu	V	Tensile strength (MN m^{-2})	Yield stress (MN m^{-2}) Thickness of specimen (mm) Up to 12	12–25	25–40	40–50	Charpy V-notch impact test Temp (°C)	Energy min. av. (J)	Max. thickness (mm)
WR50A1	0.12 max	0.25–0.75	0.60 max	0.05	0.07–0.15	0.30–1.25	0.25–0.55		480	345	325	325		–	–	–
WR50A										345				0	27	12
WR50B1	0.10–0.19	0.15–0.50	0.90–1.25	0.05	0.04 max	0.40–0.70	0.25–0.40	0.02–0.10	480	345	345	345	340	–	–	–
WR50B														0	27	50
WR50C1	0.10–0.19	0.15–0.50	0.90–1.25	0.05	0.04 max	0.04–0.70	0.25–0.40	0.02–0.10	480	345	345	345	340[1]	-15	27	50
WR50C	0.10–0.22	0.15–0.50	0.90–1.45	0.05	0.04 max	0.40–0.70	0.25–0.40	0.02–0.10								

[1] WR50C only: up to 63 mm. Minimum yield stress for sections> 63 mm negotiable.

petroleum and coal tar pitch wrapping tapes and polymer wrapping materials. Corrosion at breaks in the coatings is prevented by cathodic protection (BS 5943).

Concrete Reinforcement

Concrete has low tensile and bending strengths and a high compressive strength. Steel reinforcement overcomes the deficiencies in the tensile and bending strengths.

The reinforcing steel must have adequate tensile properties and form a strong bond with the concrete since the concrete transmits load to the steel by shearing stresses. The bond is purely mechanical and arises from surface roughness and friction. Mild steel with a maximum carbon content of 0.25/0.40 per cent is suitable and is supplied in three conditions. These are hot rolled (BS 4449), cold rolled (BS 4461) and hard drawn (BS 4482) which give yield strengths between 250 and 485 MN m^{-2}. The tensile strength of the hot-rolled grade must be 15 per cent greater than the yield strength and for the other grades the requirement is a 10 per cent difference. Reinforcing steels are supplied as plain, indented or twisted round or square bars, in a variety of sectional shapes in straight lengths or bent shapes and as woven or electrically welded mesh. Protection against corrosion is provided by the highly alkaline environment of the Portland cement hydrates within the concrete. Carbonation, that is, the reaction of the hydrates with carbon dioxide can, however, break down this protection if it penetrates as far as the steel.

Prestressing Steels

With pretensioned concrete the strained steel members attempt to shorten when the straining device is removed. The shortening is resisted by the concrete and a condition is established of compressive stress in the concrete balanced by tensile stress in the steel. With post-tensioning the same condition is effected by threading steel members through ducts in the concrete after which the steel is elastically strained and anchored to the concrete. When the composite is stressed in service the compressive stress in the concrete must be overcome before the concrete is subjected to a damaging tensile stress.

Prestressing steels must have a high yield strength in tension so that a high elastic strain can be induced in them. A stress approaching the yield strength must be supported at an elongation of 1.5/2.0 per cent without the steel suffering creep relaxation. In addition pretensioning steels must form a good bond with the concrete.

0.6/0.9 per cent carbon, 0.5/0.9 per cent manganese steel (BS 2691) is suitable for pretensioning. It has better tensile properties than mild steel owing to the higher carbon content (figure 5.1) and can be further enhanced by cold working to give a specified characteristic strength of 1470/1720 MN m^{-2} and a minimum 0.2 per cent proof stress of 1250/1460 MN m^{-2}. High and low relaxation grades are produced which are required to give maximum relaxations of 8.5 and 3.0 per cent respectively for an initial stress equal to 70 per cent of the characteristic strength.

A number of steels are used for post-tensioning. BS 3617 covers stranded wire ropes manufactured from cold drawn 0.6/0.9 per cent carbon, 0.5/0.9 per cent manganese steel wire with a tensile strength of 1735/1850 MN m^{-2} and 0.1 per cent proof stress greater than 70 per cent of the tensile strength. Cold-worked rods with a tensile strength of 1030/1110 MN m^{-2} and a 0.2 per cent proof stress of 870/950 MN m^{-2} are produced in 0.5/0.6 per cent carbon, 0.7/1.0 per cent manganese, 1.5/2.0 per cent silicon steel, 250 A53 (BS 970 : Part 5). When exceptional corrosion resistance is essential stainless steel is specified. This was considered necessary for the post-tensioning bars of the concrete foundations of the York Minster structural restoration. Dowrick and Beckmann (1971) report that a precipitation hardening stainless steel, Firth Vickers FV 520 with a typical composition of 0.07 per cent carbon, 16 per cent chromium, 6 per cent nickel, 1.5 per cent copper, 1.5 per cent molybdenum, 0.3 per cent titanium, was chosen. Following ageing at 550 °C for 2 hours a minimum 0.1 per cent proof stress of 800 MN m^{-2} was obtained in the 32 mm diameter bars used for the post-tensioning.

Bridge Wire

The mechanical properties required for suspension-bridge cables are high strength, toughness and fatigue resistance to withstand the fluctuations of stress caused by loading of the bridge and climatic effects. Corrosion resistance is also important. Suitable cables are produced from heavily galvanised, cold-drawn 0.75/0.85 per cent carbon, 0.5/0.7 per cent manganese steel with a minimum tensile strength of 1600 MN m^{-2}.

Cladding Steels

Steels are gaining favour for external walls and curtain walls, and for covering internal and external structural columns. The cladding is not a stressed part of the structure but despite this it must be strong and rigid so that cost and weight can be reduced by the use of thin sheet. The steel selected must have ductility to permit forming of the thin sheet, corrosion resistance and a pleasing appearance. These requirements are satisfied by stainless steels and mild steel if it is protected against corrosion.

Two austenitic stainless steels containing 18 per cent chromium, 10 per cent nickel with or without 3 per cent molybdenum, and a 17 per cent chromium ferritic steel are normally specified. For external use additional corrosion resistance is required and the molybdenum-containing austenitic steel is essential. Protective treatment is not required and this partly offsets the high material cost. A range of attractive finishes from a dull matt to a bright polish are produced. The mild steel sheet is protected with coatings of vitreous enamel or PVC. Vitreous enamel is fused at 800 °C on to both sides of the 1.0/1.6 mm sheet to give complete encasement in a chemically inert glass. A wide range of durable colours may be incorporated in the enamel, singly or as designs and murals. The enamel is hard and exposure of the steel by abrasion is unlikely to occur. Galvanising of the steel prior to coating with PVC is essential for corrosion resistance because PVC coatings have poor scratch resistance and tend to be porous. The coatings may be applied at the works by painting or lamination and on site by painting. Treatment at the works gives superior coatings at a smaller over-all cost. The colours available are limited because of the need to ensure good fire and weathering properties.

High Tensile Bolt Steels

Bolts should have adequate tensile strength and toughness and should not slacken owing to relaxation when used at elevated temperatures. Fatigue strength is required when loading is likely to be variable and cases have been brought to the authors' attention of the importance of corrosion fatigue resistance when bolts are used in structures subjected to varying stresses by tidal waters. A variety of steels are suitable and typical compositions and properties are summarised in table 5.3. The toughest condition is produced in the steels by oil-quenching from 820/880 °C and tempering at 550/700 °C but the strength is low. Tempering at lower temperatures down to a specified minimum of 430 °C may be used to give high strengths with poorer toughness. Owing to the effect of hardenability (see section 3.3) the properties obtained depend on the composition of the steel and the size of the section treated. In general the higher the alloy content the greater the hardenability and the larger the section that can be effectively heat-treated or the better the strength that can be produced in a section of a given size. Thus BS 970 : Part 2 specifies properties for a range of limiting ruling sections and the extremities of the ranges for each steel are included in table 5.3. By choosing the correct steel composition and heat treatment it is possible to develop the required properties in bolts of appropriate size.

Railway Steels

Principles discussed earlier can be applied to the selection of steels for railway structures, for example, railway bridges and signal gantries are constructed in structural steels and galvanised wire is used for signal operation. Rails can be usefully considered in more detail. They need to be strong and hard to resist the wearing of the wheels. Fatigue strength is required because the dynamic part of the load borne by the rails is high and fatigue is a common cause of failure. Fatigue cracks are initiated at the bolt hole nearest to the rail end in bolted rails and on the running surface of welded rails. The cracks in welded rails develop when there is temporary loss of adhesion between a driving wheel and the rail. This causes local heating and the subsequent rapid cooling causes the formation of martensite in which cracks are initiated by the stresses imposed by the following wheels. Both types of crack propagate slowly under the cyclic loading to give fatigue failure. The required strength and hardness are provided by hot-rolled 0.45/0.6 per cent carbon, 0.95/1.25 per cent manganese normal grade steel or the alternative weather-resistant grades with increased carbon and/or manganese content (BS 11). Periodic tests are made for fatigue cracks and when they are found the rail is replaced.

TABLE 5.3

High tensile bolt steels

Steel type	Chemical composition (%) C	Mn	Ni	Cr	Mo	Limiting ruling section (mm)	Minimum tensile strength (MN m^{-2})	Minimum yield stress (MN m^{-2})	Minimum Izod impact value (J)	Minimum elongation (%)
Medium C	0.3/0.45	0.7/1.0				63.5	617.8	432.4	33.4	22
						19.0	695.0	494.2	27.1	20
*530 M40	0.3/0.45	0.6/0.95		0.85/1.15		100	695.0	565.1	54.2	22
						28.5	849.5	679.6	54.2	18
*640 M40	0.3/0.4	0.6/0.9	1.0/1.5	0.45/0.75		150	695.0	535.1	54.2	22
						28.5	926.1	741.3	47.5	17
*709 M40	0.35/0.45	0.5/0.8		0.9/1.5	0.2/0.4	150	695.0	535.1	54.2	22
						25	1235.6	1050.2	13.5	10
*816 M40	0.35/0.45	0.4/0.8	1.2/1.6	0.9/1.4	0.1/0.2	150	772.2	586.9	54.2	20
						28.5	1091.1	895.8	40.7	15

*BS 970: Part 2: 1970

Steel Forgings

The applications considered above involve the use of wrought products with uniform cross-sections produced by rolling or drawing. Such sections make up the bulk of steel usage in civil engineering but complex shapes are also required. These may be produced by forging or casting. Forging involves hot shaping between dies and the deformation sequence is chosen so that adequate deformation is given to all parts of the component and the metal flow direction is controlled to give the optimum fibre structure and properties. In appropriate cases further enhancement of the properties is achieved by heat treatment. A limitation on the shape of forgings arises from the need to remove them from the dies. Plain carbon steels are adequate for most civil engineering forgings and BS 29 covers recommended forging steels in the compositional range 0.6 per cent carbon maximum, 0.3/1.3 per cent manganese and tensile strength range 485/695 MN m^{-2}. Alloy forging steels with tensile strengths in the range 600/1250 MN m^{-2} are covered in BS 4670.

5.2 Steel and Iron Castings

Casting may provide an alternative to forging for the production of complex shapes. A casting technique is likely to be selected when the number of components is so small that the cost of forging dies is not warranted, the shape is too complex for forging or components with adequate properties can be produced more cheaply. In general, forgings have better properties than castings because the properties of castings are determined by the solidification characteristics of the casting alloys and any heat treatment that may be possible. Thus castings may show gas and shrinkage porosity and compositional segregation because these cannot be removed by working, as is the case with forgings. When a casting alloy is selected, castability must be considered as well as mechanical properties and in this respect consultation with the founder could be beneficial. He will give advice concerning the best design, casting technique and alloy which will help to ensure that castings with optimum properties, soundness, surface finish and machinability are produced at minimum cost.

Cast Steels

BS 3100 covers seventeen grades of steels for general engineering castings. Many of these are unlikely to be of interest to civil engineers since they are for service at elevated temperatures, in severe corrosive environments and where special magnetic properties are required. The remaining grades, BS 3100, give a range of compositions and properties corresponding to those of the plain carbon and alloy forging steels.

Hadfields manganese steel castings, BS 3100, have exceptional resistance to impact and abrasion of the type experienced at rail crossovers. The steel contains 1.0/1.25 per cent carbon, 11 per cent manganese and water-quenching from above 1000 °C gives an austenitic structure because the high manganese content causes the pearlite and martensite transformations to be suppressed. The austenite has excellent toughness and impact loading during service heavily work hardens the surface so that it is very wear resistant.

Cast Irons

The effect of graphite flakes in ordinary grey irons, mentioned in section 3.4, limits their tensile strength to 120/175 MN m^{-2} together with almost zero impact and percentage elongation values. Despite this they are applied where the strength and toughness requirements are not stringent but where advantage can be taken of their beneficial properties. These are good compressive strength, castability and machinability, better corrosion resistance than mild steel and reasonable fatigue strength due to effective damping of vibration by the mixed graphite - metal structure. They are covered in BS 1452 together with improved grades which give tensile strengths up to 425 MN m^{-2}. Typical applications are spun and vertically cast pipes and fittings, manhole covers and frames, gulley drains, rainwater wear, components of bridge bearings and screw piles.

Cast irons for use where strength and toughness are required have their graphite form modified (see section 3.4). Malleable cast irons with tensile strengths of 290/410 MN m^{-2}, 0.5 per cent proof stresses of 170/250 MN m^{-2} and impact values of 9/18 J are suitable for applications as pipes, components of conveyor systems, insulation caps and cable supports in electric power systems and guard rails. The properties of spheroidal graphite cast irons vary depending on whether they are used cast or annealed and fall within the following ranges, tensile strength 415/725 MN m^{-2}, yield strength 310/540 MN m^{-2}, impact value 7.5/16 J. Typical applications are pipes and fittings, nuts and bolts for buried pipes to avoid galvanic corrosion, bridge-bearing blocks, manhole covers and frames and comminutor drums for sewage.

5.3 Nonferrous Metals

Because the initial cost of nonferrous metals is generally greater than that of ordinary ferrous alloys they are only selected when utilisation of their special properties reduces the cost difference or the properties are essential to the application. Cost difference may be reduced by the utilisation of superior working properties, low specific gravity or corrosion resistance, an example being the competition of aluminium alloys with structural steels discussed below. Essential properties may be electrical conductivity or corrosion resistance, illustrated by the electrical applications of copper.

Civil engineers may be concerned with the erection of the shells of large buildings but infrequently with the fitting-out stage. Although the latter is considered to be building and outside the scope of this discussion, nevertheless brief references to metallic building materials are included.

TABLE 5.4

Summary of BS 1470 for wrought aluminium sheet materials

Material [1]	Composition (%) Al	Cu	Mg	Si	Mn	Selected conditions [2]	0.2% proof stress min. N mm^{-2} [3]	Tensile strength N mm^{-2} [3] Min.	Max.
S1	99.99					O	–	–	65
						H8	–	100	–
S1A	99.8					O	–	–	90
						H8	–	125	–
S1B	99.5					O	–	55	95
						H8	–	135	–
S1C	99.0					O	–	70	105
						H8	–	140	–
NS3					0.8 – 1.5	O	–	90	130
						H8	–	175	–
NS4			1.7 – 2.4			O	60	160	200
						H6	175	225	275
NS5			3.1 – 3.9			O	85	215	275
						H4	225	275	325
NS8			4.0 – 4.9		0.5 – 1.0	O	125	275	350
						H4	270	345	405
HS15		3.9 – 5.0	0.2 – 0.8	0.5 – 0.9	0.4 – 1.2	TB	245	385	–
						TF	345	420	440
HS30			0.5 – 1.2	0.7 – 1.3	0.4 – 1.0	TB	115	200	–
						TF	240	295	–

[1] N non-heat-treatable; H heat-treatable; S plate, sheet and strip; other wrought forms are denoted T drawn tube, F forging stock and forgings, R rivet, B bolt and screw stock, E bars, extruded round tube and sections, G wire.

[2] O annealed, H1 to H8 work hardened and partially annealed in increasing order of tensile strength; TB solution-treated and naturally aged; TF solution-treated and precipitation-treated.

[3] N mm^{-2} = MN m^{-2}.

Aluminium

The useful engineering properties of both unalloyed and alloyed aluminium are low specific gravity, resistance to corrosion, high electrical conductivity and excellent forming properties. The low strength of aluminium, 65/105 MN m^{-2}, is a disadvantage and for satisfactory service it must be supported or alloyed and mechanically or thermally treated to give improved strength.

Wrought materials

Aluminium and its alloys are available in many wrought forms which are covered by BS 1470. Table 5.4 gives a summary of the compositions and properties of these materials in the form of sheet and for selected conditions.

Super-purity 99.99 per cent aluminium is too costly for general engineering applications and commercial grades, which vary in purity from 99.0 to 99.8 per cent aluminium, are selected on the requirements of corrosion resistance, formability or tensile strength. Corrosion resistance and formability are enhanced and tensile strength is lowered by increased purity. Strengthening may be achieved by cold working and softening by annealing at various temperatures to produce conditions varying from fully work hardened to fully softened.

Alloy additions are chosen so that the strength can be enhanced without an adverse effect on the specific gravity or corrosion resistance. The alloying elements commonly added to aluminium are copper, manganese, magnesium, silicon, nickel and iron. Part of the addition forms a solid solution with the aluminium and part forms a compound with the aluminium or with one of the other alloying elements. The corrosion resistance of an alloy is good if the solid solution and associated compound have similar chemical properties. Alloys containing copper, nickel and iron have poorer corrosion resistance than those with magnesium, silicon and manganese. Wrought alloys may be capable of strengthening either by precipitation hardening or by cold working and annealing as for aluminium.

Two precipitation-hardening alloys are included in table 5.4; HS15 is of the copper - magnesium type and HS30 of the magnesium - silicon type. They are solution-treated at 490 – 500 °C, water-quenched and aged at room temperature or precipitation-treated at 100 - 150 °C. The latter gives higher yield and tensile strengths, for example, HS15 has a tensile strength of 430 MN m^{-2} compared with 385 MN m^{-2} for natural ageing and the yield strengths are 360 MN m^{-2} and 245 MN m^{-2} respectively. HS15 is stronger than HS30 but has poorer corrosion resistance because it contains copper.

The non-heat-treatable alloys N3, 4, 5 and 8 have single or mixed additions of manganese and magnesium and consequently they have good corrosion resistance. Strengths equivalent to those of the heat-treatable alloys are possible if they are heavily cold worked.

TABLE 5.5

Summary of BS 1490 for selected aluminium casting alloys

Designation LM	Composition (%) Cu	Mg	Si	Mn	Selected conditions [1]	Tensile strength min. Sand cast $N\ mm^{-2}$ [2]	Chill cast $N\ mm^{-2}$ [2]
2	0.7 – 2.5		9.0 – 11.5		M	–	150
4	2.0 – 4.0		4.0 – 6.0	0.2 – 0.6	M	140	160
					TF	230	280
5		3.0 – 6.0		0.3 – 0.7	M	140	170
6			10.0 – 13.0		M	160	190
25		0.2 – 0.45	6.5 – 7.5		M	130	160
					TF	230	280

[1] Notation for condition of alloys: M as cast; TF fully heat-treated = solution heat-treated and precipitation-treated.

[2] $N\ mm^{-2} = MN\ m^{-2}$

Casting alloys

Aluminium casting alloys are covered by BS 1490 and table 5.5 gives details of selected alloys from the twenty-one specified.

Commercial aluminium is difficult to cast owing to a large contraction during solidification and a tendency to tear when hot. Hence the elements added to casting alloys are chosen to give improvements in castability as well as strength. Silicon, copper and magnesium are beneficial in these respects. Many of the specified alloys appear to be very similar and this may be so with respect to mechanical properties but the slight differences in composition give differences in castability. Thus a large number of alloys are available to meet the varying requirements of castability and service properties. Casting alloys are divided into non-heat-treatable types and those capable of precipitation hardening.

LM25 is the most popular heat-treatable casting alloy. When fully heat-treated its tensile strength of 230/280 MN m^{-2} is adequate for most requirements and, because the alloying elements are magnesium and silicon, it has good corrosion resistance. The copper–silicon heat-treatable alloy LM4 has similar strength to LM25 and is suitable for general engineering castings requiring good castability and high strength but not corrosion resistance.

The remaining alloys listed in table 5.5 are non-heat-treatable. However, the 10/13 per cent silicon alloy LM6 can be strengthened by 'modification'. When cast normally a coarse eutectic structure of brittle silicon plates in aluminium-rich solid solution is obtained and the tensile strength is only 125 MN m^{-2}. If the alloy is chill cast and 'modified' by the addition of 0.1 per cent of sodium to the melt the eutectic structure is considerably refined and the tensile strength raised to 190 MN m^{-2}.

Applications

Wrought aluminium alloys are alternatives to steels for bridges, roof structures of buildings and reservoirs, pedestrian underpasses, etc. The decision to use aluminium alloys is usually made on the grounds of economy because the higher material cost can be offset by lower erection costs resulting from their lightness, and lower maintenance costs resulting from their corrosion resistance. There are also savings in power with movable bridges. Aluminium has a specific gravity approximately equal to one-third of that for steel and weight savings of the same order would be expected. This figure is not quite achieved in practice because

(1) aluminium has an elastic modulus about one-third of that for steel so that deeper aluminium sections are required to compensate for the greater deflection; and

(2) the welding of aluminium alloy structures is difficult (see section 5.4) and the structures are generally bolted or riveted.

Weight savings vary from 50 per cent for small structures to 65 per cent for larger structures.

Fully heat-treated aluminium, magnesium, silicon, manganese alloy H30 is the most frequently used loadbearing alloy although aluminium, copper, magnesium, silicon, manganese alloy H15 has been used. Owing to poorer corrosion resistance the latter alloy must be protected when used in severely polluted industrial atmospheres. The protection involves painting with zinc chromate primer followed by aluminium paint. Recommended bolt materials are fully heat-treated H30, fully heat-treated and anodised H15, galvanised mild steel and 0.16 per cent carbon, 11/14 per cent nickel, 11/14 per cent chromium stainless steel. Rivet materials are H30, H15, annealed aluminium, magnesium alloy N5 and steel (CP 118). The heat-treatable aluminium alloy rivets are driven in the solution-treated condition and age naturally.

Fully supported roofing sheet and flashing for which strength is unnecessary but for which corrosion resistance is important are produced in the commercial grades of aluminium. Stronger aluminium - manganese alloy N3 is specified for troughed and corrugated sheets and the aluminium, magnesium, silicon, manganese alloy H30 for sliding roof sections. Accessories with complex shapes are cast, for example, condensation under roofing sheet is prevented by ventilating cowls cast in aluminium, silicon alloy LM6M.

Anodised aluminium, manganese alloy N3 is widely used for curtain walling. A range of natural and dyed colours is available and in industrial areas these should be protected by a coating of acrylic polymer.

The resistance of aluminium alloys to dust, dirt and moisture makes them very suitable for road signs and railings on bridges and the approach roads to motorways. Rails and posts and the supports of road signs have been produced in extruded aluminium, magnesium, silicon, manganese alloy H30 and road-sign blanks in sheet of the same alloy. In the United States an aluminium, copper, magnesium, silicon alloy similar to H15 has been used for crash barriers. The alloy is clad with aluminium to give corrosion resistance and is supported on steel or timber posts.

The high electrical conductivity of aluminium is utilised in steel-cored cables for electrical power transmission and railway electrification. The steel core provides the necessary tensile, creep and fatigue strengths for long spans.

Aluminium is prominent in buildings, being used extensively in the natural and anodised conditions for external decoration. Wrought alloys have a variety of indoor applications notably for lift doors, acoustic or radiant ceilings, balustrades of moving stairways, carpet tread plates, frames of display cabinets and polished and anodised reflectors of space heaters. Castings provide matching components such as moving stairway treads in aluminium, silicon, copper alloy LM2 and door handles, shelf brackets and finger plates in aluminium, magnesium, manganese alloy LM5.

Copper

Copper is of interest to engineers because of its ease of working, high electrical conductivity and resistance to corrosive attack by the atmosphere and mildly corrosive environments.

Commercial grades of copper

Eleven grades of copper are covered by British Standards, many of which are for special purposes and not of interest to civil engineers. The important grades are as follows.

(1) Tough pitch copper (BS 1037 and 1038), which contains 0.02/0.04 per cent oxygen from the smelting process, occurring partly in solid solution and partly as small particles of copper oxide. The presence of the copper oxide makes it unsuitable for flame-welding or brazing since with heating above 400 °C hydrogen enters the copper from the atmosphere and reacts with the copper oxide to give water vapour. This collects in the copper and regions of high internal pressure develop which eventually lead to blistering. Tough pitch copper is suitable for electrical purposes.

(2) Phosphorus deoxidised copper (BS 1172). The addition of 0.04 per cent of phosphorus to tough pitch copper removes the oxygen and gives a copper which can be joined by flame techniques. Some phosphorus dissolves in the copper and this reduces the conductivity to such an extent that it is unsuitable for electrical applications.

The mechanical properties of both grades are similar. In the annealed condition copper is weak, with tensile strength 225 MN m^{-2} and 0.1 per cent proof stress 45 MN m^{-2}, but can be strengthened by cold deformation to give tensile and 0.1 per cent proof strengths of 555 MN m^{-2} and 495 MN m^{-2} respectively.

The electrical applications of copper include hard-drawn wire for power cables, lightning conductors and electrical wiring of buildings. Both grades serve as supported roofing, flashing, architectural trim, damp-proof course, rainwater goods, domestic gas, water and central-heating pipes and cisterns. Deoxidised grades are chosen whenever flame joining is involved. In plumbing systems care is taken to avoid galvanic couplings with, for example, galvanised steel cylinders and contact with acid waters which dissolve the copper and produce a green stain on sanitary equipment. Copper used for roofing and flashing develops a pleasing green patina owing to the formation of a layer of copper salts by reaction with the atmosphere.

Copper alloys

Copper can be alloyed with many metals to produce the engineering alloys covered by BS 2874. These generally retain the good corrosion resistance of copper and are stronger but have poorer conductivity. Brasses and tin bronzes are the only alloys used to any extent in civil engineering and the present discussion is limited to these.

Brasses and bronzes are the result of alloying with zinc and tin respectively. Increasing amounts of both elements give first an α solid solution and then two-phase alloys consisting of α plus a compound, as shown in figure 5.2. The α-alloys are not heat-treatable but can be strengthened by cold working and annealing treatments. The two-phase alloys cannot be cold worked because the compounds are hard. Although they are capable of being hot worked the two-phase bronzes are normally regarded as casting alloys whereas the two-phase brasses are used in

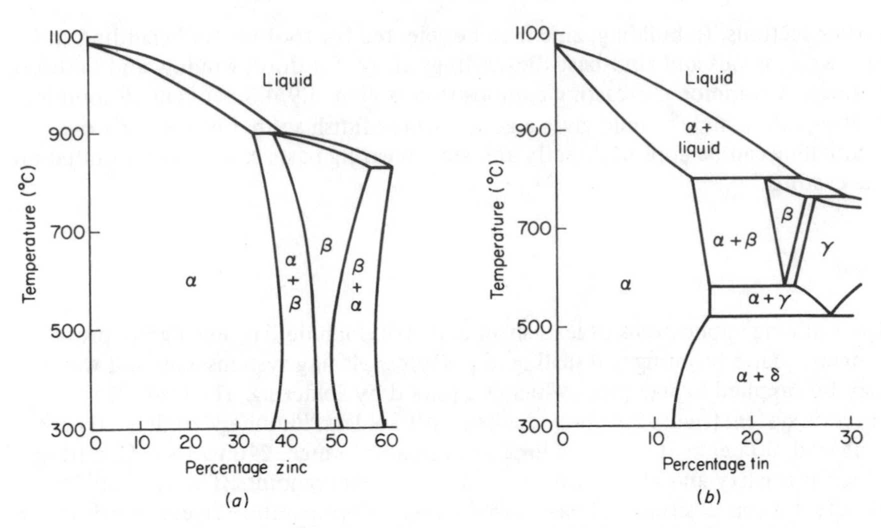

Figure 5.2 *Equilibrium diagrams for (a) copper–zinc alloys (b) copper–tin alloys*

the cast or hot-worked condition, for example, in the case of the 60 per cent copper, 40 per cent zinc, as plumbing fittings.

Toner (1969) describes the application of brasses for the curtain walling of the Polytechnic of Central London engineering and science building. By careful design and correct utilisation of the properties of copper alloys the cost was made competitive with that for alternative materials. A high tensile brass consisting of 60 per cent copper, 40 per cent zinc modified by small additions of tin, iron and aluminium was used for the window sections. The alloy additions give good corrosion resistance and enhance the tensile properties to give a tensile strength of 455/530 MN m^{-2} and a 0.1 per cent proof stress of 215/278 MN m^{-2} so that thin, rigid window sections were possible. A special jointing system was devised to permit easy, rapid brazing without distortion. The excellent working properties of the α-alloy 85 per cent copper, 15 per cent zinc were utilised to give thin sheet for pressed facias and for the outer layers of the asbestos, glass fibre, aluminium foil composite underwindow spandrel panels. Finally the development of a natural patina eliminated the need for a protective treatment and costly maintenance.

The normal composition of the two-phase bronze is copper, 10 per cent tin and if 0.3 per cent of phosphorus is added the castability and strength of the alloy is improved and a phosphor-bronze is produced. Phosphor-bronze castings are suitable for components of the expansion bearings of static bridges and the centre bearings of swing bridges. They have the required compressive and fatigue strengths but require liberal lubrication. In recent years rubber has proved a competitor to bronze for static bridge bearings because it requires less maintenance.

Zinc

The use of zinc in civil engineering is limited to the prevention of corrosion by galvanising and cathodic protection of steel, examples of which are given in earlier sections. In building, zinc may be selected for roofing, wall cladding and rain-water goods and zinc-base die-castings alloys for door, window and bathroom fittings. A common die-casting composition is zinc, 3.9/4.3 per cent aluminium. Casting into a metal mould gives a good surface finish so that plating or stove enamelling can be applied directly and steel wearing parts can be incorporated into the casting.

Lead

The building applications of lead sheet and strip include flashing, damp-proof courses, sound proofing and sealing in puttyless glazing systems. Gas and water may be supplied in lead pipes which are joined by soldering. The lead - tin eutectic system (figure 3.4) provides two soft solders. Plumbers' solder, 70 per cent lead, 30 per cent tin, has a long solidification range, 240 to 183 °C, during which it is pasty and in a suitable condition for wiping joints. Tinmans' solder is of eutectic composition and has a rapid constant temperature freeze which makes it useful for electrical work. Various combinations of lead, bismuth, tin, cadmium and indium give special-purpose solders with melting points down to 47 °C. A relevant application is the fusible plugs in automatic sprinkler systems for fire prevention. The alloy must have a constant melting point so that the plug does not yield at too low a temperature but releases quickly at the dangerous temperature of 74 °C.

5.4 Welding

Metals may be joined by bolting, riveting, adhesion bonding, soldering, brazing and welding, some of which may be further subdivided. Detailed discussion is limited here to the metallurgical principles of fusion-welding because of its importance in the fabrication of structures and because it is the only technique which involves major metallurgical considerations.

Bolting, riveting and adhesion bonding do not alter the metallurgical condition of the joined parts and need not be discussed except to emphasise the growing importance of adhesion bonding in engineering. Joints in aluminium window frames, lamp standards and railings on bridges and motorway approach roads are bonded with organic adhesives, which are compounded from epoxy, vinyl and phenolic resins and nitrile and Neoprene rubbers.

Applications of soldering and brazing are given in earlier paragraphs. As with adhesion bonding a joint is produced without fusion of the joined parts. However, metallurgical changes are produced by the heat used to melt the filler metal. These changes are similar to those discussed below for fusion-welding, but generally less severe because lower temperatures are involved. An arbitrary distinction may be made by using the term brazing for those processes where the filler metal has a fusion temperature greater than 500 °C and soldering for those processes where fusion takes place below 500 °C.

Welding processes may be classified broadly as follows

(1) pressure, such as fire-welding or solid-phase welding which are unsuitable for major structural work;

(2) resistance, such as the spot or seam-welding of sheet materials; and

(3) fusion, in which part of the parent metal is fused with or without filler metal and solidified to form a joint. Fusion processes which involve heating by an electric arc are used to fabricate structures. In each of these the filler electrode composition is chosen to give a joint with satisfactory mechanical and corrosion properties. Protection against oxidation and gas pick-up is achieved either by coating the electrodes with a flux to give a protective slag and atmosphere or by the use of uncoated electrodes and an inert atmosphere such as argon or a protective atmosphere such as carbon dioxide.

Metallurgical Principles of Fusion Welding

During welding a temperature gradient is established which varies from the fusion temperature at the weld metal to room temperature at some point away from the weld. This produces changes in the metallurgical condition and properties of the metal in the vicinity of the joint and consideration of these is facilitated if the affected region is divided into zones as shown in figure 5.3.

The factors that determine the quality of a fusion-welded joint are

(1) the structure and properties of the weld metal;

(2) the structure and properties of the part of the parent metal that undergoes significant metallurgical changes, known as the heat affected zone; and

(3) the cracking tendency.

These may be affected by the metallurgical characteristics of the parent and filler metals and the cooling rate following welding. The latter is high if the parent metal has a high thermal conductivity and a large mass, if the initial temperature of the plate is low and if the total heat input is small.

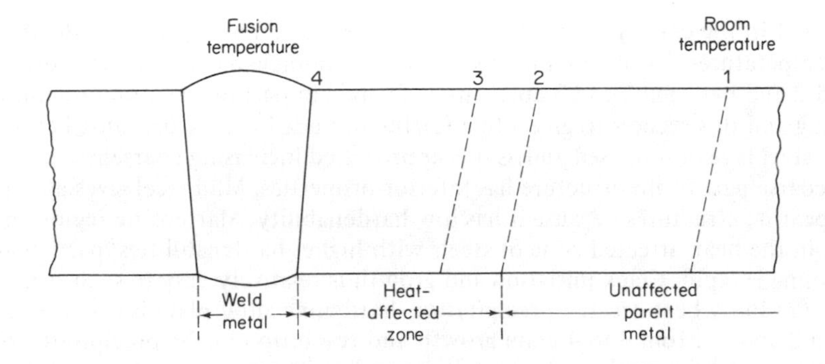

Figure 5.3 *Division of the temperature-affected part of a weld into zones which show different metallurgical effects*

Weld metal

This part of the weld is rapidly cooled from the liquid stage and can show all the features of a cast metal as follows.

(1) Owing to the moving heat source, heating and cooling are rapid and non-equilibrium conditions exist in the microstructure. Grain structures are coarse, single-phase solid solutions show compositional segregation and low carbon steels exhibit typical coarse cast structures. The inferior properties characteristic of these structures are obtained.

(2) Slag inclusions or incompletely fused oxide films may be trapped and these are sources of stress concentration at which brittle fracture or fatigue may be initiated under the appropriate conditions.

(3) Gases may be absorbed into the molten metal from the atmosphere or contaminants such as oil or grease and some may enter the heat-affected zone. Welding is unlikely to be inhibited by a soluble gas but porosity may result owing to evolution of gas as a result of the decrease in solubility which results from solidification. If the gas forms a compound it may be soluble and cause embrittlement or if insoluble form a slag or scale and give the effects described in (2).

(4) The fusion zone is a mixture of the filler and parent metals. If these have different compositions dilution will occur and variable composition is possible if the recommended welding practice with respect to type of joint, edge preparation and welding technique is not followed.

Heat-affected zone

The microstructural changes in the heat-affected zone can be considerable and they are generally accompanied by a deterioration of the mechanical properties. The changes depend on the metal, its original condition and possibly the cooling rate after welding. Those which occur in some of the materials of interest to civil engineers are discussed with reference to figure 5.3.

(1) In the case of steels, 2 and 3 of figure 5.3 can be equated to the A_1 and A_3 temperatures. Normalised mild steel shows an unchanged structure between 1 and 2 and between 2 and 3 austenite is formed in part of the structure during heating and this recools to give a fine ferrite and pearlite mixture. Just beyond 3 the steel is re-normalised and as 4 is approached increasing coarseness occurs. The coarse part of the structure has inferior properties. Mild steel gives all ferrite and pearlite structures because it has low hardenability. Martensitic regions may form in the heat-affected zone of steels with higher hardenabilities, particularly if cooling is rapid. Crack initiation and growth is relatively easy in such regions.

(2) Fully heat-treated precipitation-hardening alloy plate is overaged between 2 and 3. From 3 to 4 grain growth and resolution of the precipitated phase may occur and the final condition will depend on the rate of cooling and the ageing characteristics of the alloy. If cooling is slow the presolution-treated condition may be obtained and if cooling is rapid the alloy may be solution-treated but age during service. However, natural ageing usually occurs too

slowly to be effective and thus inferior properties result.

(3) At appropriate points on the temperature gradient between 2 and 4, cold-worked solid-solution alloy plate suffers loss of strength as a result of recovery, recrystallisation and grain growth.

Cracking tendency

Cracking may be classified as supersolidus or subsolidus. The former is obtained when the metal is partly molten and the latter when it is completely solid. Both types occur only when the metal is brittle and the thermal stresses due to contraction exceed the fracture strength of the metal. The brittleness which leads to supersolidus cracking arises when the growing crystals join to form a weak solid skeleton which has intergranular liquid films owing to the presence of low melting point eutectics or compounds. Subsolidus embrittlement may arise from brittle grain boundary films owing to melting in the heat-affected zone, brittle phases such as martensite or the effects of absorbed gases. With these conditions the subsolidus cracking tendency may be increased by the presence of stress concentrators.

Weldability

The weldability of an alloy is good if a relatively simple welding technique can produce a sound weld with little deterioration of the properties of the parent metal.

Structural steels

Welding is not difficult because the steels have satisfactory physical characteristics and the important metallurgical factors can be controlled. Any oxides formed are readily fluxed and heat requirements are modest owing to the low specific heats and thermal conductivities of these steels.

Weldability decreases as the carbon and alloy content of steels is increased, owing to an increasing tendency to both supersolidus and subsolidus cracking, and careful compositional control is practised to prevent cracking. Low melting point iron sulphide films can cause supersolidus cracking and are prevented by keeping the sulphur content low and by the addition of sufficient manganese. The sulphur combines with manganese in preference to iron and the resulting manganese sulphide occurs as isolated particles and not as films. The subsolidus cracking tendency is reduced by specifying carbon and alloy contents that keep the hardenabilities of the steels within safe limits. BS 4360 defines a carbon equivalent which is calculated from the percentages of the alloying elements present as

$$\text{carbon equivalent} = \text{C} + \frac{\text{Mn}}{6} + \frac{\text{Cr} + \text{Mo} + \text{V}}{5} + \frac{\text{Ni} + \text{Cu}}{15}$$

and a value of between 0.39 and 0.51 is specified for each steel which takes into account the mass to be welded as well as the composition of the steel. Control of the cooling rate may also be necessary and may be achieved by preheating the plate or the specification of a minimum size of electrode for a particular plate size and type of joint. The recommended practice is covered by BS 5135. Absorbed hydrogen increases the cracking tendency, particularly during service at low temperatures. Hydrogen pick-up during metal-arc welding with coated electrodes can be prevented by the use of electrodes with coatings that give hydrogen-free atmospheres, baking the electrodes to remove water vapour and by the avoidance of contamination by oil and grease.

Steels with optimum properties are supplied in the hot rolled or normalised condition and the changes in the heat-affected zone have little effect on the properties.

Aluminium alloys

Welding of aluminium alloys is more difficult than for structural steels because of unsatisfactory physical characteristics and difficult metallurgical effects. No colour change precedes melting owing to the low melting points of the alloys. A large heat input is required owing to their high thermal conductivities and specific heats, which result in a wide heat-affected zone and the need to preheat large sections before welding is possible. Aluminium oxide has a much higher melting point than the metal, which inhibits mixing of the parent and filler metals, and a high density, which increases the possibility of inclusions. Oxide scale can be removed by chloride and fluoride fluxes although these are not favoured because they are hygroscopic and corrosive and the completed weld requires thorough cleaning. The alternative is to remove the oxide by the action of the arc in the tungsten- and metal-inert-gas techniques.

The supersolidus cracking tendency is high because the alloys show a large thermal contraction and most have low melting point constituents. This is particularly true of the precipitation-hardening alloys. Subsolidus cracking below 200 °C is possible if the grain boundary films are brittle.

Aluminium alloys are strengthened by cold work or precipitation hardening and the changes in the heat-affected zone cause considerable deterioration of the properties of both types of alloy.

References

BS 11 : 1978 Specification for railway rails
BS 18 : Methods for tensile testing of metals; Part 1 : 1970 Non-ferrous metals; Part 2 : 1971 Steel (general); Part 3 : 1971 Steel sheet and strip (less than 3 mm and not less than 0.5 mm thick); Part 4 : 1971 Steel tubes
BS 29 : 1976 Specification for carbon steel forgings above 150 mm ruling section
BS 131 : Part 1 : 1961 The Izod impact test on metals; Part 2 : 1972 The Charpy V-notch impact test on metals
BS 240 : Part 1 : 1962 Testing of metals (Brinell hardness test)
BS 970: Part 2: 1970 Direct hardening alloy steels, including alloy steels capable of surface hardening by nitriding

BS 970: Part 5: 1972 Carbon and alloy spring steels for the manufacture of hot-formed springs
BS 1037: 1964 Fire refined tough pitch high conductivity copper
BS 1038: 1964 99.85 per cent tough pitch copper
BS 1452 : 1977 Specification for grey iron castings
BS 1470: 1972 Wrought aluminium and aluminium alloys for general engineering purposes – plate, sheet and strip
BS 1490: 1970 Aluminium and aluminium alloy ingots and castings
BS 1172: 1964 Phosphorus deoxidised non-arsenical copper for general purposes
BS 2691 : 1969 Steel wire for prestressed concrete
BS 2874: 1969 Copper and copper alloys, rods and sections (other than forging stock)
BS 3100 : 1976 Specification for steel castings for general engineering purposes; includes BS 592 Carbon steel castings for general purposes, BS 1456 1½ per cent manganese steel castings; BS 1457 Austenitic manganese steel castings; BS 1458 Alloy steel castings with higher tensile strengths; BS 1461 3 per cent chromium – molybdenum steel castings; BS 1462 5 per cent chromium – molybdenum steel castings
BS 3617: 1971 Seven-wire steel strand for prestressed concrete
BS 4360 : 1979 Specification for weldable structural steels
BS 4449 : 1978 Specification for hot rolled steel bars for the reinforcement of concrete
BS 4461 : 1978 Specification for cold worked steel bars for the reinforcement of concrete
BS 4482: 1969 Hard drawn mild steel wire for the reinforcement of concrete
BS 4486: 1969 Cold worked high tensile alloy steel bars for prestressed concrete
BS 4670: 1971 Alloy steel forgings
BS 5135 : 1974 Metal-arc welding of carbon and carbon manganese steels
BS 5493 : 1977 Code of practice for protective coating of iron and steel structures against corrosion
CP 118: 1969 The structural use of aluminium
Dowrick, D. J., and Beckmann, P., 'York Minster Structural Restoration', *Proc. Instn civ. Engrs,* suppl. vi (1971) pp. 93 - 156.
Sage, A. M., 'Factors influencing the Choice of High-strength Low-alloy Steels for High-duty Applications', *Steel Times a. Rev. Steel Ind.,* Oct. (1970) pp. 107 – 25.
Toner, D. E., 'A Lesson in Bronze', *Copper,* 3 (1969) pp. 8 - 9

Further Reading

J. P. Chilton, *Principles of Metallic Corrosion,* monographs for teachers No. 4 (Royal Institute of Chemistry, London, 1963).
J. Comrie (ed.), *Civil Engineering Reference Book* (Butterworth, London, 1961).
V. B. John, *Introduction to Engineering Materials* (Macmillan, London and Basingstoke, 1972).
V. B. John, *Understanding Phase Diagrams* (Macmillan, London and Basingstoke, 1974).
J. W. Martin, *Elementary Science of Metals* (Wykeam, London, 1969).
E. C. Rollason, *Metallurgy for Engineers* (Arnold, London, 1973).

II TIMBER

Introduction

Timber is one of the oldest known materials of construction, and skill in its use combined with understanding of its nature have together produced many satisfactory solutions to structural and design problems throughout the ages. Such solutions are diverse in range – from the simplicity of primitive huts of indigenous design, through the intricacy and craftsmanship of Gothic screens and roofs to the elegance, scale and strength of many of the glued-laminated structural forms of today. In addition to its usefulness as a material for structural and applied design, because of its relative cheapness, timber has also fulfilled a role in the capacity of temporary structure, allowing materials such as concrete, brick and stone to be erected in structural forms which would otherwise have been vulnerable in the stage before the binding medium gained strength.

As with any material, it is necessary for those using it to know its nature and limitations. With man-made materials, standards of manufacture and design codes readily guide users, specifiers and designers through the limitations of the particular material. Because timber is a natural material which is not subjected to pre-use factory processing, it is even more necessary that the limitations of use dictated by its natural form be fully understood. Relevant codes of practice and standard techniques of stress grading assist in control over the problem posed when the performance of a timber in practice is unrelated to the results of laboratory tests on standard specimens of the same type of timber.

In this part of the book the student is introduced to the subject of timber as a material inexhaustible in supply – as a result of intelligent afforestation programmes – with wide potential in a diverse range of construction fields.

6

Sources of Timber

Mature trees, of whatever type, are the source of structural timber and it is important that users of timber should have a knowledge of the nature and growth pattern of trees in order to understand the behaviour, under a variety of circumstances, of those timber elements used in the construction industry. From that understanding there should follow intelligent specification in terms of requirements of performance. (A glossary of terms relating to timber and woodwork is given in BS 565.)

6.1 Growth Structure

Basically, it is possible to consider a tree in the form of its three subsystems: the roots, the trunk, and the crown (figure 6.1). Each subsystem has a role to play in the growth pattern of the tree. These roles are quite distinct but equally important in their integrated contribution to the quality of the tree, which in turn directly dictates the quality of the marketable timber and its related cost.

The function of the *root system* is twofold; as well as absorbing moisture-containing minerals from the soil for transfer via the *trunk* to the *crown*, the spread of the roots through the soil acts as an efficient anchorage, or 'foundation', which enables the growing tree to withstand wind forces.

The trunk provides rigidity, mechanical strength and height to maintain the crown at a level above ground at which it functions efficiently, as described later. In addition, the trunk contains and protects the growth cells and provides a two-way transport system for moisture travelling up from the roots and sap travelling down from the crown.

The function of the crown is to provide as large a 'catchment area' as possible covered with leaves which contain chlorophyll. In the leaves the photosynthetic process takes place whereby light is used to break up the carbon dioxide absorbed from the air. In the resultant process, oxygen is released and the carbon combines with the sap to form sugar, cellulose and other carbohydrates on which the growth of the tree depends.

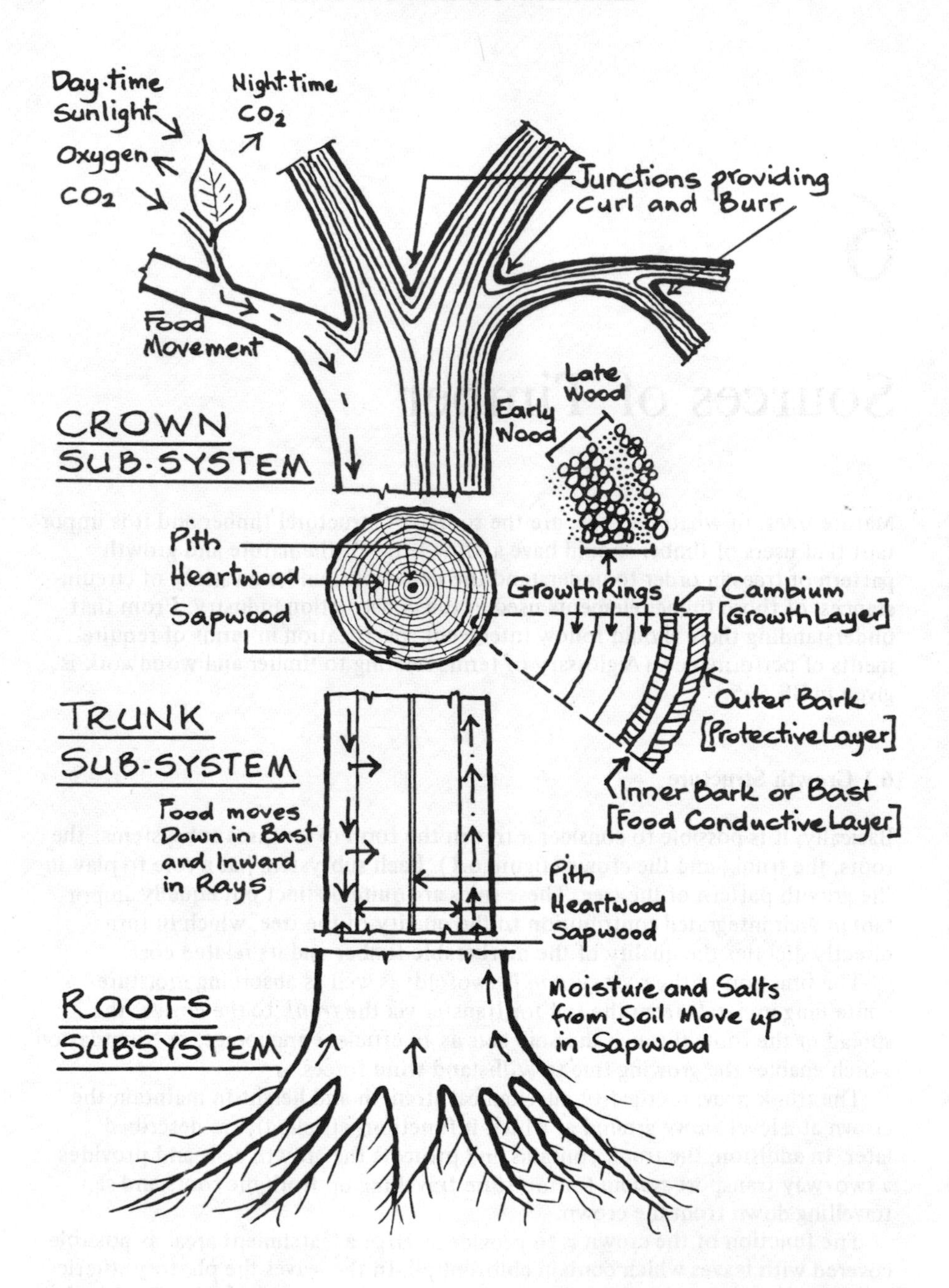

Figure 6.1 *Growth structure*

The Trunk

Although each subsystem is important, it is with the trunk of the tree that specifiers and users need to be most concerned since it is within the trunk that the marketable timber exists. Accordingly, it is worth while examining the composition of a typical tree trunk in some detail. The cross-section of a trunk provides perhaps the easiest picture of those relevant components of the tree with which users need to be concerned.

The periphery of the trunk is formed of *bark* or, more correctly, of an *outer bark* and an *inner bark* (known also as *bast*). The outer bark is rough in texture and dense enough in consistence to provide a protective 'coat' covering the vital growth and food layers immediately inside it. The inner bark is soft, moist and spongy and transports the converted sap from the leaves to the growing parts of the tree.

Between the inner bark and the actual growing timber is a thin layer of cells called the *cambium.* It is here that growth takes place by the splitting of single cells into two cells, each of which grows and splits in a process which continues throughout the growing season, eventually forming a sheath of cells which in cross-section appears as a ring, referred to as an 'annual ring'. These cells, which make up the wood tissue, or *xylem,* on the inner side of the cambium, are tubular in shape, with diameters between about 0.02 and 0.50 mm, and vary in length from about 1 mm in hardwoods to 6 mm in softwoods. They provide means of conducting and storing food and also provide mechanical support. In many species of tree, each annual ring appears to have two layers. The inner layer consists of cells with comparatively large cavities and thin walls. This cellular structure is due to a more rapid spring growth and, not illogically, is referred to as *springwood.* Later in the year, cells grow more slowly and have thicker walls and smaller cavities, resulting in a heavier, harder and stronger material called the *summerwood.* The amount of summerwood may vary in different species of tree and as a result of different weather and soil conditions. This affects the over-all density of the timber, which has a direct relationship with the strength of the timber.

Other groups of cells, known as *medullary rays,* run at right angles to the main cells from the outer layers inward. These carry food material from the inner bark to the cambium and act as storage tanks for the food material by transporting the excess towards the centre of the tree, where it is stored in cells in the inner rings which cease to function as a live part of the tree. This older timber is known as *heartwood* and is usually darker in colour as well as being drier and harder than the living layer, known as *sapwood.*

Heartwood is composed of dead tissue, its cells being completely filled, and its function is the mechanical support of the tree. Sapwood, containing more moisture, is not as strong in the 'green' state as heartwood but, after seasoning, when both heartwood and sapwood are reduced to the same moisture content, the difference in density and strength is very small. Sapwood is inferior to heartwood in respect of durability, containing starches which may attract insects and provide food for fungal growth. Sapwood, however, is very permeable and more easily impregnated with preservative and, where the conditions under which the timber must serve are such that treatment with a preservative is essential, it may be beneficial to use sapwood as a deliberate choice.

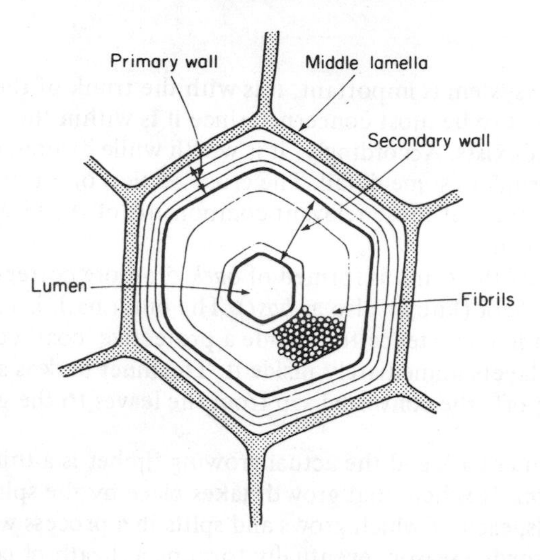

Figure 6.2 *Cell structure*

Individual cell structure

The cell walls (see figure 6.2) comprise a *primary wall* and a *secondary wall* surrounding the central cavity or *lumen.* The primary walls of adjacent cells are separated by the *middle lamella* which forms an integral part of the cell structure and consists of polyuronides. Both primary and secondary walls are composed of *fibrils* which are thin threads of cellulose whose direction varies in the different layers but which, in the secondary wall, lie principally in the direction of the cell axis. The fibrils themselves are composed of *microfibrils* which are aggregates of micelles (bundles of cellulose chain molecules). The chemical composition of the primary and secondary walls varies but includes from 45 to 60 per cent cellulose, 10 to 25 per cent hemicellulose and 20 to 35 per cent lignin. The *cellulose* which occurs in the form of long chain molecules of sugar glucose is very strong in tension and provides the basic strength of the cell walls in conjunction with the *lignin* which permeates the walls and acts as a binding medium.

Up to this stage, the description of the growth pattern relates to all types of tree. It is convenient, however, at this point to examine the two broad classifications of tree which provide timber for the construction industry. The classifications are significant in that there is a marked difference in profusion of growth as well as in cellular structure of the timber. This variation in profusion of growth is of major significance in cost aspects of the use of timber.

6.2 Tree Classification

The botanical name for those plants which grow outwardly, acquiring a new sheath of cellular tissue during each growing season is *exogens,* and this classification can be sub-divided into

(1) *Angiosperms*, or *Dicotyledons*, which have broad leaves shed in the autumn and which are normally classed as *hardwoods;*

(2) *Gymnosperms*, or *Conifers*, which have needle-like leaves, broadly evergreen, and which are generally classed as *softwoods.*

It should be noted that the terms hardwood and softwood in relation to a species of tree do not necessarily indicate relative hardness or density; balsa, for example, although soft in texture and easily worked, is a hardwood by classification whereas yew, a softwood by classification, has a density six times that of balsa. The ease with which Conifers are grown and their speed of growth make softwoods commercially much cheaper than hardwoods, which have a slower rate of growth and do not generally grow in profusion in those Northern Hemisphere countries with major construction industries. The quality of many hardwoods is not the only factor which raises their market value since scarcity and transport costs affect that value to a considerable degree. Accordingly, most of the constructional timber used in building work is of the softwood classification, ideally stress-graded and used in sizes which are suited to the task on hand, for maximum cost benefit. A similar attitude should be adopted in the specification and use of hardwoods, where cost benefit is obtained by matching the requirements for strength, quality, weathering and appearance to the particular timbers possessing those qualities.

Cellular Structure

Softwoods

Cells called *tracheids* fulfil the functions of conduction and mechanical support in softwoods. Conduction from one cell to another is by permeable areas, called *pits*, in the cell walls. The springwood with large cell cavities and thin walls readily provides conduction, while the summerwood with small cell cavities and thick walls gives mechanical support. Storage tissues are known as *parenchyma*, which remain alive for some time after their development is completed. This is a necessary feature of the growth pattern because plant food is usually stored in a form other than that required by the growing *cambium* and there is therefore a necessity for satisfactory pre-use conversion which can only take place in a living cell.

Hardwoods

Cells called *vessels* perform the function of conduction in hardwoods and in cross-section these appear as *pores*. Mechanical support is provided by cells called *fibres* which have thick walls. In softwoods the cells providing conduction and mechanical support occur principally in the springwood and summerwood respectively. In hardwoods, however, the vessels and fibres which fulfil these functions are distributed equally between these two stages of growth. Storage of food is performed by cells forming tissue called parenchyma as in softwoods.

The cellular structure of timber explains many aspects of the use of timber: for example, timber holds glue better than other nonporous materials because the glue is able to penetrate the cell cavities to form a 'key'; the preservation of wood is easily achieved by forcing a liquid into the cellular structure; and widely varying surface finishes occur because timbers with thick cell walls and small cell cavities are hard and difficult to work whereas those with thin cell walls and large cell cavities are easy to work.

6.3 Conversion of Timber

This is the term used to describe the process whereby the felled trunk is converted into marketable sizes of timber. Since the trunk is not of the same diameter over its entire length, it is not possible to obtain the same amount of marketable timber over the entire length of the trunk. Thus there will be varying lengths of different cross-section and in these lengths there may be a number of growth defects which will, in correct practice, further limit the marketable lengths or size of section, or both. Different strength characteristics are possible with varying methods of sawing the trunk (see figure 8.2). While plain sawing is the most economical run for the sawmill, it produces sections in which the angle of grain or the axis of the annual growth rings is not optimum, as might be provided by radial sawing. Similarly, figuring and appearance of timber vary with the method of sawing and, where a particular grain structure is required for the finished appearance, it is often necessary to specify this well in advance of requirements since special saw runs may be necessary. This can result in higher costs.

7

Characteristics of Timber

The principal characteristics of timber with which specifiers may be concerned are strength, durability and finished appearance. All of these are derived from natural characteristics present in the growing tree, and this chapter is essentially an orientation and amplification of the information in the previous chapter.

7.1 Strength

Strength of timber is affected by factors such as density, moisture content and grain structure as well as by the various defects discussed in chapter 8. Density is almost certainly an indication of strength: the more dense the timber the stronger it is. All timber is made up of much the same chemical constituents, but dense timber has thicker cell walls, which contribute to the strength of the timber. In softwoods in particular this occurs with a high proportion of summerwood in the growth ring, and this factor of growth ring proportion is related to the number and width of growth rings, that is, to *growth rate.* This has major strength significance and for various species there are optimum growth rates relative to strength and it has been shown that timber which has not matched or has exceeded these optima is likely to be weaker. A rough rule of thumb for most timbers is that a range of six to fifteen rings per 25 mm measured radially can be considered to indicate reasonable strength.

While it has been said that strength increases with density, this is not so with an increase in moisture content: the strength generally decreases with a rise in moisture content. Indeed, in 'green' timber in the seasoning stack this is what causes *bowing* and *springing.* There is an optimum moisture content for species, and even within species, for particular use and an excessively dried-out timber may have a lower strength. A side aspect of moisture content is the support which high moisture content might give to the germination and flourishing of fungal growth or to the attraction of insects. In each of these cases there will be an indirect reduction in strength. Grain structure and continuity are of significance in a strength context and any disruption due to growth defects or conversion and seasoning defects will induce a reduction in strength from that of 'clear' specimens of the same timber. Slope of grain in any given timber member will affect the strength properties about the most significant axis, for instance, a slope of 1 in 25 can reduce the strength to 96 per cent of that of the clear specimen, while a slope of 1 in 15 can reduce the strength to 89 per cent of that of the clear specimen. (Grading rules for sawn home-grown hardwood are given in BS 4047. Methods of test for clear plywood are published in BS 4512.)

Stress Grading

In order to design a timber structure properly the following properties of the timber should be known

(1) permissible bending stress;
(2) permissible shear stress;
(3) permissible compressive stress perpendicular to the grain;
(4) permissible compressive stress parallel to the grain; and
(5) modulus of elasticity.

Tests on large numbers of comparatively small clear specimens, that is, specimens clear of defects, under short-term loading conditions have shown that there is a significant variation in strength of apparently similar specimens. Statistical analysis of these results can be used to estimate the strength below which only 1 in 100 results, say, will fall. Factors affecting the strength of clear timber such as moisture content, size and shape of specimen and sustained loading have also been investigated and estimates of their effect on strength made. These factors together with a general factor of safety are all incorporated in reduction factors which when multiplied by the statistical 'minimum' strength give *basic stresses* which are in effect safe, or permissible, working stresses for clear timber. However, in practice, timber as commercially converted is seldom free from defects and further reductions in working stress are required. It is convenient to classify timber into various grades, each grade having associated with it a reduction factor, or *strength ratio,* which when multiplied by the basic stress for that timber gives the corresponding working stress.

Visual stress grading may be carried out using stress grading rules such as those in BS 4978 which specify the maximum size of defects acceptable in each of the different stress grades.

While visual stress grading can be carried out rapidly by an experienced grader, an inherent weakness of the method is that factors such as density, which influence strength, are completely disregarded. Fortunately it has been found that for timber containing defects there is a direct relationship between the deflection of a short length under a small transverse load and the transverse load required to cause failure. In other words there is a direct relationship between the stiffness (modulus of elasticity) and the bending strength (modulus of rupture). Thus once tests have established the relationship between modulus of elasticity and modulus of rupture for a given timber, nondestructive tests to determine the modulus of elasticity of a small length of timber may be used to estimate its strength. This has led to the development of *mechanical stress grading,* in which a machine is continuously fed with timber to which it applies a small transverse load over successive short lengths. The maximum deflection recorded for a given piece of timber, that is, the deflection associated with the minimum modulus of elasticity and hence minimum strength, is automatically compared with the maximum deflections associated with each stress grade. The end of the piece of timber is then sprayed with the colour of dye appropriate to its stress grade. Mechanical grading is more reliable than visual grading and lower factors of safety can therefore be used, resulting in an even more efficient use of timber.

7.2 Moisture Content

The moisture content of timber is the quantity of moisture contained by it expressed as a percentage of the dry weight. In 'green' timber, moisture is contained within the cells and the cell walls. The moisture may be removed from the cells without any effect other than a reduction in bulk density. The condition when all the cells are empty but the cell walls are still saturated is referred to as the *fibre saturation point,* usually between 23 and 27 per cent moisture content. Any further reduction in moisture content results in shrinkage of timber and if this reduction, or *seasoning,* is inefficient then considerable wastage of timber occurs. All timber, being hygroscopic, attempts to achieve an *equilibrium moisture content* with its environment, in use or in storage. This fact must be borne in mind when specifying moisture contents for timber in use, and especial care must be taken in storing timber which has been delivered at a low moisture content.

Timber Seasoning

Air seasoning

The timber is stacked in open-sided sheds in such a way as to promote drying without artificial assistance. The timber stack is supported about 450 mm clear of the ground and adjacent boards in each layer are kept separate; layers are provided with air space by means of spacers or sticks about 25 mm square. Hardwoods and some softwoods which dry out slowly are usually stacked in winter so that the timber will not be affected by summer heat until its moisture content has been reduced, while quick-drying softwoods may, with advantage, be stacked in spring or early summer. The advantages of air seasoning are that it is a cheap method with very little loss in quality of timber if done properly. The disadvantages are that both timber and space are immobilised for long periods, while very little control is possible, and then only by varying the size of spacers or by mobile open-slatted sides to the sheds.

Kiln drying

This method of seasoning employs a heated, ventilated and humidified 'oven'. Kiln drying must be used to reduce the moisture content below 17 per cent; temperatures much greater than atmospheric shade temperature are used, which cause the moisture in the wood to move more rapidly to the surface, whence it is removed by the circulating air. If this air were merely heated, excessive evaporation of moisture from the surface would take place – faster than moisture could move out to the surface. The outer parts of the timber would then tend to shrink peripherally so that splitting might occur. It is therefore necessary to humidify the circulating air in order to control the rate of evaporation, and this humidity is reduced as the drying proceeds. Different species of timber withstand different initial kiln temperatures and as drying proceeds these temperatures can be raised as the humidity is lowered.

Air circulation should be uniform over the face of the timber pile at which it enters, and the velocity of the air through the pile should be sufficiently high to be consistent with economic operation.

In order to assist in controlling the humidity of the circulating air, it has been found expedient to exhaust some of the air from the system from time to time and to replace it with fresh air from outside, any deficiency in water vapour being supplied by means of carefully regulated spray. The essentials of a kiln are therefore

(1) heat, under proper control and sufficient to raise the temperature to the maximum required;
(2) humidification, also under proper control and sufficient to meet all requirements;
(3) air circulation, uniform and of sufficient velocity; and
(4) air interchange, controlled at will.

Uniform air circulation is the most difficult requirement to obtain and the difference between various types of kiln lies principally in the methods adopted for circulating the air.

7.3 Timber Infestation

As well as the growth and other defects which can occur in timber, there is an associated problem of infestation by insects or fungal growth and it is necessary to protect timber against such infestation by proper specification and use, or by treatment with preservatives.

Insects

Beetles of one kind or another infest timber because the organic nature of the material is favourable to the grub's life cycle of hatching, growing and emerging. The effect is to reduce the cross-sectional area of the timber and so reduce its strength. Eggs are laid in cracks in timber, hatching out as larvae or grubs which tunnel through the timber for the whole of their growth period; the larvae develop into pupae and then into beetles, which emerge through flight holes to fly off and perpetuate the cycle. Common furniture beetles or woodworm (*Anobium punctatum*) infest both softwoods and hardwoods and are the main source of trouble in Scotland. Death watch beetles (*Xestobium ruforillosum*) prefer well-matured hardwood and are found mainly in England, seldom in Scotland. House longhorn beetles (*Hylotrupes bajulus*) can cause serious damage to sapwood of seasoned hardwoods and are generally confined to Europe and southern England. Powder post beetles (*Lyctus*) infest sapwood of newly seasoned hardwoods, softwoods being immune. Their location is mainly in timber yards in England.

Treatment of infested timber is basically that of cutting out and removal, with replacement timbers being treated with preservative. All remaining existing timber should be brush- or spray-treated with preservative or insecticide.

Fungal Growth

This may be destructive or nondestructive, and the following notes give outline descriptions of both types. Fungi are plants with no flowers or foliage and require no chlorophyll or sunlight since they are able to consume 'readymade' organic matter such as cellulose in timber.

Nondestructive fungi

These may be of the mould type which shows a black or green powder easily brushed away when dry, or of the staining type which feeds mainly on starch in sapwood cells, the cell walls being not weakened. Proprietary fungicide can usually eradicate such fungi, but this will only be effective if the moisture content of the timber is kept low.

Destructive fungi

These can be either of the dry rot or the wet rot category. The fungus producing dry rot is known as *Merulius lacrymans,* optimum conditions for its growth being a moisture content of 30 to 40 per cent, a temperature of 23 °C and a lack of ventilation. Above 26 °C and below 3 °C the fungus remains inactive and growth out of doors or in saturated timber is rare. After germination, the mould growth is sustained by tendrils, or hyphae, which are capable of travelling across brickwork and concrete, beneath plaster, through walls and along steelwork in search of nutrient material, including moisture, which it conducts back to the parent fungus. Dry rot is the most serious of all wood-rotting fungi since once established it can spread rapidly, the parent fungus 'fruiting' millions of red dust-like spores which are readily distributed by air currents, insects, vermin and other animals. The spread of fungus within a building may be extensive before any visible signs appear. Such signs include reddish dust, waving of the surface of panelling, shrinkage and malformation of the affected timber and also softening of the timber. Poor construction and detailing may encourage a constant ingress of water to brickwork, stonework or concrete in unventilated areas, thus providing a continuing source of likely germination. The use of 'green' timber with a high moisture content always creates a hazardous situation. In the final stages of attack, wood becomes dry and friable, hence the term 'dry' rot. This is a brown rot and the wood breaks into cubical pieces. *Merulius lacrymans* can cause widespread damage and may be difficult to treat satisfactorily.

Wet rot may occur in timber which is excessively wet, whether located in the interior or exterior of the building. There are many types of wet rot, some of which may show external growths and others which may only occur within the timber. Cellar fungus (*Coniophora cerebella*) is perhaps the most widely encountered. This is also a brown (but not cubical) rot and, characteristically of brown rots, it destroys the cell structure of timber by consuming only the cellulose. The decay and growth is almost completely internal with very little external evidence apart from a dark discoloration and some longitudinal surface cracking. In the final stages of attack the wood becomes very brittle and is readily powdered.

Eradication of both dry and wet rot is essentially the same. It is first necessary to eliminate all sources of moisture supporting the rot. Next, both the timber and the building should be dried out, either naturally or with the aid of dehumidifiers, by which time fungal growth should have ceased, although it is desirable that all growths be traced and destroyed. All visibly infected timber, and up to 500 mm of adjoining timber, should be cut out and burnt. Further treatment of noncombustible elements adjacent to infected timber may be carried out by sterilising with a blowlamp and then treating with a solution of 50 g sodium pentachlorophenate per litre of water. Any timber not removed should be treated with at least two brush coats of a suitable preservative. All new timber should be fully treated with preservative using one of the vacuum pressure methods.

8

Defects in Timber

Defects can occur in timber at various stages, principally during the growing period and during the conversion and seasoning process. Any of these defects can cause trouble in timber in use either by reducing its strength or marring its appearance (see table 8.1).

8.1 Natural Defects

Natural defects occur during the growing period and some of these are summarised here and in figure 8.1.

Cracks or Fissures

These may occur in various parts of the tree, and are generally given specific names to identify them more readily. 'Checks', 'shakes' and 'splits' are terms which refer to particular fibre disruptions which affect the strength of the timber by reducing the cross-sectional area of complete fibrous section. 'Heart shakes' which occur in the centre of the trunk may even indicate the presence of decay, or the beginnings of decay. 'Resin pockets' are fissures containing resin which constitute a strength defect and also interfere with decorative treatment in the finished use. BS 1186 and CP 112 limit or prohibit such fissures in timber for various uses.

Rind Gall, Burr and Curl

These are defects arising from the overgrowth of one part of a tree with another. *Rind gall* occurs where surface wounds or indentations are enclosed in subsequent growth. A *burr* is a swelling resulting from severely twisted grain caused by undeveloped buds emerging adjacent to a wound and then being enclosed by subsequent growth. A *curl* occurs at the crotch where a large branch intersects with the trunk (see figure 6.1).

In each of these three defects there is a considerable loss of strength but, interestingly, a very useful visual bonus in that all can provide very attractive grain structure, in the solid or in veneer, 'burr walnut' and 'curl oak' being examples.

TABLE 8.1
Timber deterioration sources and forms

Source	Form of deterioration
Bacteria	Discoloration of surface
Chemicals	Discoloration of surface with disintegration of cellular structure
Fire	Charring of surface of thick sections
Fungi	
Moulds	Discoloration of surface
Dry rot, Microscopic rots, Wet rots	Discoloration of surface with disintegration of cellular structure
Insect Infestation	
Marine borers, Beetles	Flight-holes at surface with tunnelling causing disintegration of structure
Mechanical	
Loading	Fracture of fibres
Abrasion, Erosion	Surface deterioration
Sunlight	Discoloration and embrittlement of surface
Water	
Flowing	Discoloration and erosion of surface
Intermittent	Expansion and contraction leading to cracking and splitting of cellular structure

Knots

These are perhaps the most obvious natural defect, and again they detract from strength by interrupting the fibre continuity. A *knot* is the part of a branch which becomes enclosed within the growing trunk. With a *live knot* there is complete continuity between the fibres of the branch and the tree. An *intergrown knot* occurs when at least 75 per cent of the cross-section of the knot perimeter fibres are continuous with those of the tree, and a *dead knot* when the proportion becomes less than 25 per cent. With an *encased knot* the entire cross-section of the knot is surrounded by bark or resin. A *pin knot* is less than 6.7 mm in diameter and a *knot cluster* consists of two or more knots around which the timber fibres are deflected. Depending on its position as visible in the converted timber, a knot may be described as an *edge knot*, a *margin knot*, a *splay knot* or an *arris knot;* in the case of an *enclosed knot,* it may not be visible at all. In every case, such knots may be of one of the previous classifications, and BS 1186 and CP 112 contain limitations on use regarding both strength and appearance.

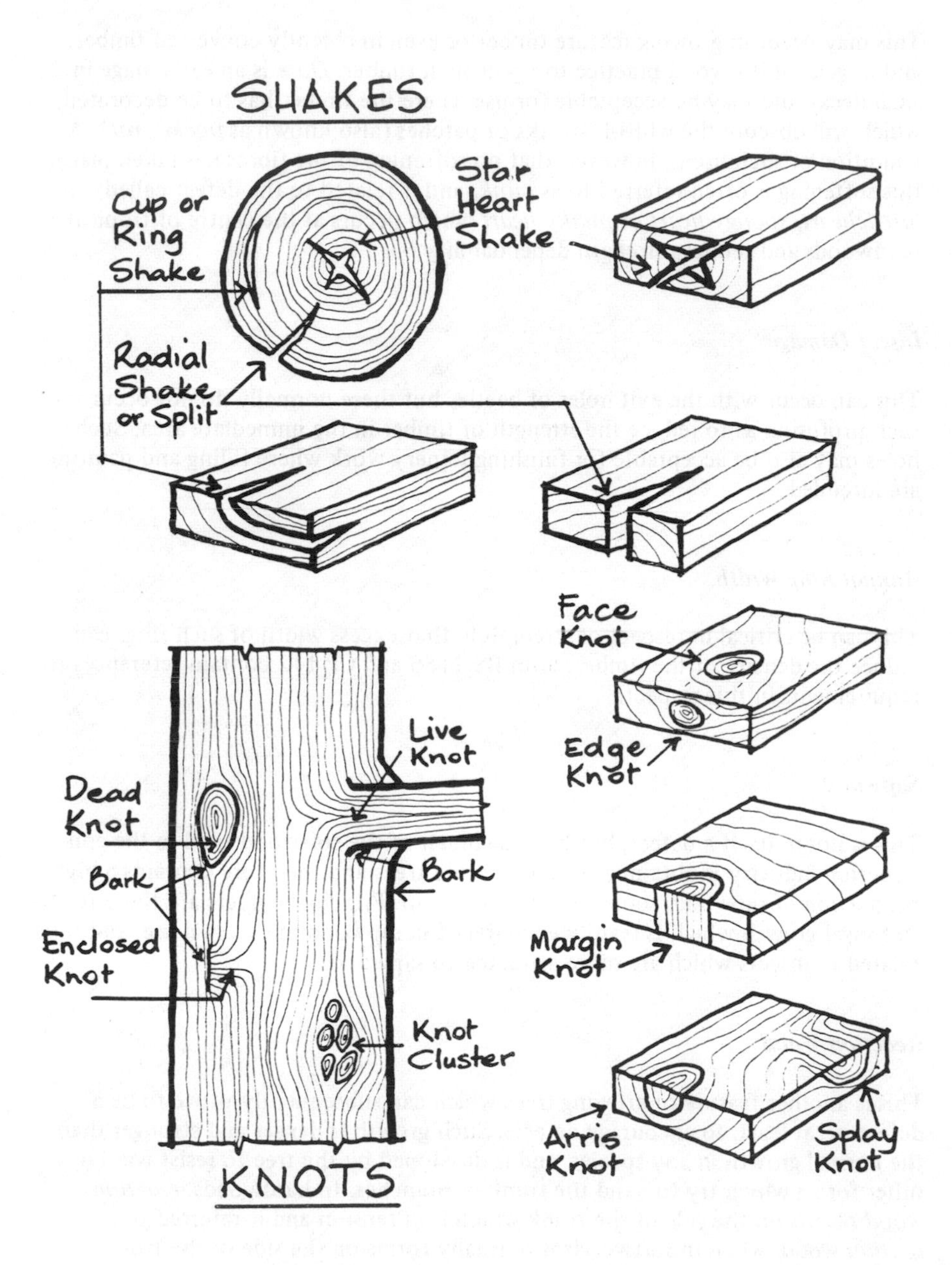

Figure 8.1 *Growth defects*

Fungal Decay

This may occur in growing mature timber or even in recently converted timber, and in general it is good practice to reject such timber. *Dote* is an early stage in such decay and may be acceptable for use where the timber has to be decorated, which will obscure the whitish streaks or patches (also known as *pocket rot*). A condition of such use is, however, that no softening of the fibres has taken place; this softening is often referred to as *punk,* and is related to the defect called *brittleheart, spongy heart* or *punky heart* which occurs at the centre of tropical hardwoods and reduces strength dependability.

Insect Damage

This can occur with the exit holes of beetles but these normally do not occur in such profusion as to reduce the strength of timber in the immediate area. Such holes may also be acceptable for finishing joinery work where filling and painting are intended.

Annual Ring Width

This can be critical in respect of strength in that excess width of such rings can reduce the density of the timber; both BS 1186 and CP 112 contain references to requirements in this respect.

Sapwood

This is not in itself a defect, but because much of the softwood used in the construction industry is obtained from immature trees and therefore contains a high proportion of sapwood, some defects may occur. Principal defects are those related to fungal growth which is readily encouraged in sapwood in external use, and those related to insects which are more attracted to sapwood.

Reaction Wood

This is another feature of growing trees which can subsequently prove to be a defect or, at least, to encourage defects. Such growth is denser and stronger than the normal growth in any species, and is developed by the tree to resist wind or other forces which try to bend the trunk or branches. In hardwoods, *reaction wood* occurs on the side of the trunk which is in tension and is referred to as *tension wood,* while in softwoods it normally forms on the side of the trunk which is in compression, and is then known as *compression wood.* Since the nature of such timber is so different from normal growth, distortion and splitting may occur in converted timber when in use, and generally such *reaction wood* is not readily workable with tools.

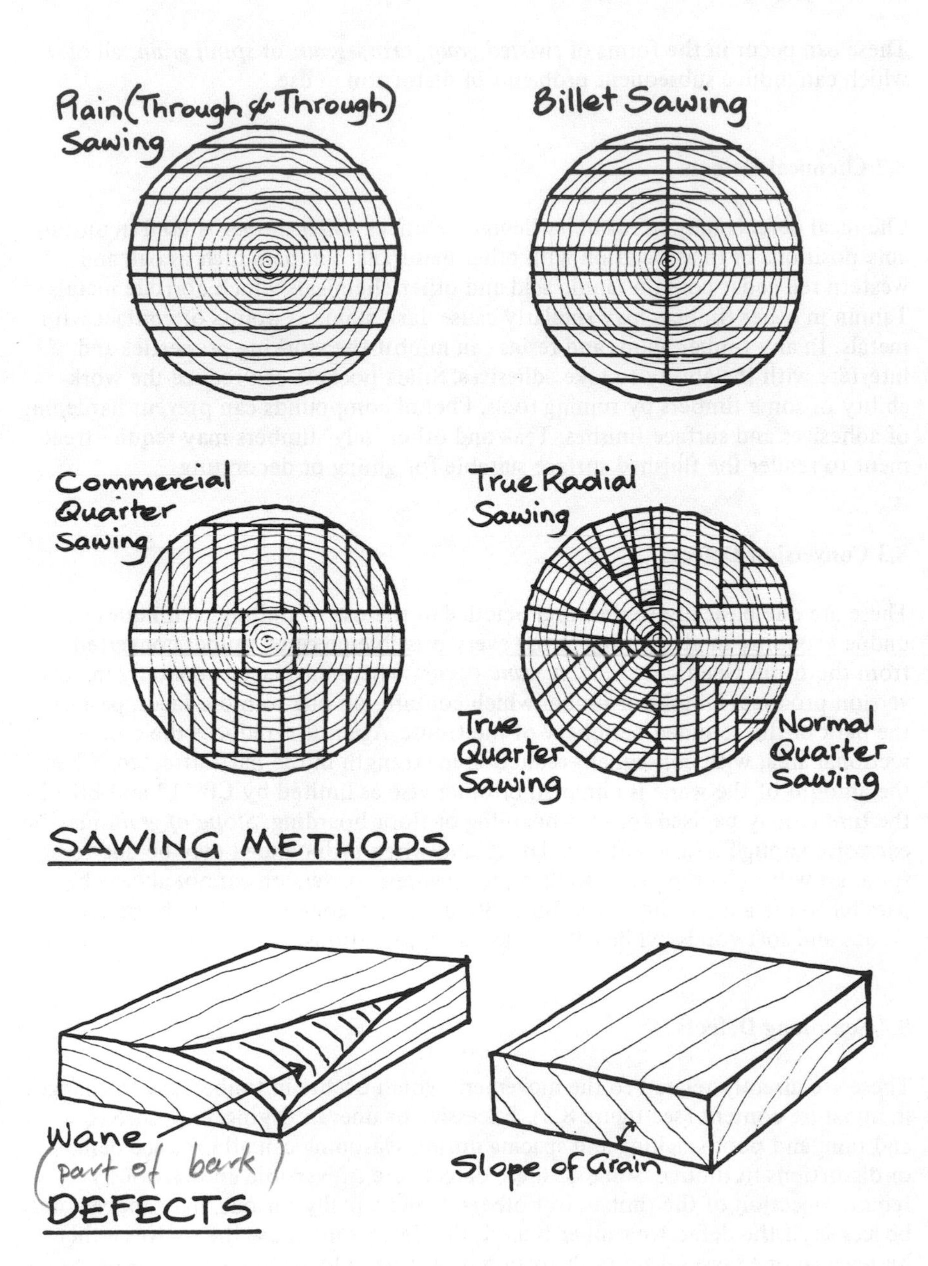

Figure 8.2 *Conversion features*

Grain Defects

These can occur in the forms of *twisted grain, cross-grain* or *spiral grain,* all of which can induce subsequent problems of distortion in use.

8.2 Chemical Defects

Chemical defects may occur in particular instances when timber is used in unsuitable positions or in association with other materials. Timbers such as oak and western red cedar contain tannic acid and other chemicals which corrode metals. Tannin in other timbers may similarly cause dark stains at points of contact with metals. In any timber, gums and resins can inhibit the working properties and interfere with the ability to take adhesives. Silica pockets can reduce the workability of some timbers by ruining tools. Phenol compounds can prevent hardening of adhesives and surface finishes. Teak and other 'oily' timbers may require treatment to render the finished surface suitable for gluing or decorating.

8.3 Conversion Defects

These are due basically to unsound practice in the use of milling techniques or to undue economy in attempting to use every possible piece of timber converted from the trunk (see figure 8.2). A *wane* occurs where misplaced economy in conversion produces lengths of timber which contain, on one or more faces, part of the bark or the rounded periphery of the trunk. Again this reduces the cross-sectional area, with consequent reduction in strength in the parts affected. When the amount of the wane is minimal, or otherwise as limited by CP 112 and BS 1297, the timber may be used for roof boarding or floor boarding. *Slope of grain* may be excessive enough to amount to a defect, and when pronounced may be due to spiral growth or to the problem that occurs when conversion cannot always be parallel to the axis of the trunk. BS 1186 limits the slope of grain in both hardwoods and softwoods, while CP 112 also sets limitations.

8.4 Seasoning Defects

These are directly related to the movement which occurs in timber due to changes in moisture content (see figure 8.3). Excessive or uneven drying, exposure to wind and rain, and poor stacking and spacing during seasoning can all produce defects or distortions in timber. Some of these defects are irreversible and obviously require rejection of the timber, but others, while equally serious, may appear to be less so; if the defective timber is used, the defects manifest themselves either by reversal or exaggeration of distortion, resulting in loosening of fixings or disruption of decoration, or both. *Checks* and *splits* similar to shakes may be induced in endwood or faces exposed in the drying stack to excessive sun or wind action. *Washboarding, honeycombing* and *case hardening* are all defects which can occur in the drying stack. *Cupping* occurs when a sawn board curves in section

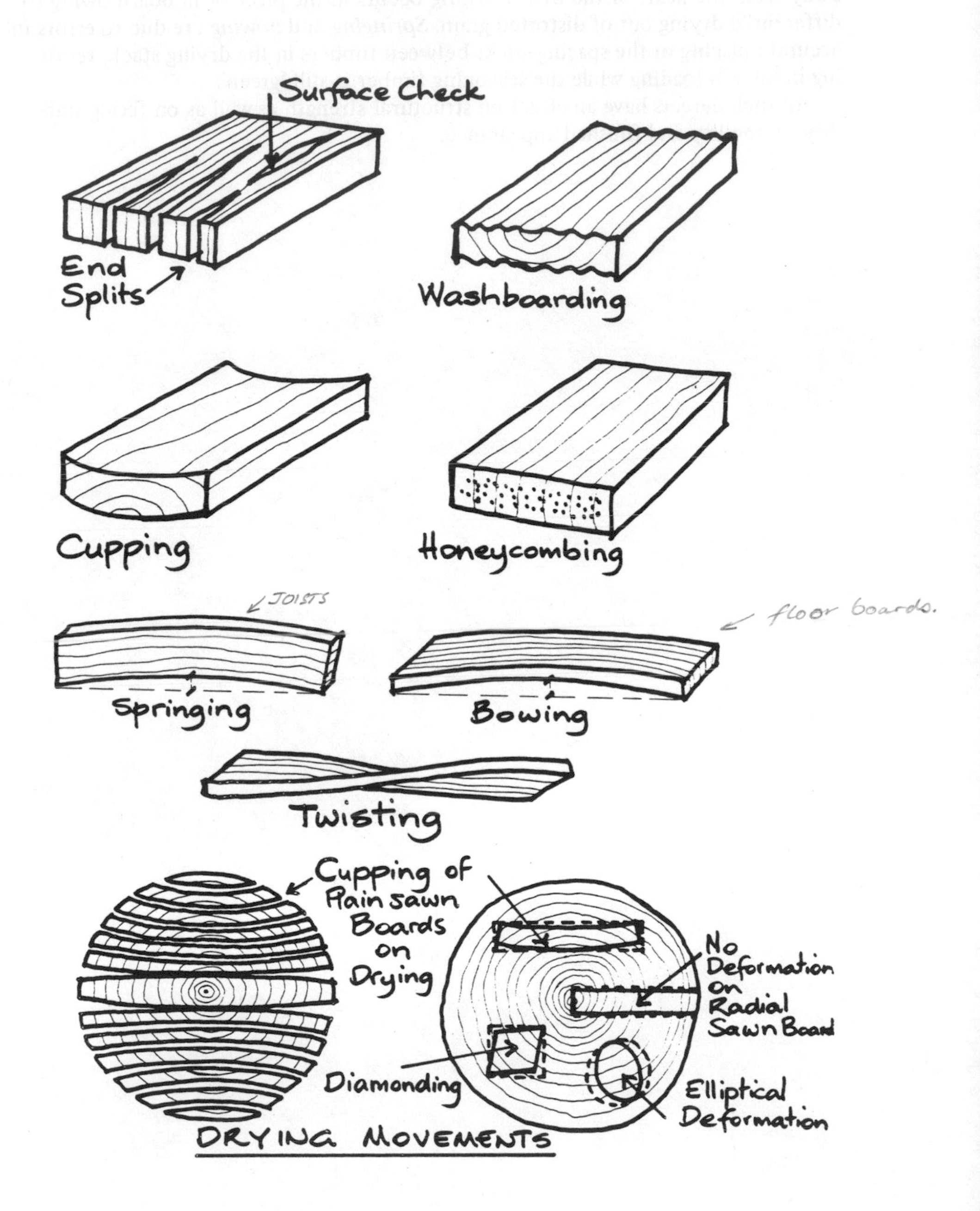

Figure 8.3 *Seasoning defects*

away from the heart of the tree. *Twisting* occurs in the piece or in board owing to differential drying out of distorted grain. *Springing* and *bowing* are due to errors in accurate placing of the spacing sticks between timbers in the drying stack, resulting in uneven loading while the seasoning timber is still 'green'.

All such defects have an effect on structural strength as well as on fixing stability, durability and finished appearance.

9

Preservation of Timber

As stated earlier and in chapter 11, not all timbers require treatment with a preservative to enable them to resist infestation and decay. Many timbers have a natural resistance owing not so much to great density and strength but often to the presence of natural oils or resins. Such timbers include cedar, greenheart, jarrah, oak and teak, but their cost may outweigh other considerations such as, for instance, appearance. This makes it necessary to use timbers not so well endowed naturally and to subject them to some form of preservative treatment which will increase their durability. A glossary of terms relating to timber preservation is given in BS 4261. The treatment of plywood with preservatives is dealt with in BS 3842.

9.1 Types of Preservative

Types are many and varied within the definition that any preservative must possess qualities which are poisonous to agents of decay. Nevertheless, for practical purposes only a few are in use because they meet all or most of the following requirements: toxicity to wood-destroying insects or fungi; permanency; economy and availability; penetrability; nonpoisonous nature to humans, plants and animals; noncorrosive nature; nonpromotion of flammability. Wood preservatives are classified in BS 1282.

Tar-oil Preservatives

These are very widely used, particularly in the form of coal-tar creosote to BS 144 for pressure impregnation and to BS 3051 for brush application, and are excellent for use on timbers which are to be located on the exterior of buildings or otherwise to be exposed to the elements (see also BS 913). Drawbacks with this type include encouragement of flammability without several months' weathering, the effect of the odour on food and the difficulty in painting treated timber, all of which restrict internal use.

Water-soluble Preservatives

These are generally odourless and nonstaining with no restrictions on decorative treatment, making timber so treated suitable for interior finishing joinery work. Because of the waterborne nature of this type of preservative it may be necessary to 're-dry' the timber to an acceptable moisture content. Normally, such preservatives do not increase the flammability of timber, and in some cases actually reduce it. The most commonly used type is that where the preservative is copper-chrome-arsenic (see BS 3452, BS 3453, BS 4072).

Organic Solvent Preservatives

These are generally poor in respect of nonflammability because of the volatile nature of the solvent (as opposed to the preservative). Such preservatives are generally noncorrosive, but may adversely affect food. Treated timber can be decorated in a variety of paints, and there is normally no necessity to 're-dry' timber after treatment. After painting, the treated timber is no more flammable than untreated timber similarly painted.

Pre-preservative Preparation

This consists basically of seasoning to a moisture content which varies with the species of tree and, moreover, with the method of impregnation. Ideally, all cutting and machining should have been completed and all timber should be clean and surface-dry. It is sometimes necessary with certain timbers to incise regularly spaced slits in the timber as a means of improving the penetration of the preservative.

9.2 Methods of Preservation

These vary from a surface treatment of notional protective value to full pressure impregnation. The following descriptions of methods are arranged in ascending order of efficiency and effectiveness.

Brush Application

This is the least effective method but is better than none and, provided the preservative is flooded over the surfaces to encourage absorption, reasonable penetration is possible in permeable timbers. Similar comment applies for application by spray.

Deluging, Dipping or Steeping

These applications are often specified for a variety of preservatives although the organic solvent type is most frequently used in one or other of the methods. Depending on the preservative used, some pre-heating may assist penetration. The above terms are used in increasing order of length of immersion, *deluging* usually taking place as part of a production line process and being equivalent to minimum *dip,* the length of which may be as short as a few seconds or as long as ten minutes or more. The term *steeping* is reserved for a period of several days' immersion which might result in deep penetration in permeable timbers.

Open Tank Application

This can be extremely efficient in ensuring penetration in permeable timbers. The method is simple and economic but provides little control. Accordingly, it is generally used to ensure maximum penetration of preservative into softwoods which have to be used for exterior work. The application takes place in a large metal tank in which the preservative can be heated by open fire or steam. The timber is kept fully immersed in the preservative, which is brought to a temperature of about 80 °C, at which it is maintained for several hours. The heat is turned off or removed and the tank and preservative are allowed to cool. During this cooling process, the air which has been expelled from the timber during heating is replaced by the preservative, thus giving good penetration and, in some timbers, complete penetration. Such intense penetration can considerably increase the weight of the timber, but the process can be continued by re-heating the preservative before removal of the timber with a view to reducing the retention of preservative within the cells while yet having deep impregnation.

Pressure Application

By far the most efficient and controllable method of preservation, pressure application is widely and economically practised. Pressure plants were originally only to be found at major timber importing ports, but are now also located at inland sawmills, which can maintain more localised control over demand and production. Basically, there are two different processes or techniques, the *full cell process* and the *empty cell process.*

In the full cell process, the timber is placed in a large enclosed pressure vessel and is subjected to a vacuum for about an hour. While the vacuum is maintained the preservative, usually preheated, is introduced into the vessel until it is filled. Pressure is then gradually increased until the required amount of preservative has been introduced into the timber, after which pressure is reduced and the vessel is emptied of preservative. A further vacuum is applied for a brief period only long enough to clean the surface of the timber.

In the empty cell process there are several methods, each varying slightly, but basically differing only from the full cell process by the absence of a preliminary vacuum period. The methods normally listed are the Rueping method, the Lowry method and the Boulton method.

The Rueping method subjects the timber in the pressure vessel to an initial pressure which is maintained while the cylinder is filled with preservative, after which pressure is increased to force the preservative into the timber. The pressure is subsequently released, allowing air which has been compressed in the inner cells of the timber to escape and in the process to expel excess preservative. A vacuum is then applied to further assist the escape of air.

The Lowry method does not use an initial period of pressure, but is otherwise the same as the Rueping method, up to the stage of a long vacuum period ensuring the escape of air and the evacuation of excess preservative.

The Boulton method is really a combination of seasoning and preservative application using a coal-tar creosote and is the main North American method for 'green' timber. The timber is immersed in hot creosote in a pressure vessel. A vacuum is then applied which has the effect of lowering the boiling point of water in the timber, so that the water boils off without damage to the timber, whose moisture content is thus reduced. The 'seasoned' timber can then be subjected to normal impregnation by one or other of the methods previously described.

In all techniques of pressure application it is important that seasoning of the timber has taken place beforehand, except in the Boulton method. This precaution ensures that during subsequent seasoning, cracks or shakes do not expose untreated timber.

10

Uses of Timber

The uses of hardwoods and softwoods are summarised according to moisture content in table 10.1. Information on plywood is given in BS 1455 and on blockboard and laminboard in BS 3444 and BS 3583.

10.1 Structural Carpentry

This term is considered to include all uses of timber in permanent carcassing or structure, and a range of such uses is given below.

Marine Work

Practice

Much of the marine work traditionally requiring timber for wharves, piers, sheet piling and cofferdams is now undertaken using precast or prestressed concrete products or both. This is due to several reasons, such as scarcity and cost of good quality timber of the oak or greenheart type, together with the advance of quality and design limits of the concrete products. Nevertheless, there are many locations where it is still an economic proposition to use timber for such constructional purposes because of the local availability of good timber and the high cost of transporting good class concrete products.

Requirements

For timber used in marine work, requirements are high density, close grain structure and a natural resistance to impact, infestation, fungal attack, salt or wave erosion and temperature variation. Where local timber does not meet all such requirements, impregnation with preservatives may provide an acceptable quality.

TABLE 10.1
Moisture content range and uses of hardwoods and softwoods

Seasoning technique	Approx. R.H. at 16°C (%)	Moisture content (%)	Construction suitability	Rot hazard
Green timber	90 ↑	27	Fibre saturation point	Wet rot and dry rot potential
		26		
	88 ↑	25	Maximum for pressure impregnation	
		24		
Obtainable in U.K. by air-seasoning		23		
		22	General carpentry: framing, sheathing, rafters, ground and upper floor joists (CP 112)	
		21		
	84	20		Germination of fungus spores unlikely
	82	19		
	80	18		
	76	17	External joinery: framing and sheathing for precision work (systems) (BS 1186 and CP 112)	
Obtainable in U.K. only by application of controlled heat and humidity	72	16		
	68	15	Internal joinery in intermittently heated areas: glued construction (BS 1186 and CP 112)	
	64	14		
	58	13		
	54	12	Internal joinery in continuously heated areas (room temp. 12 – 18°C) (CP 112)	
	48	11		
	44	10	Internal joinery in continuously heated areas (room temp. 20 – 24°C) (BS 1186)	
	38	9		
	32	8	Internal joinery located near heat source	
	26	7	Flooring over panel heating	
	20	6		
	16 ↑	5		
		4		
		3		
		2		
		1		
Kiln-dried	0	0		

Problems

Associated with marine work are problems of *impact* or *abrasion* due to movement of vessels, *infestation* by a variety of sea creatures such as marine borers (*Toredo navalis* and *Limnoria lignorum*), *fungal attack* by fungi akin to the wet rot or cellar fungus types, *erosion* by chemical action of salt or by wave action, and *dimensional movement* caused by differentials in temperature and moisture content.

Suitability of timber

For marine work timber should, as stated previously, be of high density, with close grain structure, resistance to impact and preferably having a preservative content such as that of natural oils. Generally, it is only in the hardwood classification that all these qualities are found and, traditionally, oak and greenheart are used for marine work in Britain because they combine the qualities with availability and economy. Larch, teak and jarrah are other timbers which could be used, although economy has always been affected by the cost of transport.

Heavy Constructional Work

Practice

Precast and/or prestressed concrete piles, various forms of *in situ* concrete piles and steel sheet piling have all served to reduce the amount of timber piling used in the construction industry. Timber is, however, still used for heavy constructional purposes where availability and cost of materials are favourable. Among other uses under this heading are pylons, gantries, bridges, shoring and abutments.

Requirements

In many instances of heavy work, requirements are the same as for marine work in that high density, closeness of grain and resistance to impact are all important, with the addition of resistance to acidity, alkalinity or other chemical nature of the soil with which the timber will be in contact. It is possible, of course, to augment the natural properties by appropriate preservative treatment.

Problems

In heavy constructional work there are many types of problem, complex and interrelated, varying often with the nature of the task to be performed and the location of the work. In the main, problems are of *impact* due to deliberate or accidental loading, *infestation* by airborne insects, *chemical attack* by soil, *fungal attack* by fungi of the wet-rot type and *dimensional movement* due to temperature and moisture variability.

Suitability of timber

For such work the suitability of timber depends on the requirements stated previously, together with market availability and cost in the sizes required for the work. The timbers mentioned for marine work are frequently used, namely, greenheart, jarrah, larch, oak and teak, as also are camphorwood, Douglas fir, and pitch pine, of which Douglas fir normally requires pressure impregnation with preservative (see BS 5268).

Medium/Light Constructional Work

Practice

In recent years there has been little or no diminution in the range of uses for timber in this type of constructional work. Indeed, with the development of off-site prefabrication and rationalisation of traditional building techniques, much more use has been made of the availability and ease of working and handling of the cheaper types of timber. Roof trusses, partitions, screens, floors and walls are no longer painstakingly put together on site with attendant time and material wastage, but are factory produced in large numbers, with reductions in man and machine fabrication time as well as site erection time, all of which lead to economic advantage.

Requirements

Chiefly, these are resistance to insect and fungal attack in storage and in position, together with a minimum of dimensional change due to temperature and humidity variations likely to be encountered within the building. Again, individual significance varies with the location and type of building. Natural qualities of resistance to deterioration can be augmented by techniques of application of preservative. Because the location in this type of work will almost certainly be inside a building, it may be necessary to reinforce any natural resistance to fire or rate of flame spread by impregnation with a fire-retardant chemical (see BS 476).

Problems

In medium/light constructional work, problems are principally of *dimensional movement* due to temperature and humidity variability, *fixing* using the correct methods and materials for the particular location and *infestation* and *fungal attack* as before. Another problem is often that of *market availability* of the correct type of timber in sizes and lengths suited to the purpose and of acceptable moisture content. The questions of *cost* and *site expediency* not infrequently combine with inadequacies in market availability to produce construction which is of continuing nuisance and expense during the life of the building.

Suitability of timber

For this class of work, suitability is very much related to cost as well as to fulfilling several or all of the requirements described, owing to the very great amount of timber required for the multiplicity of uses in this classification. Because of this, the timbers most frequently used are of the softwood type, principally Douglas fir, Western hemlock, European redwood and European whitewood, all of which are readily available in suitable sizes and quality at costs which are cheap by comparison with other softwoods and certainly most, if not all, of the hardwoods. All these timbers can have their resistance to infestation and fungal attack improved by preservation treatment.

10.2 Falsework Carpentry

This term is considered to include all uses of timber in section, board or plywood form erected for the purpose of facilitating other types of construction, such as shuttering for *in situ* or precast concrete work, support formwork for brick or stone arch or shell forms, or jigs for glued-laminated timber beam or shell forms. A glossary of formwork terms is given in BS 4340. For glued-laminated structural members reference should be made to BS 4169.

BS 4169 : 1970 - Glued-Laminated timber structural members
see BS - - - -

Practice

Although steel panshuttering and even some plastic shuttering have been available in recent years, by far the bulk of shuttering is still of timber. This is because of ease of working, cost and availability, and ignores consideration of misuse by contractors due to poor fixings and the excessively green state of the timber in many instances. Despite advances in the design of steel scaffolding connections and adjustable steel props, timber continues to be used for formwork for temporary support to timber shuttering – and even to plastic or metal shuttering.

Requirements

Generally for such work the principal requirements are cheapness and availability, but these are allied to requirements such as dimensional stability, ease of fixing and demounting, weight for handling and transportation and resistance to impact and abrasion. All of these should combine to allow maximum reuse of moulds and formwork. Sometimes the timber in board or plywood form may be required to have a grain or texture capable of easy transfer to the finished surface of the concrete when the shuttering is stripped.

Problems

In falsework carpentry, the problems are often specific to the location and the site environment but, in the main, are those mentioned in previous sections, namely, undue or unexpected loadings, dimensional movement, fixing and demountability, and, in the case of plywood particularly, suitability for use.

Suitability of timber

This is very much a matter of availability and cost, especially related to the repeated use aspect. Cost dictates the use of softwoods, unless there is an overriding factor such as grain for board-finished concrete in section, board or plywood form. Of this classification, Western hemlock or European whitewood are most frequently used in section or board form, while birch or gaboon species provide the plywood.

10.3 Finishing Joinery

By this term is meant all timber used in flooring, facings, skirtings, windows, doors, stairs, panelling and furniture (see CP 201, BS 459, BS 584, BS 585, BS 644).

Practice

Practice varies from country to country and with social stratum as well as class of building and cost limitation. Virtually any of the timbers previously mentioned for other purposes can be used for elements within this class of work, as well as many of the more exotic, more expensive and, not infrequently, less reliable timbers.

Requirements

These are ease of working and finishing, good grain pattern and appearance when clear-finished, dimensional stability in conditions of variability of temperature and humidity, both internal and external, resistance to infestation and fungal attack, and availability and cost.

Problems

These occur principally because of internal and external environmental variations affecting timber with possibly an excess, or at least a mis-match, of moisture content. They also arise in particular locations with infestation and fungal attack, with many timbers requiring preservative treatment to protect them. Inadequacy of finishing paint or varnish can be a problem in timber elements exposed to driving wind and rain or to excessive variations in temperature or humidity. Poor factory priming treatment of timber elements such as windows, doors or screens can produce trouble in the event of inadequate site storage or unduly long exposure to wind and rain before final paintwork is applied.

Suitability of timber

This is as complex as the range of costs and acceptability of performance in various building situations might indicate. It is very much a case of having to pay for what is required if, for instance, appearance is an overriding consideration. This limits the choice, which is further reduced according to the availability of the timber in the required dimensions. If preservative treatment or kiln drying or both have to be employed then costs can rise to a point where a compromise might have to be made by relaxing one or perhaps more of the requirements.

11

Types of Timber

The descriptions of types of timber given in this chapter are general, but in the range given provide a useful basis for comparison. (The nomenclature of commercial timbers, including sources of supply, is covered in BS 881 and BS 589.)

Afrormosia

A hardwood from West AFrica, afrormosia has a density about 700 kg m^{-3} in a seasoned condition. It resembles teak, but is finer textured, slightly heavier, lacks the oily nature of teak, is stronger and harder and very resistant to fungal attack. It is widely used for flooring, shopfitting and high quality joinery.

Agba

A hardwood from West Africa, agba has a density about 500 kg m^{-3}. It is very resistant to decay and is a strong-structured timber, eminently suitable for uses such as flooring, shopfitting and general joinery work.

Balsa

A hardwood from the West Indies, Central and Southern America, balsa has a density from 110 to 160 kg m^{-3}, which makes it the lightest and softest timber used commercially. It is often impregnated with hot paraffin solution to avoid undue absorption of moisture. Its special qualities tend to limit its use to purposes such as theatre and film sets, sub-flooring and landing stages.

Beech

A hardwood grown in Britain and throughout Europe, beech has a seasoned density of 720 kg m^{-3}. Although stronger than oak, it is not so resistant to decay and must be treated with preservative for use in the open. It is used principally for furniture but also for flooring.

Birch

A hardwood native to Britain and Europe, birch has a seasoned density of about 670 kg m^{-3}. It has a tendency to knots, is only of moderate strength, is subject to attack by furniture beetle, and requires treatment with preservative for use in the open. It is principally used in the construction industry as a source for plywood.

Camphorwood

A hardwood grown in Borneo and Malaya, camphorwood has a density of about 780 kg m^{-3} and is moderately hard. When seasoned it is harder, stronger and stiffer than teak, but is not durable in contact with the ground or with water. It is also readily subject to attack by beetles. It is used for flooring, general construction work and exterior joinery.

Cedar

Western red cedar is a softwood grown principally in North America and has a density of about 470 kg m^{-3}. Its strength properties are poor, being about 30 per cent below those of Baltic redwood, and it is not used for major structural work, or where it would be subject to hard wear. It is, however, exceptionally resistant to all forms of decay and can be used out of doors without need of preservative treatment. It is best used for general joinery work and finishings with care being taken to use copper or galvanised nails, since steel nails will be corroded by chemicals in the timber.

Fir

Douglas fir or Oregon pine as it is often called is a softwood native to North America with density about 530 kg m^{-3}. It is very straight grained and its strength properties are higher than Baltic redwood, as well as being more difficult to work. With only moderate resistance to decay, it may require preservative treatment for certain uses. Principal uses are for heavy, medium and light carpentry work and as a source of plywood.

Gaboon

A hardwood from West Africa, gaboon has a density of about 400 kg m^{-3}. It has only moderate strength and its most important use in the construction industry is as a source of plywood.

Greenheart

A hardwood from Guyana, greenheart has a density from 990 to 1090 kg m^{-3}. It is hard, strong and free from knots and other defects. It is almost immune from fungal attack and highly resistant to marine borers and these properties make it suitable for use in marine work and heavy construction work.

Gurjun

A hardwood from India and Burma, gurjun has a seasoned density of about 740 kg m^{-3}. It has good strength properties and fair resistance to decay and is used in general construction work and for flooring.

Hemlock

Western hemlock is a softwood grown throughout most of North America with a density about 480 kg m^{-3}. Originally, it had poor performance in construction usage, but improved seasoning methods have now made it a very reliable timber for general carcassing work, flooring and finishing joinery.

Iroko

A hardwood grown throughout Africa, iroko has a density about 640 kg m^{-3}. It is prone to hardened deposit in the cells and this has produced a reputation as a bad timber for sawing. Equivalent in strength to oak, it is reasonably resistant to decay and can be used for good class finishing joinery and for exterior work as well as for flooring.

Jarrah

A hardwood grown mainly in Australia, jarrah has a density about 900 kg m^{-3}. Gum veins and interlocking grain make seasoning and working somewhat tricky. It is very resistant to decay and to infestation and these properties make it eminently suitable for heavy construction work and for marine work.

Keruing

A hardwood grown extensively throughout the Far East, keruing, when seasoned, has a density from 640 to 910 kg m^{-3}. Strength figures are much the same as for teak, but the sapwood requires preservative treatment for use externally to resist insect attack although it is resistant to fungal growth.

Larch

A softwood common to Britain and Europe, larch has a density of about 590 kg m^{-3}. It is about 60 per cent harder than Baltic redwood and its heartwood is very resistant to decay. Its principal advantage as a constructional timber is that it can be used for a variety of purposes at different stages in its growth from fencing to flooring and marine work.

Mahogany

African mahogany is a hardwood available from all parts of Africa and has a density about 560 kg m^{-3}. It has an interlocking grain which gives beautiful figuring, but can be a problem in working and in its tendency to warp. It is only moderately resistant to decay and is liable to infestation by furniture beetles. It is used generally for high quality joinery and as a source of plywood.

Maple

A hardwood grown chiefly in North America, maple has a density about 750 kg m^{-3}. It is the heaviest and hardest of the North American commercial timbers, and possesses good strength properties so that it is particularly resistant to abrasion. Its principal uses are for flooring and high quality joinery particularly in panelling.

Oak

A hardwood grown in Britain and most European countries, oak has a density from 740 to 770 kg m^{-3}. English oak is one of the most useful hardwoods in the world, possessing not only high strength properties but also great durability for both internal and external uses. The sapwood is, however, vulnerable to the death watch beetle and care must be taken to exclude sapwood from structural use. Generally oak can be used for marine work and heavy constructional work as well as for all types of interior and exterior finishing joinery.

Pine: parana

Parana pine is a softwood grown in Central and South America and has a density from 480 to 640 kg m^{-3}. Strength properties vary with source but it is generally stronger and harder than Baltic redwood. Preservative treatment is required for exterior use, but otherwise the timber is eminently suitable for finishing joinery work.

Pine: pitch

Pitch pine is a softwood, now in short supply, grown in North America and with a density about 660 kg m^{-3}. It is very resinous, hard and tough but with a tendency to split in nailing. In its structural grades it can be used for heavy constructional, piling and marine work. Lower grades, which are less liable to splitting and which take paint better, are used in general joinery work.

Pine : Scots

A softwood, Scots pine is the only pine native to Britain, but is also found in most parts of Europe. When imported from the Baltic, the timber is referred to as redwood. It has an average density of about 530 kg m^{-3} and its strength properties are similar to those of Baltic redwood. Pressure creosoting is advisable for external use, and knots may be a problem. Much of the home-grown timber is too thin and knotty for use in good class joinery, but the better specimens provide good timber for medium to light constructional work.

Redwood

European redwood is a softwood grown extensively in Europe and Northern Asia, and it is similar to the Scots pine tree, with a density slightly more, about 540 kg m^{-3} It is strong for its weight and is eminently suited to general construction usage. Although the heartwood is durable, it is necessary to treat the sapwood with preservative before external use. Knots are a problem, but otherwise the timber is suitable for all general work from rough carcassing to finishing joinery.

Sapele

A hardwood of the mahogany family, sapele is grown in West Africa and has a density about 640 kg m^{-3} which makes it harder and heavier than ordinary African mahogany. Its principal characteristic is a marked and regular striped effect, and it is almost as hard as oak. It is used for general finishing joinery and as a source of plywood.

Spruce

Canadian spruce is a softwood grown over most of North America and has a density about 450 kg m^{-3}. The strength is good for a lightweight timber and it compares favourably with Baltic redwood. It is not resistant to decay and does not readily take treatment by preservative, so its use is limited to general carcassing work and, in the better grades, for general joinery. The Sitka spruce is similar in character, and has recently been planted extensively in Britain.

Teak

A hardwood of which the best qualities are grown in India and Burma, teak has a density from 640 to 720 kg m^{-3}. It is the most valuable of hardwoods because of high strength, durability, resistance to decay, resistance to fire and resistance even to acids. It is used extensively in the construction industry for purposes as varied as piling, heavy construction work, all general interior and exterior joinery, laboratory bench tops and fitments and kitchen work tops and fitments.

Whitewood

European whitewood is a softwood grown in Britain and in Northern Europe. With a density about 430 kg m^{-3}, this species is very similar to its relative the Sitka spruce and knots are a problem. Strength properties are only adequate for general construction work and preservative treatment is necessary for exterior use. Normally, it is only used for interior joinery, but in Scotland it is used for all general building purposes instead of redwood for cost reasons.

References

BS 144: 1973 Coal tar creosote for the preservation of timber
BS 459: Part 1 : 1954 Panelled and glazed wood doors; Part 2: 1962 Flush doors; Part 3: 1951 Fire-check flush doors and wood and metal frames (half-hour and one-hour types); Part 4 : 1965 Matchboarded doors
BS 476: Part 3 : 1975 External fire exposure roof test; Part 4 : 1970 Non-combustibility test for materials; Part 5: 1968 Ignitability test for materials; Part 6: 1968 Fire propagation test for materials; Part 7: 1971 Surface spread of flame tests for materials; Part 8: 1972 Test methods and criteria for the fire resistance of elements of building construction
BS 565: 1972 Glossary of terms relating to timber and woodwork
BS 584: 1967 Wood trim (softwood)
BS 585: 1972 Wood stairs
BS 644: Part 1 : 1951 Wood casement windows; Part 2: 1958 Wood double hung sash windows; Part 3: 1951 Wood double hung sash and case-windows (Scottish type)
BS 881, 589: 1974 Nomenclature of commercial timbers, including sources of supply
BS 913: 1973 Wood preservation by means of pressure creosoting
BS 1186: Part 1: 1971 Quality of timber; Part 2: 1971 Quality of workmanship
BS 1282: 1975 Guide to the choice, use and application of wood preservatives
BS 1297: 1970 Grading and sizing of softwood flooring
BS 1455: 1972 Plywood manufactured from tropical hardwoods
BS 3051: 1972 Coal tar creosotes for wood preservation (other than creosotes to BS 144)
BS 3444: 1972 Blockboard and laminboard
BS 3452: 1962 Copper/chrome water-borne wood preservatives and their application

BS 3453 : 1962 Fluoride/arsenate/chromate/dinitrophenol water-borne wood preservatives and their application
BS 3583 : 1963 Information about blockboard and laminboard
BS 3842 : 1965 Treatment of plywood with preservatives
BS 4047 : 1966 Grading rules for sawn home-grown hardwood
BS 4072 : 1974 Wood preservation by means of water-borne copper/chrome/ arsenic compositions
BS 4169 : 1970 Glued-laminated timber structural members
BS 4261 : 1968 Glossary of terms relating to timber preservation
BS 4340 : 1968 Glossary of formwork terms
BS 4512 : 1969 Methods of test for clear plywood
BS 4978 : 1973 Timber grades for structural use
BS 5268 : Part 5 : 1977 Preservative treatments for constructional timber
CP 112 : Part 2 : 1971 The structural use of timber * see now BS 5268
CP 201 : Part 2 : 1972 Wood flooring (board, strip, block and mosaic)

Further Reading

Architects' Journal Information Library (March 1967, January 1974) (Architectural Press, London).
L. G. Booth and P. O. Reece, *The Structural Use of Timber* (Spon, London, 1967).
H. E. Desch, *Timber, Its Structural Properties* (Macmillan, London and Basingstoke, 1968).
A. B. Emary, *Carpentry, Joinery and Machine Woodworking* (Macmillan, London and Basingstoke, 1974).
A. Everett, *Mitchell's Building Construction : Materials* (Batsford, London, 1974).
A. G. Geeson, *Building Science,* vol. 2 *Materials* (English Universities Press, London, 1952).
W. B. McKay, *Building Construction* (Longman, London, 1971).
F. D. Silvester, *Mechanical Properties of Timber* (Pergamon, Oxford, 1967).
Specification (two volumes, revised annually) (Architectural Press, London).
J. G. Sunley, 'Working Stresses for Structural Timber', *Forest Prod. Res. Bull.*, 47 (1965)
(Also various publications by the Building Research Establishment, Department of the Environment, Princes Risborough Laboratory, British Wood Manufacturing Association and the Trades Research and Development Association.)

see also
BS 5268 : .. Code of practice for the structural use of timber ~ in various parts
this code replaces CP112: Part 2

BS 5268 : Part 2 : 1984 Permissible Stress design matls & workmanship

III CONCRETE

Introduction

Concrete is a man-made composite the major constituent of which is natural aggregate, such as gravel and sand or crushed rock. Alternatively artificial aggregates, for example, blastfurnace slag, expanded clay, broken brick and steel shot may be used where appropriate. The other principal constituent of concrete is the binding medium used to bind the aggregate particles together to form a hard composite material. The most commonly used binding medium is the product formed by a chemical reaction between cement and water. Other binding mediums are used on a much smaller scale for special concretes in which the cement and water of normal concretes are replaced either wholly or in part by epoxide or polyester resins. These *polymer concretes* known as resin-based or resin-additive concretes respectively are costly and generally not suitable for use where fire-resistant properties are required but they are useful for repair work and other special applications. Resin-based concretes have been used, for example, for precast chemical-resistant pipes and lightweight drainage channels. This section deals only with *normal concretes* in which cement and water form the binding medium.

In its hardened state concrete is a rock-like material with a high compressive strength. By virtue of the ease with which fresh concrete in its plastic state may be moulded into virtually any shape it may be used to advantage architecturally or solely for decorative purposes. Special surface finishes, for example, exposed aggregate, can also be used to great effect.

Normal concrete has a comparatively low tensile strength and for structural applications it is normal practice either to incorporate steel bars to resist any tensile forces (reinforced concrete) or to apply compressive forces to the concrete to counteract these tensile forces (prestressed concrete). Concrete is also used in conjunction with other materials, for example, it may form the compression flange of a box section the remainder of which is steel (composite construction). Concrete is used structurally in buildings for foundations, columns, beams and slabs, in shell structures, bridges, sewage-treatment works, railway sleepers, roads, cooling towers, dams, chimneys, harbours, off-shore structures, coastal protection works and so on. It is used also for a wide range of precast concrete products which includes concrete blocks, cladding panels, pipes and lamp standards.

The impact strength, as well as the tensile strength, of normal concretes is low and this can be improved by the introduction of randomly orientated fibres into the concrete mix. Steel, polypropylene, asbestos and glass fibres have all been used with some success in precast products, for example, pipes, building panels and piles. Steel fibres also increase the flexural strength, or modulus of rupture, of concrete and this particular type of *fibre-reinforced concrete* has been used in ground paving slabs for roads where flexural and impact strength are both important. Fibre-reinforced concretes are however essentially special-purpose concretes and for most purposes the normal concretes described in this book are used.

In addition to its potential from aesthetic considerations, concrete requires little maintenance and has good fire resistance. Concrete has other properties which may on occasions be considered less desirable, for example, the time-dependent deformations associated with drying shrinkage and other related phenomena. However, if the effects of environmental conditions, creep, shrinkage and loading on the dimensional changes of concrete structures and structural elements are fully appreciated and catered for at the design stage no subsequent difficulties in this respect should arise.

A true appreciation of the relevant properties of any material is necessary if a satisfactory end product is to be obtained and concrete, in this respect, is no different from other materials.

12

Constituent Materials

Concrete is composed mainly of three materials, namely, cement, water and aggregate, and an additional material, known as an admixture, is sometimes added to modify certain of its properties. Cement is the chemically active constituent but its reactivity is only brought into effect on mixing with water. The aggregate plays no part in chemical reactions but its usefulness arises because it is an economical filler material with good resistance to volume changes which take place within the concrete after mixing, and it improves the durability of the concrete.

A typical structure of hardened concrete and the proportions of the constituent materials encountered in most concrete mixes are shown in figure 12.1. In a properly proportioned and compacted concrete the voids are usually less than 2 per cent. The properties of concrete in its fresh and hardened state can show large variations depending on the type, quality and proportions of the constituents, and from the discussion to follow students should endeavour to appreciate the significance of those properties of the constituent materials which affect concrete behaviour.

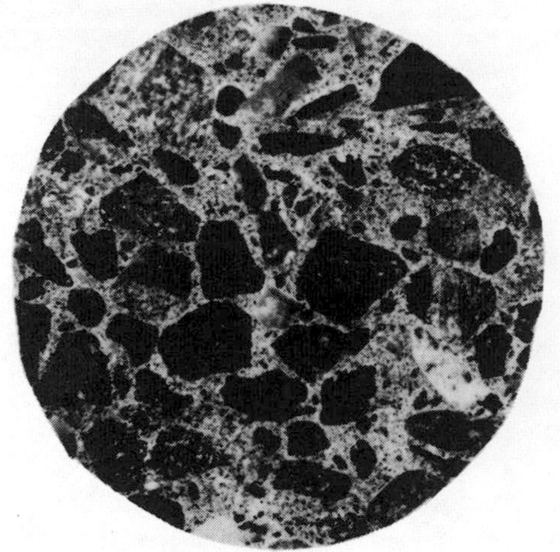

Figure 12.1 *Composition of concrete*

12.1 Cement

The different cements used for making concrete are finely ground powders and all have the important property that when mixed with water a chemical reaction (hydration) takes place which, in time, produces a very hard and strong binding medium for the aggregate particles. In the early stages of hydration, while in its plastic stage, cement mortar gives to the fresh concrete its cohesive properties.

The different types of cement and the related British Standards, in which certain physical and chemical requirements are specified, are given in figure 12.2. Of these, ordinary Portland cement is the most widely used, the others being used where concretes with special properties are required.

Portland Cement

Portland cement was developed in 1824 and derives its name from Portland limestone in Dorset because of its close resemblance to this rock after hydration has taken place. The basic raw materials used in the manufacture of Portland cements are calcium carbonate, found in calcareous rocks such as limestone or chalk, and silica, alumina and iron oxide found in argillaceous rocks such as clay or shale. Marl, which is a mixture of calcareous and argillaceous materials, can also be used.

Manufacture

Cement is prepared by first intimately grinding and mixing the raw constituents in certain proportions, burning this mixture at a very high temperature to produce clinker, and then grinding it into powder form. Since the clinker is formed by diffusion between the solid particles, intimate mixing of the ingredients is essential if a uniform cement is to be produced. This mixing may be in a dry or wet state depending on the hardness of the available rock.

The wet process is used, in general, for the softer materials such as chalk or clay. Water is added to the proportioned mixture of crushed chalk and clay to produce a slurry which is eventually led off to a kiln. This is a steel cylinder, with a refractory lining, which is slightly inclined to the horizontal and rotates continuously about its own axis. It is usually fired by pulverised coal, although gas or oil may also be used. It may be as large as 3.5 m in diameter and 150 m long and handle up to 700 t of cement in a day. The slurry is fed in at the upper end of the kiln and the clinker is discharged at the lower end where fuel is injected. With its temperature increasing progressively, the slurry undergoes a number of changes as it travels down the kiln. At 100 °C the water is driven off, at about 850 °C carbon dioxide is given off and at about 1400 °C incipient fusion takes place in the firing zone where calcium silicates and calcium aluminates are formed in the resulting clinker. The clinker is allowed to cool and then ground, with 1 to 5 per cent gypsum, to the required fineness. Different types of Portland cement are obtained by varying the proportions of the raw materials, the temperature of burning and the fineness of grinding. Gypsum is added to control the setting of the cement, which would otherwise set much too quickly for general use. Certain additives may also be introduced for producing special cements, for example, calcium chloride is added in the manufacture of extra-rapid-hardening cement.

The dry or semi-dry process is used for the harder rocks such as limestone and shale. The constituent materials are crushed into powder form and, with a minimum amount of water, passed into an inclined rotating nodulising pan where nodules are formed. These are known as *raw meal.* This is fed into a kiln and thereafter the manufacturing process is similar to the wet process although a much shorter length of kiln is used.

The grinding of the clinker produces a cement powder which is still hot and this hot cement is usually allowed to cool before it leaves the cement works.

Basic Characteristics of Portland Cements

Differences in the behaviour of the various Portland cements are determined by their chemical composition and fineness. The effect of these on the physical properties of cement mortars and concrete are considered here.

Chemical composition

As a result of the chemical changes which take place within the kiln several compounds are formed in the resulting cement although only four (see table 12.1) are generally considered to be important. A direct determination of the actual proportion of these principal compounds is a very tedious process and it is more usual to calculate these from the properties of their oxide constituents, which can be determined more easily. A typical calculation using Bouge's method is shown in table 12.2.

The two silicates, C_3S and C_2S, which are the most stable of these compounds, together form 70 to 80 per cent of the constituents in the cement and contribute most to the physical properties of concrete. When cement comes into contact with water, C_3S begins to hydrate rapidly, generating a considerable amount of heat and making a significant contribution to the development of the early strength, particularly during the first 14 days. In contrast C_2S, which hydrates slowly and is mainly responsible for the development in strength after about 7 days, may be active for a considerable period of time. It is generally believed that cements rich in C_2S result in a greater resistance to chemical attack and a smaller drying shrinkage than do other Portland cements. It may be noted from table 12.2 that the C_3S and C_2S

TABLE 12.1
Main chemical compounds of Portland cements

Name of compounds	Chemical composition	Usual abbreviation
Tricalcium silicate	$3\ CaO.\ SiO_2$	C_3S
Dicalcium silicate	$2\ CaO.\ SiO_2$	C_2S
Tricalcium aluminate	$3\ CaO.\ Al_2O_3$	C_3A
Tetracalcium aluminoferrite	$4\ CaO.\ Al_2O_3.Fe_2O_3$	C_4AF

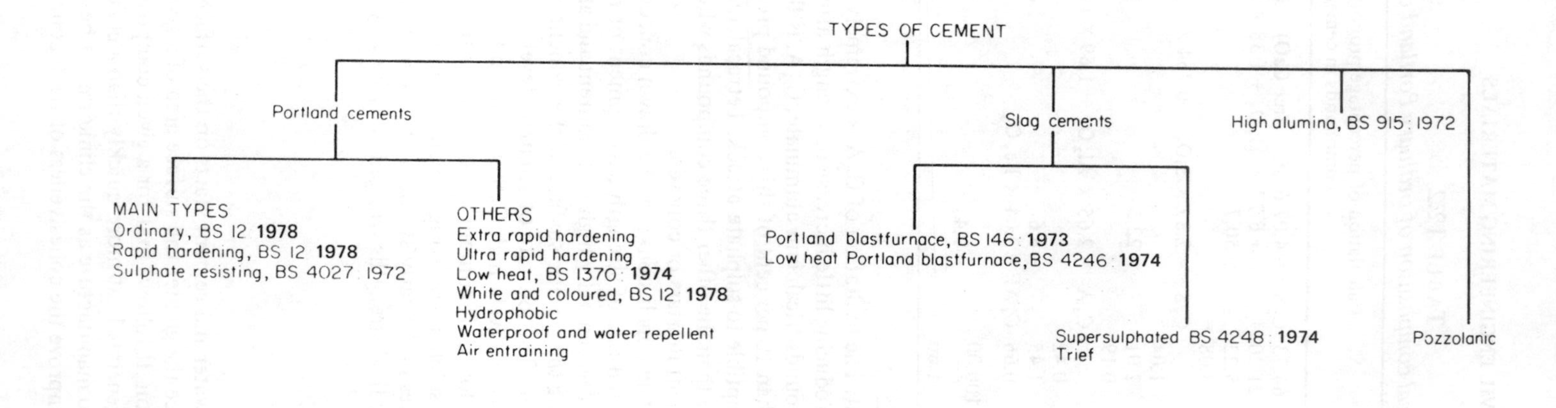

Figure 12.2 *Different types of cement used for making concrete*

TABLE 12.2
A typical chemical composition of ordinary Portland cement

Oxide composition (per cent)		Calculation of percentage proportion of main compounds in cement	
Lime, CaO	64.73	C_3S	$= 4.07\ (CaO - \text{free } CaO) - (7.60 \times SiO_2$
Silica, SiO_2	21.20		$+ 6.72 \times Al_2O_3 + 1.38 \times Fe_2O_3 + 2.85 \times SO_3)$
Alumina, Al_2O_3	5.22		$= 50.7$
Iron oxide, Fe_2O_3	3.08		
		C_2S	$= 2.87 \times SiO_2 - 0.754 \times C_3S$
Magnesia, MgO	1.04		
Sulphur trioxide, SO_3	2.01		$= 22.5$
Soda, Na_2O	0.19		
		C_3A	$= 2.65 \times Al_2O_3 - 1.69 \times Fe_2O_3$
Potash, K_2O	0.42		
Loss of ignition	1.45		$= 8.6$
Insoluble residue	0.66	C_4AF	$= 3.04 \times Fe_2O_3$
	100.00		$= 9.4$
Free lime, CaO	1.60		

contents are interdependent. The hydration of C_3A is extremely exothermic and takes place very quickly, producing little increase in strength after about 24 hours. Of the four principal compounds tricalcium aluminate, C_3A, is the least stable and cements containing more than 10 per cent of this compound produce concretes which are particularly susceptible to sulphate attack. Tetracalcium aluminoferrite, C_4AF, is of less importance than the other three compounds when considering the properties of hardened cement mortars or concrete.

From the foregoing, certain conclusions may be drawn concerning the nature of various cements. The increased rate of strength development of rapid-hardening Portland cement arises from its generally high C_3S content and also from its increased fineness which, by increasing the specific surface of the cement, increases the rate at which hydration can occur. The low rate of strength development of low-heat Portland cement is due to its relatively high C_2S content and low C_3A and C_3S contents. An exceptionally low C_3A content contributes to the increased resistance to sulphate attack of sulphate-resisting cement. It should be noted that while there can be large differences in the early strength of concretes made with different Portland cements, their final strengths will generally be very much the same (see chapter 14).

Fineness

The reaction between the water and cement starts on the surface of the cement particles and in consequence the greater the surface area of a given volume of cement the greater the hydration. It follows that for a given composition, a fine cement will develop strength and generate heat more quickly than a coarse cement. It will, of course, also cost more to manufacture as the clinker must be more finely ground. Fine cements, in general, improve the cohesiveness of fresh concrete and can be effective in reducing the risk of bleeding (see chapter 13), but they increase the tendency for shrinkage cracking.

The British Standards listed in figure 12.2 specify the fineness requirements for different cements. Several methods are available for measuring the fineness of cement, for example BS 12 prescribes a permeability method which is a measure of the resistance of a layer of cement to the passage of air. The measured fineness is an over-all value known as specific surface and is expressed in square centimetres per gram.

Hydration

The chemical combination of cement and water, known as hydration, produces a very hard and strong binding medium for the aggregate particles in concrete and is accompanied by the liberation of heat, normally expressed as calories per gram. The rate of hydration depends on the relative properties of silicate and aluminate compounds, the cement fineness and the ambient conditions (particularly temperature and moisture). The time taken by the main constituents of cement to attain 80 per cent hydration is given in table 12.3. Factors affecting the rate of hydration have a similar effect on the liberation of heat. It can be seen from table 12.4 that the heat associated with the hydration of each of the principal compounds of cement is very different and in consequence cements having different compositions also have different heat characteristics (see figure 12.3).

TABLE 12.3

Time taken to achieve 80 per cent hydration of the main compounds of Portland cement, based on Goetz (1969)

Chemical compounds	Time (days)
C_3S	10
C_2S	100
C_3A	6
C_4AF	50

TABLE 12.4

Heat of hydration of the main chemical compounds of Portland cement, based on Goetz (1969)

Chemical compounds	Heat of hydration (cal g^{-1})
C_3S	120
C_2S	60
C_3A	525
C_4AF	100

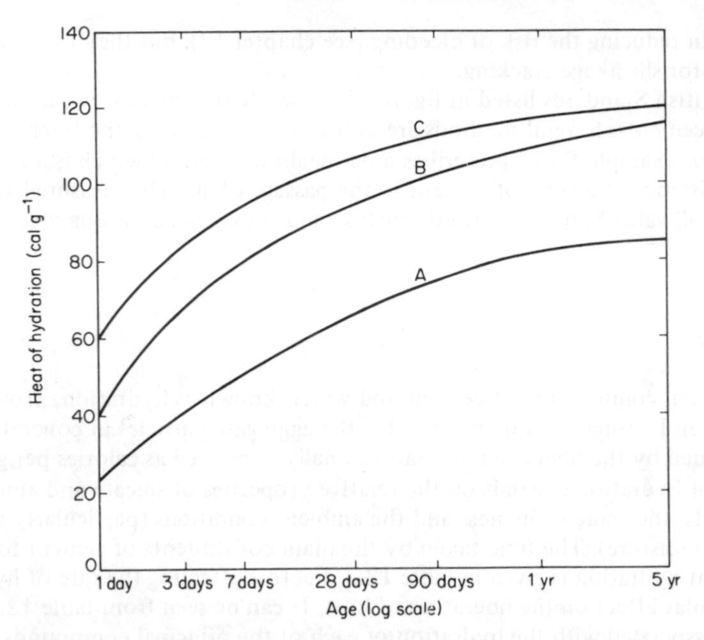

Figure 12.3 *Typical results for the heat evolution at 20 °C of different Portland cements : (A) low heat, (B) ordinary and (C) rapid hardening, based on Lea (1970)*

Concrete is a poor conductor of heat and the heat generated during hydration can have undesirable effects on the properties of the hardened concrete as a result of microcracking of the binding medium. The possible advantages associated with the increased rate of hydration may in these circumstances be outweighed by the loss in durability of the concrete resulting from the microcracking. Other factors which affect the temperature of the concrete are the size of the structure, the ambient conditions, the type of formwork and the rate at which concrete is placed. It should be noted that it is the rate at which heat is generated and not the total liberated heat which in practice affects the rise in temperature. The heat characteristics must be considered when determining the suitability of a cement for a given job.

Setting and hardening

Setting and hardening of the cement paste are the main physical characteristics associated with hydration of the cement. Hydration results in the formation of a gel around each of the cement particles and in time these layers of gel grow to the extent that they come into contact with each other. At this stage the cement paste begins to lose its fluidity. The beginning of a noticeable stiffening in the cement paste is known as the *initial* set. Further stiffening occurs as the volume of gel increases and the stage at which this is complete and the final hardening process, responsible for its strength, commences is known as the *final* set. The time from the addition of the water to the initial and final set are known as the setting times

TABLE 12.5
Typical initial and final setting times for the main Portland cements

Cements	Typical results		British Standard requirements	
	Initial setting time (min)	Final setting time (min)	Initial setting time not less than (min)	Final setting time not more than (h)
Ordinary	162	214	45	10
Rapid hardening	97	149	45	10
Sulphate resisting	154	202	45	10
Low heat	157	253	60	10

(BS 12) and the specific requirements in this respect for the different cements are given in the appropriate British Standards. The setting times for some of the more important Portland cements are given in table 12.5. In practice, when mixes have a higher water content than that used in the standard tests, the cement paste takes a correspondingly longer time to set. Setting time is affected by cement composition and fineness, and also, through its influence on the rate of hydration, by the ambient temperature.

Two further phenomena are a *flash set* and a *false set*. The former takes place in cement with insufficient gypsum to control the rapid reaction of C_3A with water, this reaction generating a considerable amount of heat and causing the cement to stiffen within a few minutes after mixing. This can only be overcome by adding more water and reagitating the mix. The addition of water results in a reduction in strength. A false set also produces a rapid stiffening of the paste but is not accompanied by excessive heat. In this case remixing the paste without further addition of water causes it to regain its plasticity and its subsequent setting and hardening characteristics are quite normal. False set is thought to be the result of intergrinding gypsum with very hot clinker in the final stages of the manufacture of cement.

Soundness

An excessive change in volume, particularly expansion, of a cement paste after setting indicates that the cement is unsound and not suitable for the manufacture of concrete. In general, the effects of using unsound cement may not be apparent for some considerable period of time, but usually manifest themselves in cracking and disintegration of the surface of the concrete. One of the methods for testing the soundness of cements is that developed by Le Chatelier (BS 12: Part 2).

Types of Cement

The different types of cement are shown in figure 12.2 and their main properties are summarised in table 12.6. A brief description of typical properties of each

TABLE 12.6

Main properties of different cements

Type of cement	Rate of strength development	Rate of heat evolution	Drying shrinkage	Resistance to sulphate attack
Portland cements				
Ordinary	medium	medium	medium	low
Rapid hardening	high	high	medium	low
Sulphate resisting	low to medium	low to medium	medium	high
Extra rapid hardening	high to very high	high to very high	medium to high	low
Ultra rapid hardening	high to very high	high to very high	high	low
Low heat	low	low	medium	medium to high
White and coloured	medium	medium	medium	low
Hydrophobic	medium	medium	medium	low
Waterproof and water repellent	medium	medium	medium	low
Air entraining	medium	medium	medium	low
Slag cements				
Portland blastfurnace	low to medium	low to medium	above medium	medium
Low heat Portland blastfurnace	low	low	above medium	medium to high
Supersulphated	medium	very low	medium	high to very high
*High-alumina cement**	very high	very high	medium	very high
Pozzolanic cement	low	low to medium	medium to high	high

*Subject to *conversion* in some environments.

type of cement is given here. For more detailed information the reader is referred to Neville (1968) and Orchard (1968).

Portland cements

Ordinary Portland cement has a medium rate of hardening, making it suitable for most concrete work. It has, however, a low resistance to chemical attack. *Rapid-hardening Portland cement* is in many ways similar to ordinary Portland cement but produces a much higher early strength. The increased rate of hydration is accompanied by a high rate of heat development which makes it unsuitable for large masses of concrete, although this may be used to advantage in cold weather. *Low-heat Portland cement* has a limited use but is suitable for very large structures, such as concrete dams, where the use of ordinary cement would result in unacceptably large temperature gradients within the concrete. Its slow rate of hydration is accompanied by a much slower rate of increase in strength than for ordinary Portland cement although its final strength is very similar. Its resistance to chemical attack is greater than that of ordinary Portland cement. *Sulphate-resisting Portland cement,* except for its high resistance to sulphate attack, has principal properties similar to those of ordinary Portland cement. Calcium chloride should not be used with this cement as it reduces its resistance to sulphate attack. *Extra-rapid-hardening Portland cement* is used when very high early strength is required or for concreting in cold conditions. Because of its rapid setting and hardening properties the concrete should be placed and compacted within about 30 minutes of mixing. Since the cement contains approximately 2 per cent calcium chloride dry storage is essential. Its use in reinforced or prestressed concrete is not recommended (CP 110: Part 1). *Ultra-high early-strength Portland cement,* apart from its much greater fineness and larger gypsum content, is similar in composition to ordinary Portland cement. Although the early development in strength is considerably higher than with rapid-hardening cement there is little increase after 28 days. It is suitable for reinforced and prestressed concrete work. *White and coloured Portland cements* are similar in basic properties to ordinary Portland cement. White cement requires special manufacturing methods, using raw materials containing less than 1 per cent iron oxide. Coloured cements are produced by intergrinding a chemically inert pigment with ordinary clinker. Because of its inert characteristics, the presence of a pigment slightly reduces the concrete strength. These cements are used for architectural purposes. *Hydrophobic Portland cement,* owing to the presence of a water-repellent film around its grain, can be stored under unfavourable conditions of humidity for a long period of time without any significant deterioration. The protective coating is broken off during mixing and normal hydration then takes place. *Waterproof and water-repellent Portland cements* produce a more impermeable fully compacted concrete than ordinary Portland cement. *Air-entraining Portland cement* produces concrete with a greater resistance to frost attack. The cement is produced by intergrinding an air-entraining agent with ordinary clinker during manufacture. However, in practice, it is more advantageous to add an air-entraining agent during mixing since its quantity can be varied to meet particular requirements.

Slag cements

Different types of slag cement can be produced by intergrinding varying proportions of granulated blastfurnace slag with activators such as ordinary Portland cement and gypsum. The chemical composition of the granulated slag is similar to that of Portland cement but the proportions are different. One general requirement is that the slag used must have a high lime content. Of the four slag cements listed in figure 12.2 only Portland blastfurnace cement has been used to any great extent.

Portland blastfurnace cement is produced by mixing up to 65 per cent granulated blastfurnace slag with ordinary Portland cement. The basic characteristics are similar to those of ordinary Portland cement although the rate of hydration is lower. It is particularly suited for structures involving large masses of concrete. Its resistance to chemical attack, particularly seawater, is somewhat better than that of ordinary Portland cement. *Low-heat Portland blastfurnace cement,* except for its greater slag content, is similar to Portland blastfurnace cement and can therefore be effectively employed where a control in the rise of temperature is the main requirement. It has a greater resistance to chemical attack than Portland blastfurnace and ordinary Portland cements. *Supersulphated cement* contains up to 85 per cent slag, 10 to 15 per cent gypsum and a small percentage of ordinary Portland cement. Its resistance to chemical attack is similar to that of sulphate-resisting Portland cement. Because of its low heat properties it can also be used for mass concrete work. In order to prevent the friable and dusty surface appearance often associated with this cement, careful initial curing is required.

High-alumina cement

This cement, which is manufactured by melting a mixture of limestone, chalk and bauxite (aluminium ore) at about 1450 °C and then grinding the cold mass, has different composition and properties from those of Portland cements. The high proportion of aluminate, about 40 per cent, brings about a very high early strength and consequently it often becomes necessary to keep concrete, in which this cement is used, continuously wet for at least 24 hours to avoid damage from the associated heat of hydration. Control of temperature is particularly critical as, above 25 °C, a substantial reduction in strength may occur owing to the conversion of the hydrated cement to a more porous form. This can result in some circumstances in concrete strength after a period of time being considerably lower than its strength at 24 hours. The degree of conversion is also very sensitive to the water-cement ratio, which should not normally exceed 0.4. The cement can be beneficially employed for concreting in winter conditions. It has a wide application in refractory concrete. Its resistance to chemical attack is greater than that of Portland cements although in its converted, more porous, state it is very susceptible to alkali and sulphate attack.

Pozzolanic cement

Pozzolanic cement is made by grinding together up to 40 per cent of a pozzolanic material with ordinary Portland cement. Pozzolanic materials combine with the lime released during the setting and hardening of ordinary Portland cement and form cementitious materials. Pozzolanas occur naturally, for example, volcanic ash, or are obtained in other forms such as pulverised fuel ash. The rate of gain in strength and liberation of heat is slower than for ordinary Portland cement, and this can be useful for mass concrete work. Like sulphate-resisting Portland cement, it has a high resistance to chemical attack.

12.2 Aggregate

Aggregate is much cheaper than cement and maximum economy is obtained by using as much aggregate as possible in concrete. Its use also considerably improves both the volume stability and the durability of the resulting concrete. The commonly held view that aggregate is a completely inert filler in concrete is not true, its physical characteristics and in some cases its chemical composition affecting to a varying degree the properties of concrete in both its plastic and hardened states.

Basic Characteristics of Aggregate

The criterion for a good aggregate is that it should produce the desired properties in both the fresh and hardened concrete. In testing aggregates it is important that a truly representative sample is used. The procedure for obtaining such a test sample is described in BS 812.

Physical properties

The properties of the aggregate known to have a significant effect on concrete behaviour are its strength, deformation, durability, toughness, hardness, volume change, porosity, specific gravity and chemical reactivity.

The *strength* of an aggregate limits the attainable strength of concrete only when its compressive strength is less than or of the same order as the design strength of concrete. In practice the majority of rock aggregates used are usually considerably stronger than concrete. While the strength of concrete does not normally exceed 80 N mm^{-2} and is generally between 30 and 50 N mm^{-2} the strength of the aggregates commonly used is in the range 70 to 350 N mm^{-2}. In general, igneous rocks are very much stronger than sedimentary and metamorphic rocks. Because of the irregular size and shape of aggregate particles a direct measurement of their strength properties is not possible. These are normally assessed from compressive strength tests on cylindrical specimens taken from the parent rock and from crushing value tests on the bulk aggregate (BS 812). For weaker materials, that is, those with crushing values greater than 30, the crushing value may be unreliable and the load required to produce 10 per cent fines in the crushing test should be used. The results of these tests for the strength properties of aggregates are only a guide to aggregate quality, however, which may also be assessed from the intensity of aggregate fracturing in ruptured concrete specimens.

The *deformation* characteristics of an aggregate are seldom considered in assessing its suitability for concrete work although they can easily be determined from compression tests on specimens from the parent rock. In general, the modulus of elasticity of concrete increases with increasing aggregate modulus. The deformation characteristics of the aggregate also play an important part in the creep and shrinkage properties of concrete as the restraint afforded by the aggregate to the creep and shrinkage of the cement paste depends on their relative moduli of elasticity.

A commonly used definition for aggregate *toughness* is its resistance to failure by impact and this is normally determined from the aggregate impact test (BS 812). Since the apparatus is portable, cheap, simple to operate and rapid in application it can be used in the field for quality control purposes. *Hardness* is the resistance of an aggregate to wear and is normally determined by an abrasion test (BS 812). Toughness and hardness properties of an aggregate are particularly important for concrete used in road pavements.

Volume changes due to moisture movements in aggregates derived from sandstones, greywackes and some basalts may result in considerable shrinkage of the concrete. If the concrete is restrained this produces internal tensile stresses, possible tensile cracking and subsequent deterioration of the concrete. If the coefficient of thermal expansion of an aggregate differs considerably from that of the cement paste this too may adversely affect the concrete performance.

Aggregate *porosity* is an important property since it affects the behaviour of both freshly mixed and hardened concrete through its effect on the strength, water absorption and permeability of the aggregate. An aggregate with high porosity will tend to produce a less durable concrete, particularly when subjected to freezing and thawing, than an aggregate with low porosity. Direct measurement of porosity is difficult and in practice a related property, namely, water absorption, is measured. The water absorption is defined as the weight of water absorbed by a dry aggregate in reaching a saturated surface-dry state and is expressed as a percentage of the weight of the dry aggregate. In general, sedimentary rock materials have the highest absorption values. A gravel aggregate with a similar petrology to that of a crushed-rock aggregate will absorb more water because of its greater weathering. It should be noted that, for a given rock type, absorption can vary depending on the way in which it is measured and also on the size of the aggregate particles.

The total moisture content, that is, the absorbed moisture plus the free or surface water, of aggregates used for making concrete varies considerably. The water added at the mixer must be adjusted to take account of this if the free water content is to be kept constant and the required workability and strength of concrete maintained. Concrete mix proportions are normally based on the weight of the aggregates in their saturated surface-dry condition and any change in their moisture content must be reflected in adjustments to the weights of the aggregates used in the mix. Several methods for the determination of moisture content and absorption are described in BS 812.

The *specific gravity* of a material is the ratio of its unit weight to that of water. Since aggregates incorporate pores the value of specific gravity varies depending on the extent to which the pores contain (absorbed) water when the value is determined (BS 812). For the purposes of mix design the specific gravity on a 'saturated and surface-dry' basis is used. This is given by $A/(A - B)$ where A is the weight of the saturated surface-dry sample in air and B the weight of the saturated sample in water. The specific gravity of most natural aggregates falls within the range 2.5 - 3.0. For artificial aggregates the specific gravity varies over a much wider range. Aggregate is the major constituent of concrete and as such its specific gravity is an important factor affecting the density of the resulting concrete.

Shape and surface texture

Aggregate shape and surface texture can affect the properties of concrete in both its plastic and hardened states. These external characteristics may be assessed by observation of the aggregate particles and classification of their particle shape and texture in accordance with tables 12.7 and 12.8 from BS 812. The classification is somewhat subjective, however, and the particle shape may also be assessed by direct measurement of the aggregate particles to determine the flakiness, elongation and angularity.

TABLE 12.7
Shape of aggregates, BS 812

Classification	Description
Rounded	Fully water-worn or completely shaped by attrition
Irregular	Naturally irregular, or partly shaped by attrition and having rounded edges
Angular	Possessing well-defined edges formed at the intersection of roughly planar faces
Flaky	Material of which the thickness is small relative to the other two dimensions
Elongated	Material, usually angular, in which the length is considerably larger than the other two dimensions
Flaky and elongated	Material having the length considerably larger than the width, and the width considerably larger than the thickness

TABLE 12.8
Surface texture of aggregates, BS 812

Surface texture	Characteristics
Glassy	Conchoidal fracture
Smooth	Water-worn, or smooth due to fracture of laminated or fine-grained rock
Granular	Fracture showing more or less uniform rounded grains
Rough	Rough fracture of fine or medium-grained rock containing no easily visible crystalline constituents
Crystalline	Containing easily visible crystalline constituents
Honeycombed	With visible pores and cavities

The angularity is expressed in terms of the angularity number. This is the difference between the solid volume of rounded aggregate particles after compaction in a standard cylinder, expressed as a percentage of the volume of the cylinder, and the solid volume of the particular aggregate being investigated when compacted in a similar manner. It is a measure of the increased voids in the compacted non-rounded aggregate, expressed as a percentage of the total volume. The angularity number ranges from zero for a perfectly rounded aggregate to about 12. Hughes (1966) proposed an alternative method for determining shape based on an *angularity factor.* This is defined as the ratio of the solid volume of loose single-size spheres to the solid volume of a single-size aggregate of similar nominal dimensions placed under identical conditions. The method has been shown to be suitable for both fine and coarse aggregates.

Grading

The grading of an aggregate defines the proportions of particles of different size in the aggregate. The size of the aggregate particles normally used in concrete varies from 37.5 to 0.15 mm. BS 882 places aggregates in three main categories: *fine aggregate* or *sand* containing particles the majority of which are smaller than 5.00 mm, *coarse aggregate* containing particles the majority of which are larger than 5.00 mm and *all-in aggregate* comprising both fine and coarse aggregate.

The grading of an aggregate can have a considerable effect on the workability and stability of a concrete mix (see chapter 13) and is a most important factor in concrete mix design. In the United Kingdom, the grading of natural aggregates is generally required to be within the limits specified in BS 882. The fine aggregates are generally required to conform to any one of four standard grading zones, all of which are suitable for concrete provided suitable mix proportions are used.

A sieve analysis is used for determining the grading of an aggregate (BS 812). The aggregate sample must be air-dried and the weight of material retained on each sieve should not exceed the specified maximum values as overloading will give erroneous results. Sieving may be performed by hand or machine and the results of a typical analysis are shown in table 12.9 from which it will also be seen

TABLE 12.9
A typical example of calculation for aggregate grading

BS 410 sieve size (mm)	Weight of aggregate retained (g)	Percentage retained	Cumulative percentage retained	Cumulative percentage passing
5.00	0	0	0	100
2.36	32	16	16	84
1.18	40	20	36	64
0.60	42	21	57	43
0.30	46	23	80	20
0.15	32	16	96	4
Pan	8	4	100	
Total	200			

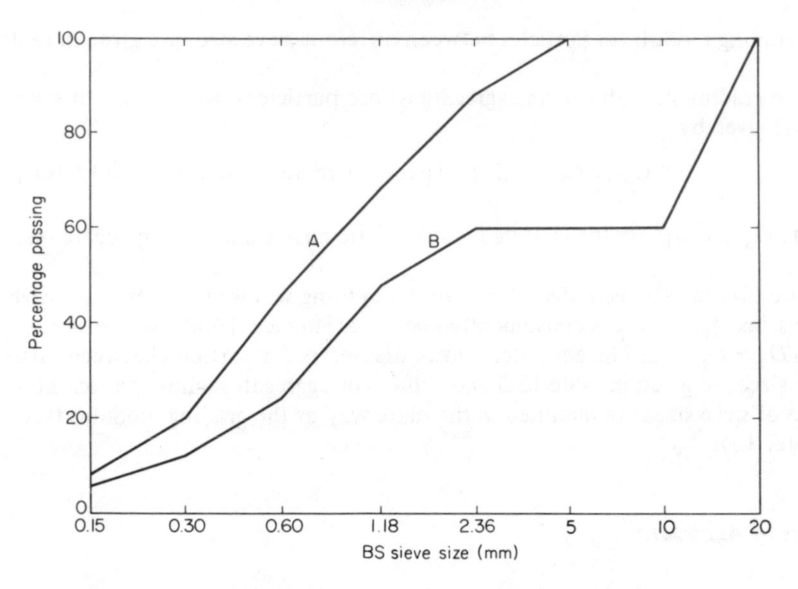

Figure 12.4 *Typical aggregate grading curves : (A) continuously graded (see table 12.9) and (B) gap-graded aggregates*

that successive sieve sizes decrease by a factor of about two. A convenient visual assessment of the particle size distribution can be obtained from a grading chart (figure 12.4). Curve A represents a continuously graded aggregate and curve B, in which two of the sizes are missing, represents a gap-graded aggregate.

In practice, fine and coarse aggregates are batched separately, their proportions being governed largely by their respective gradings. One method for determining the required aggregate contents (see chapter 15) uses the grading modulus and the equivalent mean diameter of both fine and coarse aggregate.

The *grading modulus* of an aggregate is the mean specific surface of spheres which pass through the same sieve size as the actual aggregate particles and whose size distribution, or grading, corresponds to that of those aggregate particles.

For spheres with constant diameter D, the grading modulus G and the specific surface of a sphere are identical, that is, $G = (\pi D^2)/(\pi D^3/6) = 6/D$.

Consider now the grading modulus of aggregate particles lying wholly between successive sieve sizes D_1 and D_2 corresponding to the smallest and largest diameters of the associated spheres. If, between successive sieve sizes, the proportion by volume of particles of diameter D is assumed to be inversely proportional to D_1 then the grading modulus is given by

$$G = \frac{\int_{D_1}^{D_2} \left(\frac{6}{D} \times \frac{1}{D}\right) \mathrm{d}D}{\int_{D_1}^{D_2} \frac{1}{D} \mathrm{d}D} = \frac{6\left(\frac{1}{D_1} - \frac{1}{D_2}\right)}{\log_e \left(\frac{D_2}{D_1}\right)}$$

The grading moduli for particles between different sieve sizes are given in table 15.3.

The grading modulus of an aggregate whose particles cover a range of sieve sizes is given by

$$G_a \text{ or } G_b = \Sigma\ [G\,(\text{per cent retained on each sieve}) / 100]$$

where G_a and G_b are the grading moduli of the coarse and fine aggregate respectively.

The *equivalent mean diameter* of particles lying wholly between successive sieve sizes D_1 and D_2 is conveniently assumed (Hughes, 1960) to be given by $D = (D_1 + D_2) / 2$. The equivalent mean diameters for particles between different sieve sizes are given in table 15.3 and values for aggregates whose particles cover a range of sieve sizes are obtained in the same way as the grading modulus (see chapter 15).

Types of Aggregate

In the previous sections discussion has been mainly confined to rock aggregates. Although other types of aggregate are used for making concrete their contribution is very small in comparison with rock aggregates. The general classifications of aggregates and the related British Standards are shown in figure 12.5. For a more detailed discussion the reader is referred to Taylor (1965), Short and Kinniburgh (1968) and BRE Digests 111 and 123.

Heavyweight aggregate

Heavyweight aggregates provide an effective and economical use of concrete for radiation shielding by giving the necessary protection against X-rays, gamma-rays and neutrons. The effectiveness of heavyweight concrete, with a density from 4000 to 5500 kg m^{-3}, depends on the aggregate type, the dimensions and the degree of compaction. It is frequently difficult with heavyweight aggregates to obtain a mix which is both workable and not prone to segregation.

Normal aggregate

These aggregates are suitable for most purposes and produce concrete with a density in the range 2300 to 2500 kg m^{-3}. Rock aggregates are obtained by crushing quarried rock to the required particle size or by extracting the sand and gravel deposits formed by alluvial or glacial action. Some sands and gravels are also obtained by dredging from sea and river beds. Aggregates, in particular sands and gravels, should be washed to remove impurities such as clay and silt. In the case of river and marine aggregates the chloride content should generally be less than 1 per cent if these are to be used for structural concrete.

The properties of rock aggregates depend on their composition, grain size and

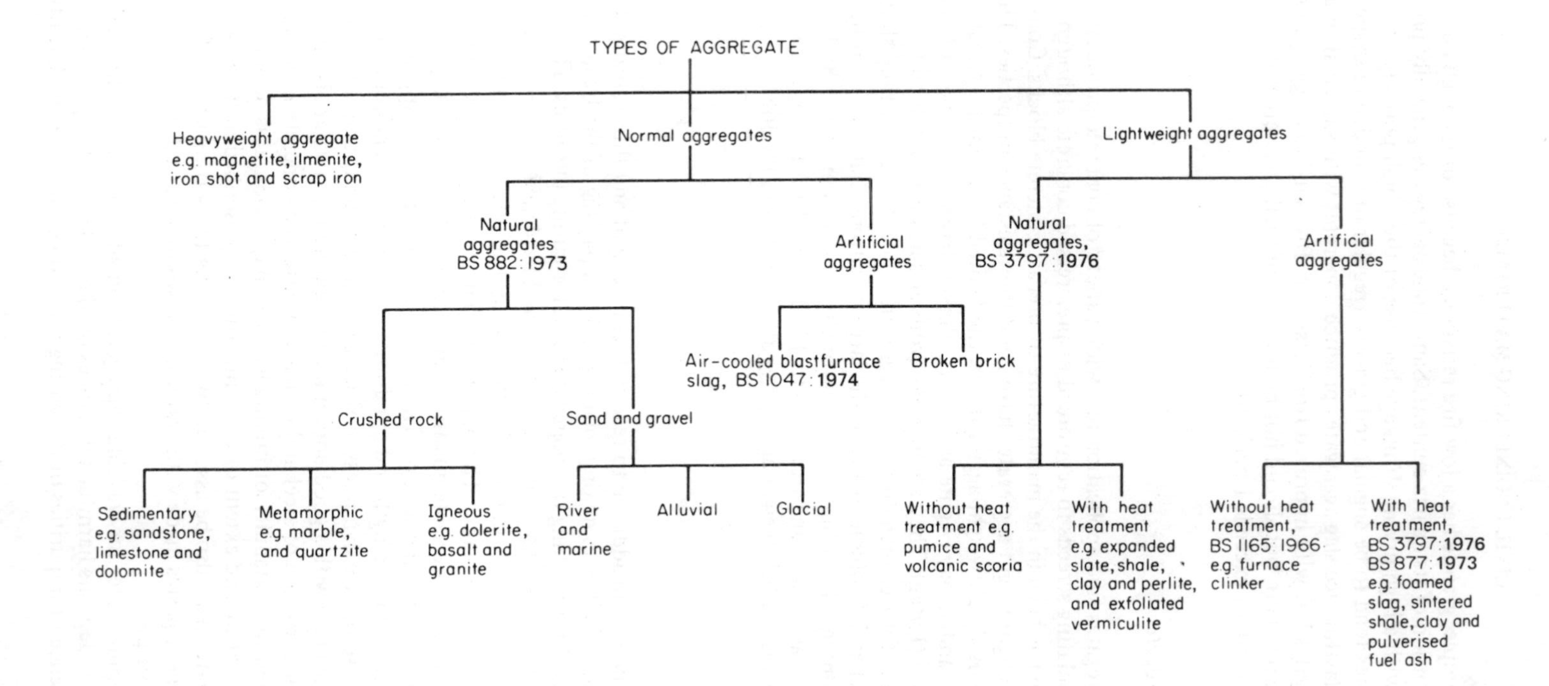

Figure 12.5 *A classification of aggregates used for making concrete*

texture. For example, granite has a low fire resistance because of the high coefficient of expansion of its quartz content. Sandstone aggregates generally produce concretes with a high drying shrinkage because of their high porosity. Crushed aggregates tend to be angular and gravels irregular or rounded in shape.

Air-cooled blastfurnace slag aggregates produce concretes with similar strength to natural aggregates but with improved fire resistance. Broken-brick aggregates are also very fire resistant, but should not be used for normal concrete if its soluble sulphate content exceeds 1 per cent.

Lightweight aggregate

Lightweight aggregates find application in a wide variety of concrete products ranging from insulating screeds to reinforced or prestressed concrete although their greatest use has been in the manufacture of precast concrete blocks. Concretes made with lightweight aggregates have good fire resistance properties. The most commonly used lightweight aggregates in the United Kingdom are expanded slate (Solite), expanded clay (Aglite and Leca), clinker, foamed slag and sintered pulverised fuel ash (Lytag). They are highly porous and absorb considerably greater quantities of water than do normal aggregates. For this reason they should normally be batched by volume owing to the large variations that can occur in their moisture content. Their bulk density normally ranges from 350 to 850 kg m^{-3} for coarse aggregates and from 750 to 1100 kg m^{-3} for fine aggregates. The methods for testing lightweight aggregates are described in BS 3681.

12.3 Water

Water used in concrete, in addition to reacting with cement and thus causing it to set and harden, also facilitates mixing, placing and compacting of the fresh concrete. It is also used for washing the aggregates and for curing purposes. The effect of water content on the properties of fresh and hardened concrete is discussed in chapters 13 and 14. In general water fit for drinking, such as tap water, is acceptable for mixing concrete. The impurities that are likely to have an adverse effect when present in appreciable quantities include silt, clay, acids, alkalis and other salts, organic matter and sewage. The use of seawater does not appear to have any adverse effect on the strength and durability of Portland cement concrete but it is known to cause surface dampness, efflorescence and staining and should be avoided where concrete with a good appearance is required. Seawater also increases the risk of corrosion of steel and its use in reinforced concrete is not recommended. When the suitability of mixing water is in question, it is desirable to test for both the nature and extent of contamination as prescribed in BS 3148. The quality of water may also be assessed by comparing the setting time and soundness of cement pastes made with water of known quality and the water whose quality is suspect.

The use of impure water for washing aggregates can adversely affect strength and durability if it deposits harmful substances on the surface of the particles. In general, the presence of impurities in the curing water does not have any harmful effects, although it may spoil the appearance of concrete. Water containing appreciable amounts of acid or organic materials should be avoided.

12.4 Admixtures

Admixtures are substances introduced into concrete mixes in order to alter or improve the properties of the fresh or hardened concrete or both. In general, these changes are effected through the influence of the admixtures on hydration, liberation of heat, formation of pores and the development of the gel structure. Concrete admixtures should only be considered for use when the required modifications cannot be made by varying the composition and proportion of the basic constituent materials, or when the admixtures can produce the required effects more economically.

Since admixtures may also have detrimental effects, their suitability for a particular concrete should be carefully evaluated before use, based on a knowledge of their main active ingredients, on available performance data and on trial mixes. The specific effects of an admixture generally vary with the type of cement, cement - water ratio, ambient conditions (particularly temperature) and its dosage. Since the quantity of admixture used is both small and critical the required dose must be carefully determined and administered. Where related British Standards exist the admixtures should comply with their specifications. It should be remembered that admixtures are not intended to replace good concreting practice and should not be used indiscriminately.

Types of Admixture

Several hundred proprietary admixtures are available and since a great many usually contain several chemicals intended simultaneously to change several properties of concrete, they are not easy to classify. Moreover, as many of the individual constituents and their proportions are not widely known the selection of an admixture must frequently be based on the information provided by the suppliers. Of the different types of admixture listed in table 12.10 the air-entraining, accelerating, retarding and water-reducing admixtures are most commonly used.

Air-entraining agents

These are probably the most important group of admixtures. They improve the durability of concrete, in particular its resistance to the effects of frost and deicing salts. The entrainment of air in the form of very small and stable bubbles can be achieved by using foaming agents based on natural resins, animal or vegetable fats and synthetic detergents which promote the formation of air bubbles during mixing or by using gas-producing chemicals such as zinc or aluminium powder which react with cement to produce gas bubbles. The first method is generally more effective and is the most widely used. The beneficial effects of entrained air are produced in two ways: first, by disrupting the continuity of

TABLE 12.10
Types of admixture

Type	Air entrainers	Accelerators		Retarders	Water reducers	Pigments	Pozzolanas	Pore fillers	Water repellers
		Setting	Setting and hardening						
Principal constituents	Foaming agents e.g. wood resins synthetic detergent or gas generators, e.g. zinc powder	Highly alkaline solutions, e.g. aluminium chloride	Calcium chloride	Lignosulphonic or hydroxylated-carboxylic acids with cellulose or starch	Lignosulphonic or hydroxylated-carboxylic acids	Natural and manufactured pigments	Pumicite and pulverised fuel ash	Kaoline, bentonite and rock flour	Metallic soaps or mineral and vegetable oil derivatives
Action with concrete	Formation of small stable discrete bubbles	Very rapid setting	Increased rate of setting and hardening	Delayed onset of setting and hardening	Increased workability or decreased water content	Coloured rendering	Combine with free lime. Reduced rate of hydration	Increase cohesiveness and workability	Decreased permeability
Main uses or effects	Increase resistance to frost action. Used in road and runway construction	Used for emergency repair work	High heat evolution. Used for concreting in winter conditions and early strength for rapid removal of formwork	Used for concreting in hot weather	Used to facilitate placing or to give higher strength and durability	Used for architectural purposes	Low heat evolution. Increased durability. Used in mass concrete work	Used with poorly graded aggregates	Prevent absorption of rainwater
Adverse effects	Some reduction in strength	Reduction in strength	Increased drying shrinkage. Decreased resistance to sulphate attack. Danger of metal corrosion	Increased bleeding with some types. Increased drying shrinkage with others	As for retarders	—	—	May result in some reduction in strength	—

capillary pores and thus reducing the permeability of concrete, and second, by reducing the internal stresses caused by the expansion of water on freezing.

Air-entraining agents also improve the workability and cohesiveness of fresh concrete and tend to reduce bleeding and segregation. However, entrained air results in some reduction in concrete strength. Since improvements in workability can permit a reduction in the water content the loss in strength can be minimised. The amount of entrained air is dependent on the type of cement, mix proportions and ambient temperature and it should therefore only be used when adequate supervision is assured. A method for determining the amount of air entrained in concrete is described in BS 1881: Part 2.

Accelerating agents

These can be divided into two groups, namely, setting accelerators and setting and hardening accelerators. The first of these are alkaline solutions which can considerably reduce the setting time and are particularly suitable for repair work involving water leakage. Because of their adverse effect on subsequent strength development these admixtures should not be used where the final concrete strength is an important consideration. Setting and hardening accelerators increase the rate of both setting and early strength development. The most common admixture for this purpose is calcium chloride which should comply with BS 3587. Since its use may result in several adverse effects such as increased drying shrinkage, reduced resistance to sulphate attack and increased risk of corrosion of steel reinforcement, it should only be used with extreme caution and in accordance with any relevant specifications. It may usefully be employed for concreting in winter conditions, for emergency repair work or where early removal of formwork is required.

Retarders

Most admixtures in this group are based on lignosulphonic or hydroxylated-carboxylic acids and their salts with cellulose or starch. They are used mainly in hot countries where high temperatures can reduce the normal setting and hardening times. A slightly reduced water content may be used when using these retarding agents, with a corresponding increase in final concrete strength. The lignin-based retarders result in some air-entrainment and tend to increase cohesiveness and reduce bleeding although drying shrinkage may be increased. The hydroxy-carboxylic retarders, however, tend to increase bleeding.

Water reducers or plasticisers

These admixtures are also based on lignosulphonic and hydroxylated-carboxylic acids. Their effect is thought to be due to an increased dispersion of cement particles causing a reduction in the viscosity of the concrete. They are used to increase workability and are normally employed with harsh mixes or where placement is difficult. They can also be used to increase strength and durability since for a given workability less water is necessary.

Pigments

Colouring pigments are normally used for architectural purposes and the best effect is produced when they are interground with the cement clinker rather than when added during mixing. Pigments used for this purpose should conform to BS 1014.

Pozzolanas

The most commonly used pozzolanas are pumicite and pulverised fuel ash. Because of their reaction with lime, which is liberated during the hydration of cement, these materials can improve the durability of concrete. Since they retard the rate of setting and hardening but have no long term effect on strength (provided proper curing is maintained) they can be used in mass concrete work. Pulverised fuel ash can also be used as a replacement for sand (up to 20 per cent) in harsh mixes to improve workability. Pulverised ash should conform to BS 3892 and be used in accordance with CP 110: Part 1.

Pore fillers

These are chemically inactive finely ground materials such as bentonite, kaoline or rock flour. These admixtures are thought to improve workability, stability and impermeability of concrete.

Water-repelling agents

These are the least effective of all admixtures and are based on metallic soaps or vegetable or mineral oils. Their use gives a slight temporary reduction in concrete permeability.

13

Properties of Fresh Concrete

Fresh concrete is a mixture of water, cement, aggregate and admixture (if any). After mixing, operations such as transporting, placing, compacting and finishing of fresh concrete can all considerably affect the properties of hardened concrete. It is important that the constituent materials remain uniformly distributed within the concrete mass during the various stages of its handling and that full compaction is achieved. When either of these conditions is not satisfied the properties of the resulting hardened concrete, for example, strength and durability, are adversely affected.

The characteristics of fresh concrete which affect full compaction are its consistency, mobility and compactability. In concrete practice these are often collectively known as workability. The ability of concrete to maintain its uniformity is governed by its stability, which depends on its consistency and its cohesiveness. Since the methods employed for conveying, placing and consolidating a concrete mix, as well as the nature of the section to be cast, may vary from job to job it follows that the corresponding workability and stability requirements will also vary. The assessment of the suitability of a fresh concrete for a particular job will always to some extent remain a matter of personal judgement.

In spite of its importance, the behaviour of plastic concrete often tends to be overlooked. It is recommended that students should learn to appreciate the significance of the various characteristics of concrete in its plastic state and know how these may alter during operations involved in casting a concrete structure.

13.1 Workability

Workability of concrete has never been precisely defined. For practical purposes it generally implies the ease with which a concrete mix can be handled from the mixer to its finally compacted shape. The three main characteristics of the property are consistency, mobility and compactability. Consistency is a measure of wetness or fluidity. Mobility defines the ease with which a mix can flow into and completely fill the formwork or mould. Compactability is the ease with which a given mix can be fully compacted, all the trapped air being removed. In this context the required workability of a mix depends not only on the characteristics and relative proportions of the constituent materials but also on (1) the methods employed for conveyance and compaction, (2) the size, shape and surface roughness of formwork or moulds and (3) the quantity and spacing of reinforcement.

Another commonly accepted definition of workability is related to the amount of useful internal work necessary to produce full compaction. It should be appreciated that the necessary work again depends on the nature of the section being cast. Measurement of internal work presents many difficulties and several methods have been developed for this purpose but none gives an absolute measure of workability.

The tests commonly used for measuring workability do not measure the individual characteristics (consistency, mobility and compactability) of workability. However, they do provide useful and practical guidance on the workability of a mix. Workability affects the quality of concrete and has a direct bearing on cost so that, for example, an unworkable concrete mix requires more time and labour for full compaction. It is most important that a realistic assessment is made of the workability required for given site conditions before any decision is taken regarding suitable concrete mix proportions.

13.2 Measurement of Workability

Three tests widely used for measuring workability are the slump, compacting factor and V - B consistometer tests (figure 13.1). These are standard tests in the United Kingdom and are described in detail in (BS 1881 : Part 2). Their use is also recommended in CP 110: Part 1. It is important to note that there is no single relationship between the slump, compacting factor and V - B results for different concretes. In the following sections the salient features of these tests together with their merits and limitations are discussed.

Slump Test

This test was developed by Chapman in the United States in 1913. A 300 mm high concrete cone, prepared under standard conditions (BS 1881:Part 2) is allowed to subside and the slump or reduction in height of the cone is taken to be a measure of workability. The apparatus is inexpensive, portable and robust and is the simplest of all the methods employed for measuring workability. It is not surprising that, in spite of its several limitations, the slump test has retained its popularity.

The test primarily measures the consistency of plastic concrete and although it is difficult to see any significant relationship between slump and workability as defined previously, it is suitable for detecting changes in workability. For example, an increase in the water content or deficiency in the proportion of fine aggregate results in an increase in slump. Although the test is suitable for quality-control purposes it should be remembered that it is generally considered to be unsuitable for mix design since concretes requiring varying amounts of work for compaction can have similar numerical values of slump. The sensitivity and reliability of the test for detecting variation in mixes of different workabilities is largely dependent

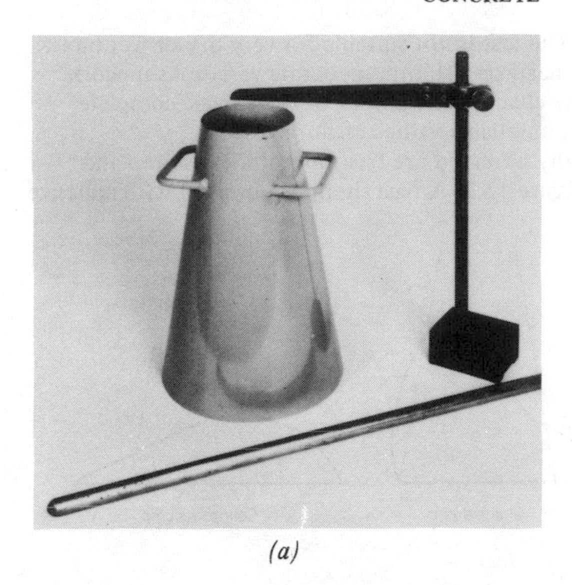

(a)

(b)

(c)

Figure 13.1 *Apparatus for workability measurement* : *(a) slump cone, (b) compacting factor and (c) V–B consistometer*

on its sensitivity to consistency. The test is not suitable for very dry or wet mixes. For very dry mixes, with zero or near-zero slump, moderate variations in workability do not result in measurable changes in slump. For wet mixes, complete collapse of the concrete produces unreliable values of slump.

The three types of slump usually observed are true slump, shear slump and collapse slump, as illustrated in figure 13.2. A true slump is observed with cohesive

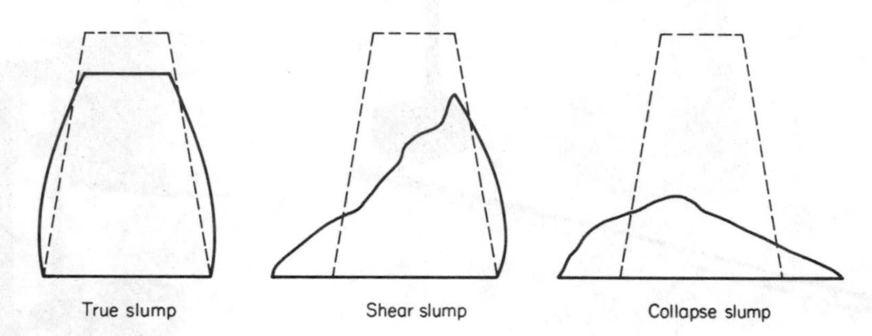

Figure 13.2 *Three main types of slump*

and rich mixes for which the slump is generally sensitive to variations in workability. A collapse slump is usually associated with very wet mixes and is generally indicative of poor quality concrete and most frequently results from segregation of its constituent materials. Shear slump occurs more often in leaner mixes than in rich ones and indicates a lack of cohesion which is generally associated with harsh mixes (low mortar content). Whenever a shear slump is obtained the test should be repeated and, if persistent, this fact should be recorded together with test results, because widely different values of slump can be obtained depending on whether the slump is of true or shear form.

The standard slump apparatus is only suitable for concretes in which the maximum aggregate size does not exceed 37.5 mm. It should be noted that the value of slump changes with time after mixing owing to normal hydration processes and evaporation of some of the free water, and it is desirable therefore that tests are performed within a fixed period of time.

Compacting Factor Test

This test, developed in the United Kingdom by Glanville *et al.* (1947), measures the degree of compaction for a standard amount of work and thus offers a direct and reasonably reliable assessment of the workability of concrete as previously defined. The apparatus is a relatively simple mechanical contrivance (figure 13.1) and is fully described in BS 1881: Part 2. The test requires measurement of the weights of the partially and fully compacted concrete and the ratio of the partially compacted weight to the fully compacted weight, which is always less than 1, is known as the compacting factor. For the normal range of concretes the compacting factor lies between 0.80 and 0.92. The test is particularly useful for drier mixes for which the slump test is not satisfactory. The sensitivity of the compacting factor is reduced outside the normal range of workabilities and is generally unsatisfactory for compacting factors greater than 0.92.

It should also be appreciated that, strictly speaking, some of the basic assumptions of the test are not correct. The work done to overcome surface friction of the measuring cylinder probably varies with the characteristics of the mix. It has been shown by Cusens (1956) that for concretes with very low workability the actual work required to obtain full compaction depends on the richness of a mix while the compacting factor remains sensibly unaffected. Thus it follows that the generally held belief that concretes with the same compacting factor require the same amount of work for full compaction cannot always be justified. One further point to note is that the procedure for placing concrete in the measuring cylinder bears no resemblance to methods commonly employed on the site. As in the slump test, the measurement of compacting factor must be made within a certain specified period. The standard apparatus is suitable for concrete with a maximum aggregate size of up to 37.5 mm.

V - B Consistometer Test

This test was developed in Sweden by Bährner (1940) (see figure 13.1). Although generally regarded as a test primarily used in research its potential is now more widely acknowledged in industry and the test is gradually being accepted. In this test (BS 1881: Part 2) the time taken to transform, by means of vibration, a standard cone of concrete to a compacted flat cylindrical mass is recorded. This is known as the V - B time, in seconds, and is stated to the nearest 0.5 s. Unlike the two previous tests, the treatment of concrete in this test is comparable to the method of compacting concrete in practice. Moreover, the test is sensitive to change in consistency, mobility and compactability, and therefore a reasonable correlation between the test results and site assessment of workability can be expected.

The test is suitable for a wide range of mixes and, unlike the slump and compacting factor tests, it is sensitive to variations in workability of very dry and also air-entrained concretes. It is also more sensitive to variation in aggregate characteristics such as shape and surface texture. The reproducibility of results is good. As for other tests its accuracy tends to decrease with increasing maximum size of aggregate; above 19.0 mm the test results become somewhat unreliable. For concretes requiring very little vibration for compaction the V - B time is only about 3 s. Such results are likely to be less reliable than for larger V - B times because of the difficulty in estimating the time of the end point (concrete in contact with the whole of the underside of the plastic disc). At the other end of the workability range, such as with very dry mixes, the recorded V - B times are likely to be in excess of their *true* workability since prolonged vibration is required to remove the entrapped air bubbles under the transparent disc. To overcome this difficulty an automatic device which records the vertical settlement of the disc with respect to time can be attached to the apparatus. This recording device can also assist in eliminating human error in judging the end point. The apparatus for the V – B test is more expensive than that for the slump and compacting factor tests, requiring an electric power supply and greater experience in handling; all these factors make it more suitable for the precast concrete industry and ready-mixed concrete plants than for general site use.

13.3 Factors affecting Workability

Various factors known to influence the workability of a freshly mixed concrete are shown in figure 13.3. From the following discussion it will be apparent that a change in workability associated with the constituent materials is mainly affected by water content and specific surface of cement and aggregate.

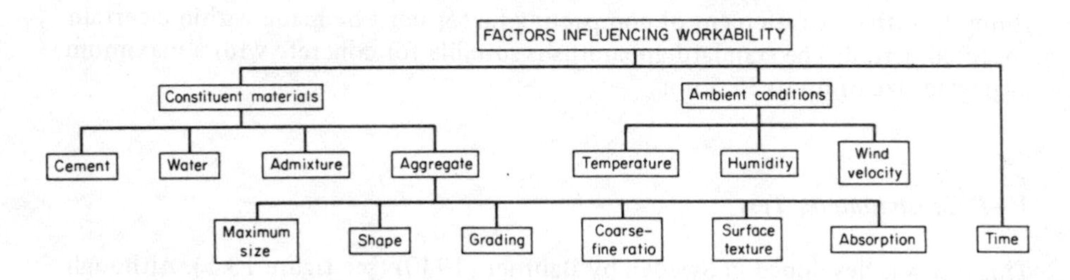

Figure 13.3 *Factors affecting workability of fresh concrete*

Cement and Water

Typical relationships between the cement – water ratio (by volume) and the volume fraction of cement for different workabilities are shown in figure 15.5. The change in workability for a given change in cement – water ratio is greater when the water content is changed than when only the cement content is changed. In general the effect of the cement content is greater for richer mixes. Hughes (1971) has shown that similar linear relationships exist irrespective of the properties of the constituent materials.

For a given mix, the workability of the concrete decreases as the fineness of the cement increases as a result of the increased specific surface, this effect being more marked in rich mixtures. It should also be noted that the finer cements improve the cohesiveness of a mix. With the exception of gypsum, the composition of cement has no apparent effect on workability. Unstable gypsum is responsible for *false set,* which can impair workability unless prolonged mixing or remixing of the fresh concrete is carried out. Variations in quality of water suitable for making concrete have no significant effect on workability.

Admixtures

The principal admixtures affecting improvement in the workability of concrete are water-reducing and air-entraining agents. The extent of the increase in workability is dependent on the type and amount of admixture used and the general characteristics of the fresh concrete.

Workability admixtures are used to increase workability while the mix proportions are kept constant or to reduce the water content while maintaining constant workability. The former results in a slight reduction in concrete strength.

Air-entraining agents are by far the most commonly used workability admixtures because they also improve both the cohesiveness of the plastic concrete and the frost resistance of the resulting hardened concrete. Two points of practical importance concerning air-entrained concrete are that for a given amount of entrained air, the increase in workability tends to be smaller for concretes containing rounded aggregates or low cement - water ratios (by volume) and, in general, the rate of increase in workability tends to decrease with increasing air content. However, as a guide it may be assumed that every 1 per cent increase in air content will increase the compacting factor by 0.01 and reduce the V - B time by 10 per cent.

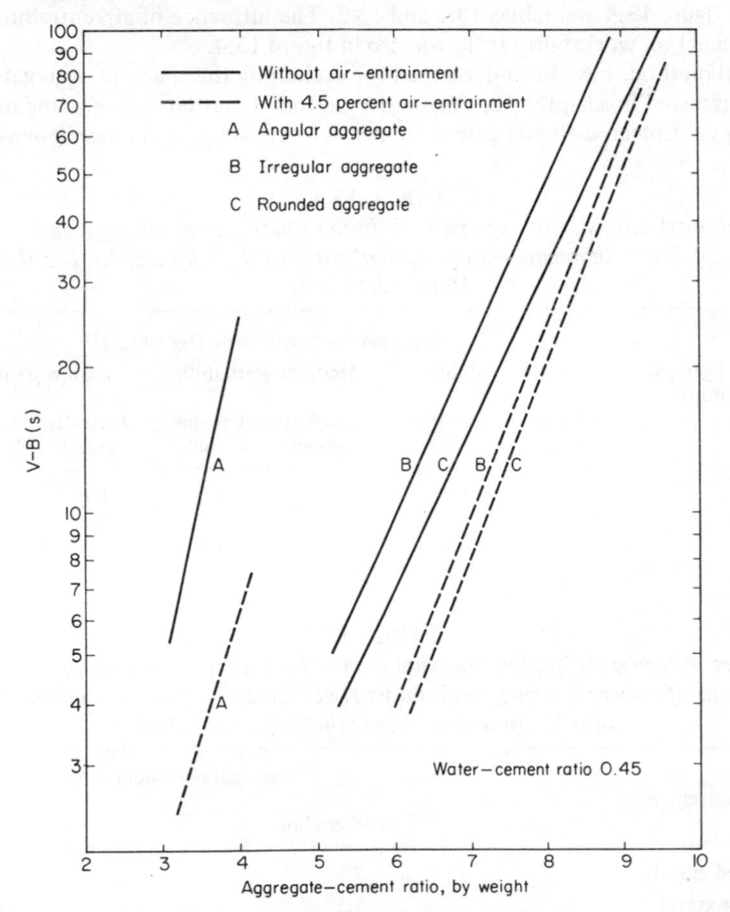

Figure 13.4 *Effect of aggregate shape on aggregate - cement ratio of concretes for different workabilities, based on Cornelius (1970)*

Aggregate

For given cement, water and aggregate contents, the workability of concrete is mainly influenced by the total surface area of the aggregate. The surface area is governed by the maximum size, grading and shape of the aggregate. Workability decreases as the specific surface increases, since this requires a greater proportion of cement paste to wet the aggregate particles, thus leaving a smaller amount of paste for lubrication. It follows that, all other conditions being equal, the workability will be increased when the maximum size of aggregate increases, the aggregate particles become rounded or the overall grading becomes coarser. However, the magnitude of this change in workability depends on the mix proportions, the effect of the aggregate being negligible for very rich mixes (aggregate - cement ratios approaching 2). The practical significance of this is that for a given workability and cement - water ratio the amount of aggregate which can be used in a mix varies depending on the shape, maximum size and grading of the aggregate, as shown in figure 13.4 and tables 13.1 and 13.2. The influence of air-entrainment (4.5 per cent) on workability is shown also in figure 13.4.

Several methods have been developed for evaluating the shape of aggregate, a subject discussed in chapter 12. Angularity factors together with grading modulus and equivalent mean diameter provide a means of considering the respective effects

TABLE 13.1

Effect of maximum size of aggregate of similar grading zone on aggregate - cement ratio of concrete having water - cement ratio of 0.55 by weight, based on McIntosh (1964)

Maximum aggregate size (mm)	Aggregate - cement ratio (by weight)					
	Low workability		Medium workability		High workability	
	Irregular gravel	Crushed rock	Irregular gravel	Crushed rock	Irregular gravel	Crushed rock
9.5	5.3	4.8	4.7	4.2	4.4	3.7
19.0	6.2	5.5	5.4	4.7	4.9	4.4
37.5	7.6	6.4	6.5	5.5	5.9	5.2

TABLE 13.2

Effect of aggregate grading (maximum size 19.0 mm) on aggregate - cement ratio of concrete having medium workability and water - cement ratio of 0.55 by weight, based on McIntosh (1964)

Type of aggregate	Aggregate - cement ratio	
	Coarse grading	Fine grading
Rounded gravel	7.3	6.3
Irregular gravel	5.5	5.1
Crushed rock	4.7	4.3

of shape, size and grading of aggregate (see chapter 15). Since the strength of a fully compacted concrete, for given materials and cement - water ratio, is not dependent on the ratio of coarse to fine aggregate, maximum economy can be obtained by using the coarse aggregate content producing the maximum workability for a given cement content (Hughes, 1960) (see figure 13.5). The use of optimum coarse aggregate content in concrete mix design is described in chapter 15. It should be noted that it is the volume fraction of an aggregate, rather than its weight, which is important.

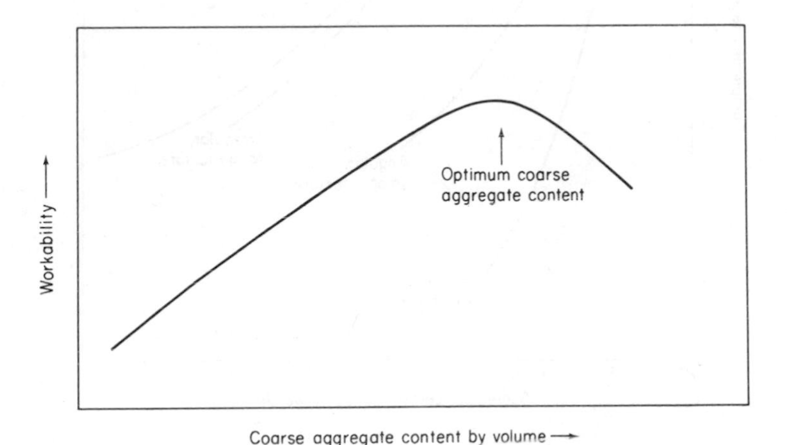

Figure 13.5 *A typical relationship between workability and coarse aggregate content of concrete, based on Hughes (1960)*

The effect of surface texture on workability is shown in figure 13.6. It can be seen that aggregates with a smooth texture result in higher workabilities than aggregates with a rough texture. Absorption characteristics of aggregate also affect workability where dry or partially dry aggregates are used. In such a case workability drops, the extent of the reduction being dependent on the aggregate content and its absorption capacity.

Ambient Conditions

Environmental factors that may cause a reduction in workability are temperature, humidity and wind velocity. For a given concrete, changes in workability are governed by the rate of hydration of the cement and the rate of evaporation of water. Therefore both the time interval from the commencement of mixing to compaction and the conditions of exposure influence the reduction in workability. An increase in the temperature speeds up the rate at which water is used for hydration as well as its loss through evaporation. Likewise wind velocity and humidity influence the workability as they affect the rate of evaporation. It is worth remembering that in practice these factors depend on weather conditions and cannot be controlled.

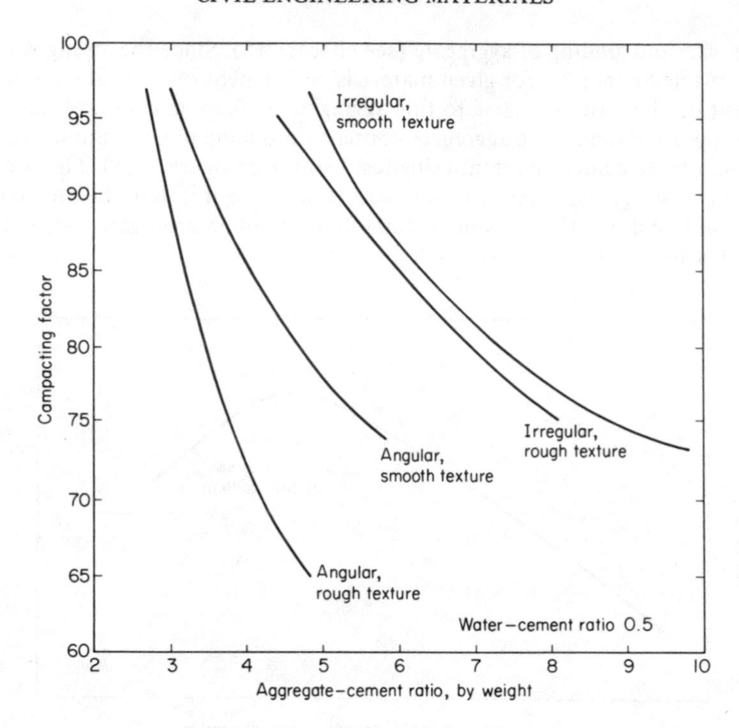

Figure 13.6 *Effect of aggregate surface texture on aggregate - cement ratio of concretes for different workabilities, based on Cornelius (1970)*

Time

The time that elapses between mixing of concrete and its final compaction depends on the general conditions of work such as the distance between the mixer and the point of placing, site procedures and general management. The associated reduction in workability is a direct result of loss of free water with time through evaporation, aggregate absorption and initial hydration of the cement. The rate of loss of workability is affected by certain characteristics of the constituent materials, for example, hydration and heat development characteristics of the cement, initial moisture content and porosity of the aggregate, as well as the ambient conditions.

For a given concrete and set of ambient conditions, the rate of loss of workability with time depends on the conditions of handling. Where concrete remains undisturbed after mixing until it is placed, the loss of workability during the first hour can be substantial, the rate of loss of workability decreasing with time as illustrated by curve A in figure 13.7. On the other hand, if it is continuously agitated, as in the case of ready-mixed concrete, the loss of workability is reduced, particularly during the first hour or so (see curve B in figure 13.7). However, prolonged agitation during transportation may increase the fineness of the solid particles through abrasion and produce a further reduction in workability. For

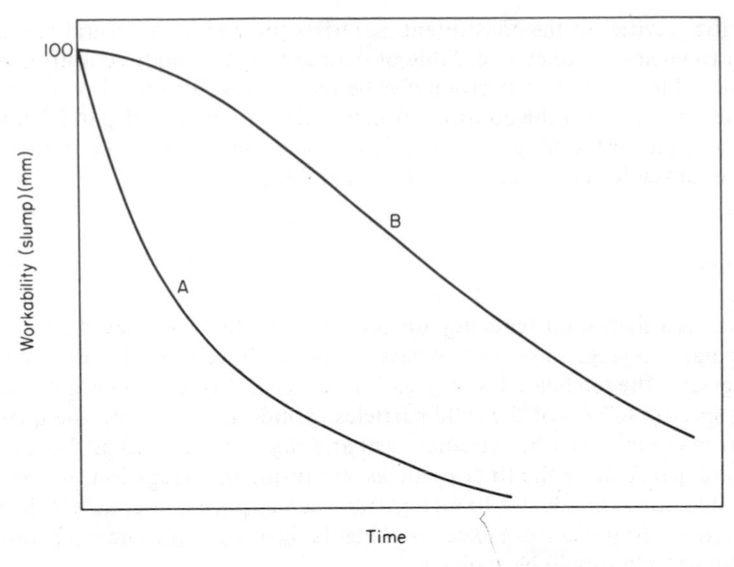

Figure 13.7 *Loss of workability of concrete with time : (A) no agitation and (B) continuously agitated after mixing*

concretes continuously agitated and undisturbed during transportation, the time intervals permitted (BS 1926) between the commencement of mixing and delivery on site are 2 hours and 1 hour respectively.

For practical purposes, loss of workability assumes importance when concrete becomes so unworkable that it cannot be effectively compacted, with the result that its strength and other properties become adversely affected. Corrective measures frequently taken to ensure that concrete at the time of placing has the desired workability are either an initial increase in the water content or an increase in the water content with further mixing shortly before the concrete is discharged. When this results in a water content greater than that originally intended, some reduction in strength and durability of the hardened concrete is to be expected unless the cement content is increased accordingly. This important fact is frequently overlooked on site. It should be recalled that the loss of workability varies with the mix, the ambient conditions, the handling conditions and the delivery time. No restriction on delivery time is given in CP 110: Part 1 but the concrete must be capable of being placed and effectively compacted without the addition of further water. For detailed information on the use of ready-mixed concrete the reader is advised to consult the work of Dewar (1973).

13.4 Stability

Apart from being sufficiently workable, fresh concrete should have a composition such that its constituent materials remain uniformly distributed in the concrete during both the period between mixing and compaction and the period following compaction before the concrete stiffens. Because of differences in the particle size and specific gravities of the constituent materials there exists a natural tendency for them to separate. Concrete capable of maintaining the required uniformity is said to be stable and most cohesive mixes belong to this category. For an unstable mix the extent to which the constituent materials will separate depends on the methods of transportation, placing and compaction. The two most common features of an unstable concrete are segregation and bleeding.

Segregation

When there is a significant tendency for the large and fine particles in a mix to become separated segregation is said to have occurred. In general, the less cohesive the mix the greater the tendency for segregation to occur. Segregation is governed by the total specific surface of the solid particles including cement and the quantity of mortar in the mix. Harsh, extremely wet and dry mixes as well as those deficient in sand, particularly the finer particles, are prone to segregation. As far as possible, conditions conducive to segregation such as jolting of concrete during transportation, dropping from excessive heights during placing and over-vibration during compaction should be avoided.

Blemishes, sand streaks, porous layers and honeycombing are a direct result of segregation. These features are not only unsightly but also adversely affect strength, durability and other properties of the hardened concrete. It is important to realise that the effects of segregation may not be indicated by the routine strength tests on control specimens since the conditions of placing and compaction of the specimens differ from those in the actual structure. There are no specific rules for suspecting possible segregation but after some experience of mixing and handling concrete it is not difficult to recognise mixes where this is likely to occur. For example, if a handful of concrete is squeezed in the hand and then released so that it lies in the palm, a cohesive concrete will be seen to retain its shape. A concrete which does not retain its shape under these conditions may well be prone to segregation and this is particularly so for wet mixes.

Bleeding

During compaction and until the cement paste has hardened there is a natural tendency for the solid particles, depending on size and specific gravity, to exhibit a downward movement. Where the consistency of a mix is such that it is unable to hold all its water some of it is gradually displaced and rises to the surface, and some may also leak through the joints of the formwork. Separation of water from a mix in this manner is known as bleeding. While some of the water reaches the top surface some may become trapped under the larger particles and under the reinforcing bars. The resulting variations in the effective water content within a concrete mass produce corresponding changes in its properties. For example, the strength of the concrete immediately underneath the reinforcing bars and coarse aggregate particles may be much less than the average strength and the resistance to percolation of water in these areas is reduced. In general, the concrete strength tends to increase with depth below the top surface. The water which reaches the top surface presents the most serious practical problems. If it is not removed, the concrete at and near the top surface will be much weaker and less durable than the remainder of the concrete. This can be particularly troublesome in slabs which have a large surface area. On the other hand, removal of the surface water will unduly delay the finishing operation on the site.

The risk of bleeding increases when concrete is compacted by vibration although this may be minimised by using a correctly designed mix and ensuring that the concrete is not over-vibrated. Rich mixes tend to bleed less than lean mixes. The type of cement employed is also important, the tendency for bleeding to occur decreasing as the fineness of the cement or its alkaline and tricalcium aluminate (C_3A) content increases. Air-entrainment provides another very effective means of controlling bleeding in, for example, wet lean mixes where both segregation and bleeding are frequently troublesome.

14

Properties of Hardened Concrete

The properties of fresh concrete are important only in the first few hours of its history whereas the properties of hardened concrete assume an importance which is retained for the remainder of the life of the concrete. The important properties of hardened concrete are strength, deformation under load, durability, permeability and shrinkage. In general, strength is considered to be the most important property and the quality of concrete is often judged by its strength. There are, however, many occasions when other properties are more important, for example, low permeability and low shrinkage are required for water-retaining structures. Although in most cases an improvement in strength results in an improvement of the other properties of concrete there are exceptions. For example, increasing the cement content of a mix improves strength but results in higher shrinkage which in extreme cases can adversely affect durability and permeability. One of the primary objectives of this chapter is to help the reader to understand the factors which affect each of the important properties of hardened concrete.

Since the properties of concrete change with age and environment it is not possible to attribute absolute values to any of them. Laboratory tests give only an indication of the properties which concrete may have in the actual structure as the quality of the concrete in the structure depends on the workmanship on site. For these reasons it is important to be able to judge the quality of concrete *in situ*. The direct method of testing drilled cores is expensive and limited because the removal of too many cores weakens a structure. Nondestructive tests have therefore been developed for the assessment of concrete quality. The limitations and applications of nondestructive testing together with a brief description of the techniques is given at the end of this chapter.

14.1 Strength

The strength of concrete is defined as the maximum load (stress) it can carry. As the strength of concrete increases its other properties usually improve and since the tests for strength, particularly in compression, are relatively simple to perform concrete compressive strength is commonly used in the construction industry for the purpose of specification and quality control. Concrete is a comparatively brittle material which is relatively weak in tension.

Compressive Strength

The compressive strength of concrete is taken as the maximum compressive load it can carry per unit area. Concrete strengths of up to 80 N mm^{-2} can be achieved by selective use of the type of cement, mix proportions, method of compaction and curing conditions.

Concrete structures, except for road pavements, are normally designed on the basis that concrete is capable of resisting only compression, the tension being carried by steel reinforcement. In the United Kingdom a 150 mm cube is commonly used for determining the compressive strength. The standard method described in BS 1881 : Part 3 requires that the test specimen should be cured in water at 20° ± 1 °C and crushed immediately after it has been removed from the curing tank.

Tensile Strength

The tensile strength of concrete is of importance in the design of concrete roads and runways. For example, its flexural strength or modulus of rupture (tensile strength in bending) is utilised for distributing the concentrated loads over a wider area of road pavement. Concrete members are also required to withstand tensile stresses resulting from any restraint to contraction due to drying or temperature variation.

Unlike metals, it is difficult to measure concrete strength in direct tension and indirect methods have been developed for assessing this property. Of these the split cylinder test is the simplest and most widely used. This test is fully described in BS 1881 : Part 4 and entails diametrically loading a cylinder in compression along its entire length. This form of loading induces tensile stresses over the loaded diametrical plane and the cylinder splits along the loaded diameter. The magnitude of the induced tensile stress f_{ct} at failure is given by

$$f_{ct} = \frac{2F}{\pi l d}$$

where F is the maximum applied load and l and d are the cylinder length and diameter respectively.

The flexural strength of concrete is another indirect tensile value which is also commonly used (BS 1881 : Part 4). In this test a simply supported plain concrete beam is loaded at its third points, the resulting bending moments inducing compressive and tensile stresses in the top and bottom of the beam respectively. The beam fails in tension and the flexural strength (modulus of rupture) f_{cr} is defined by

$$f_{cr} = \frac{FL}{bd^2}$$

where F is the maximum applied load, L the distance between the supports, and b and d are the beam breadth and depth respectively at the section at which failure occurs.

The tensile strength of concrete is usually taken to be about one-tenth of its compressive strength. This may vary, however, depending on the method used for measuring tensile strength and the type of concrete. In general the direct tensile strength and the split cylinder tensile strength vary from 5 to 13 per cent and the flexural strength from 11 to 23 per cent of the concrete cube compressive strength. In each case, as the strength increases the percentage decreases. As a guide, the modulus of rupture may be taken as $0.7\sqrt{(\text{cube strength})}$ N mm^{-2} and the direct tensile strength as $0.45\sqrt{(\text{cube strength})}$ N mm^{-2} although, where possible, values based on tests using the actual concrete in question should be obtained.

14.2 Factors influencing Strength

Several factors which affect the strength of concrete are shown in figure 14.1. In this section their influence is discussed with particular reference to compressive strength. In general, tensile strength is affected in a similar manner.

Influence of the Constituent Materials

Cement

The influence of cement on concrete strength, for given mix proportions, is determined by its fineness and chemical composition through the processes of hydration (see chapter 12). The gain in concrete strength as the fineness of its cement particles increases is shown in figure 14.2. The gain in strength is most marked at early ages and after 28 days the relative gain in strength is much reduced. At some later age the strength of concrete made with fine cements may not be very different from that made with normal cement (300 m^2 kg^{-1}).

The role of the chemical composition of cement in the development of concrete strength can best be appreciated by studying table 14.1 and figures 14.3 and 14.4. It is apparent that cements containing a relatively high percentage of tricalcium silicate (C_3S) gain strength much more rapidly than those rich in dicalcium silicate (C_2S), as shown in figure 14.3; however, at later ages the difference in the corresponding strength values is small. In fact there is a tendency for concretes made with low-heat cements eventually to develop slightly higher strengths (figure 14.4). This is possibly due to the formation of a better quality gel structure in the course of hydration.

Water

A concrete mix containing the minimum amount of water required for complete hydration of its cement, if it could be fully compacted, would develop the maximum attainable strength at any given age. A water - cement ratio of approximately 0.25 (by weight) is required for full hydration of the cement but with this water

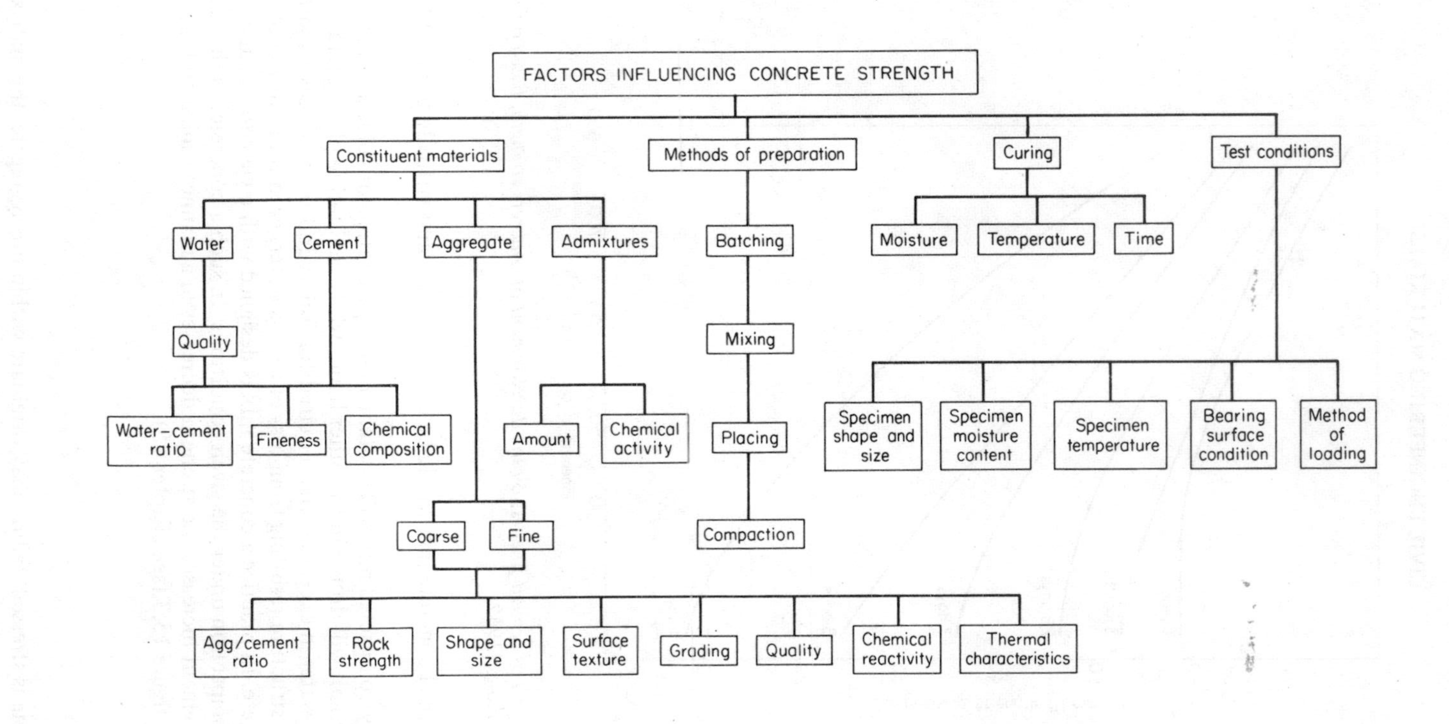

Figure 14.1 *Factors affecting strength of concrete*

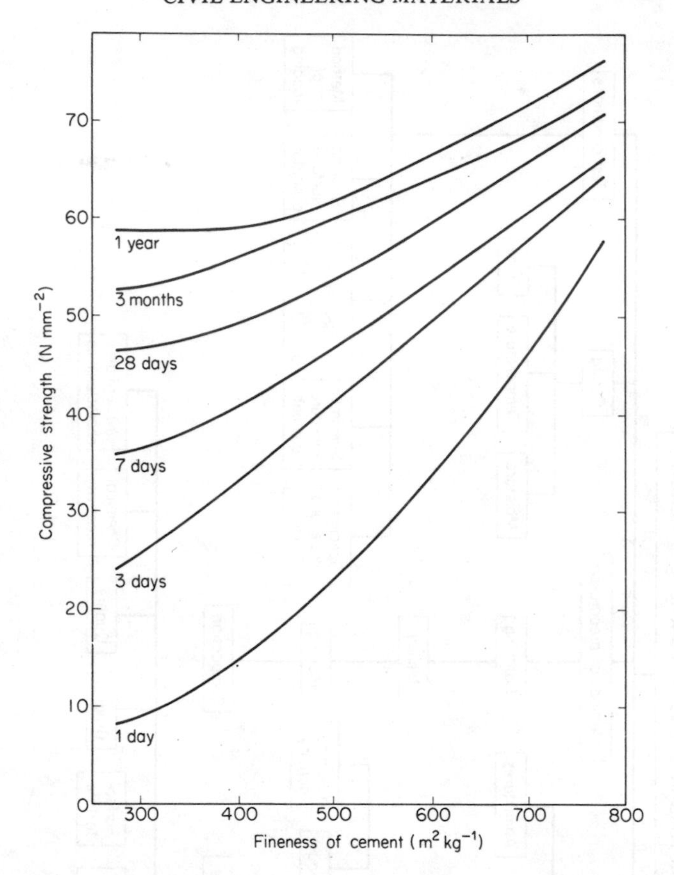

Figure 14.2 *Effect of cement fineness on the development of concrete strength, based on Bennett and Collings (1969)*

content a normal concrete mix would be extremely dry and virtually impossible to compact.

A partially compacted mix will contain a large percentage of voids and the concrete strength will drop. On the other hand, while facilitating placing and compaction, water in excess of that required for full hydration produces a somewhat porous structure resulting from loss of excess water, even for a fully compacted concrete. In practice a concrete mix is designed with a view to obtaining maximum compaction under the given conditions. In such a concrete, as the ratio of water to cement increases the strength decreases in a manner similar to that illustrated in figure 15.3 (see chapter 15).

Aggregate

When concrete is stressed, failure may originate within the aggregate, the matrix or at the aggregate - matrix interface; or any combination of these may occur. In general the aggregates are stronger than the concrete itself and in such cases the aggregate strength has little effect on the strength of concrete.

The bond (aggregate - matrix interface) is an important factor determining concrete strength. Bond strength is influenced by the shape of the aggregate, its surface texture and cleanliness. A smooth rounded aggregate will result in a

TABLE 14.1
Chemical composition of various Portland cements with similar fineness

Cement	Compound composition (per cent)			
	C_3S	C_2S	C_3A	C_4AF
A	55	16	12	8
B	50	21	11	9
C	43	33	5	11
D	35	41	6	12

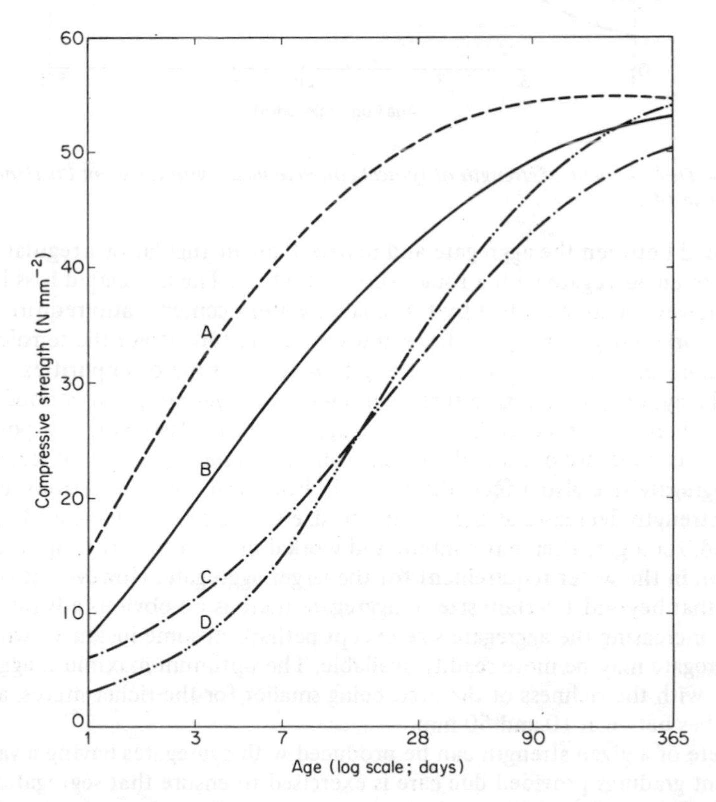

Figure 14.3 *Development of strength of typical concrete made with different Portland cements (see table 14.1)*

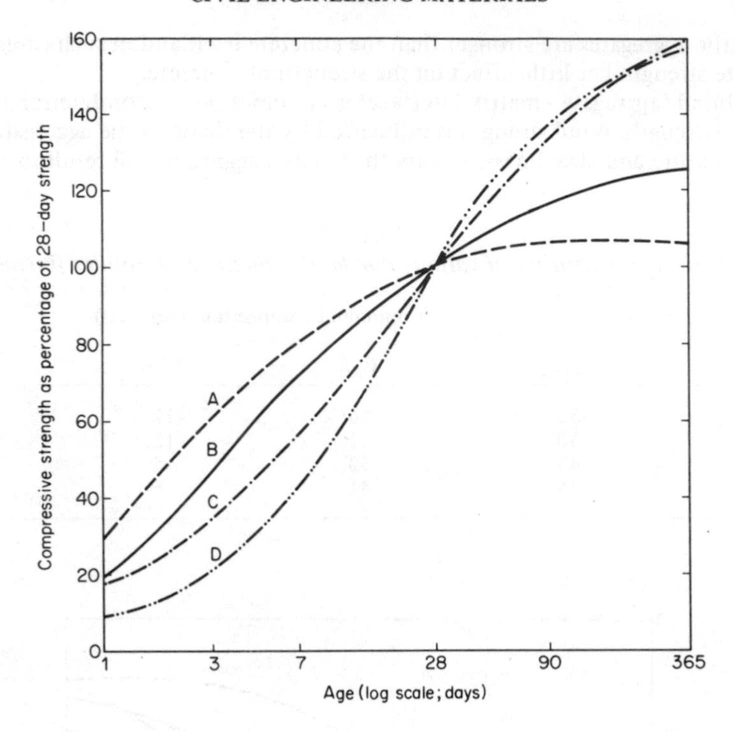

Figure 14.4 *Development of strength of typical concrete made with different Portland cements (see table 14.1)*

weaker bond between the aggregate and matrix than an angular or irregular aggregate or an aggregate with a rough surface texture. The associated loss in strength however may be offset by the smaller water-cement ratio required for the same workability. Aggregate shape and surface texture affect the tensile strength more than the compressive strength. A fine coating of impurities, such as silt and clay, on the aggregate surface hinders the development of a good bond. A weathered and decomposed layer on the aggregate can also result in a poor bond as this layer can readily become detached from the sound aggregate beneath.

The aggregate size also affects the strength. For given mix proportions, the concrete strength decreases as the maximum size of aggregate is increased. On the other hand, for a given cement content and workability this effect is opposed by a reduction in the water requirement for the larger aggregate. However, it is probable that beyond a certain size of aggregate there is no obvious advantage in further increasing the aggregate size except perhaps in some instances when larger aggregate may be more readily available. The optimum maximum aggregate size varies with the richness of the mix, being smaller for the richer mixes, and generally lies between 10 and 50 mm.

Concrete of a given strength can be produced with aggregates having a variety of different gradings provided due care is exercised to ensure that segregation does not occur. The suitability of a grading to some extent depends on the shape and texture of the aggregate. Aggregates which react with the alkaline content of a cement adversely affect concrete strength although this is rarely a problem in the United Kingdom.

Admixtures

Several types of admixture and their usage have been discussed in chapter 12. The two kinds of admixture most widely used are accelerators and air-entraining agents. Calcium chloride is the most commonly used accelerator; it increases the rate of development of concrete strength, particularly at early ages, and consequently is frequently used for concreting in winter. Figure 14.5 shows the relative gain in strength, for a particular concrete, with dosages of calcium chloride up to 2 per cent (maximum permitted) by weight of cement. Because the effectiveness varies with the type of cement, curing conditions (particularly temperature) and design strength, an accurate estimate of the increase in strength with calcium chloride is not possible.

When using an air-entraining agent, for a constant water - cement ratio, the greater the percentage of air entrained the greater the loss in strength. However, since the water - cement ratio can be reduced while maintaining the required workability, there is generally no significant change in concrete strength for the usual range of air content.

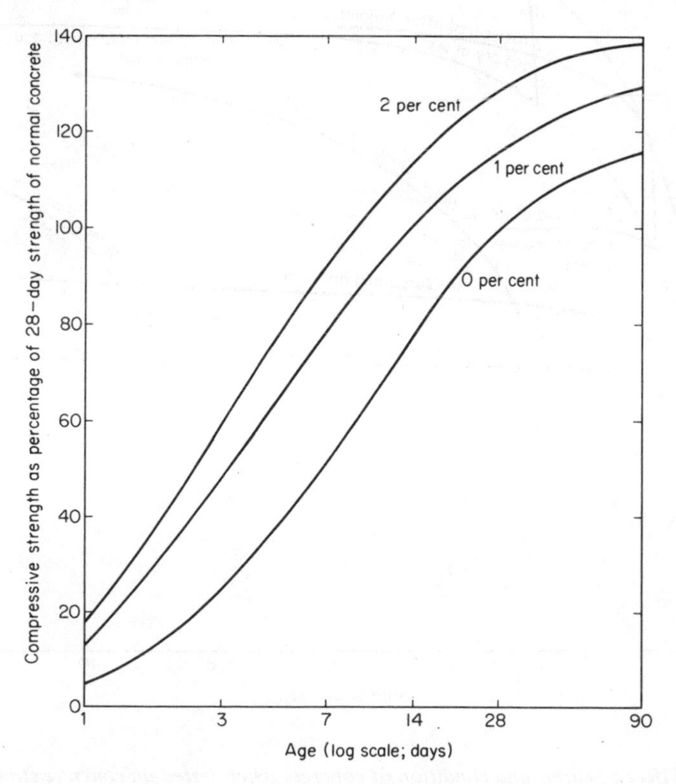

Figure 14.5 *Gain in strength of concrete containing different dosages of calcium chloride, based on Akroyd (1962)*

Influence of the Methods of Preparation

When concrete materials are not adequately mixed into a consistent and homogeneous mass, some poor quality concrete is inevitably the result. Even when a concrete is adequately mixed care must be taken during placing and compaction to minimise the probability of the occurrence of bleeding, segregation and honeycombing all of which can result in patches of poor quality concrete. A properly designed concrete mix is one that does not demand the impossible from site operatives before it can be fully compacted in its final location. If full compaction is not achieved the resulting voids produce a marked reduction in concrete strength.

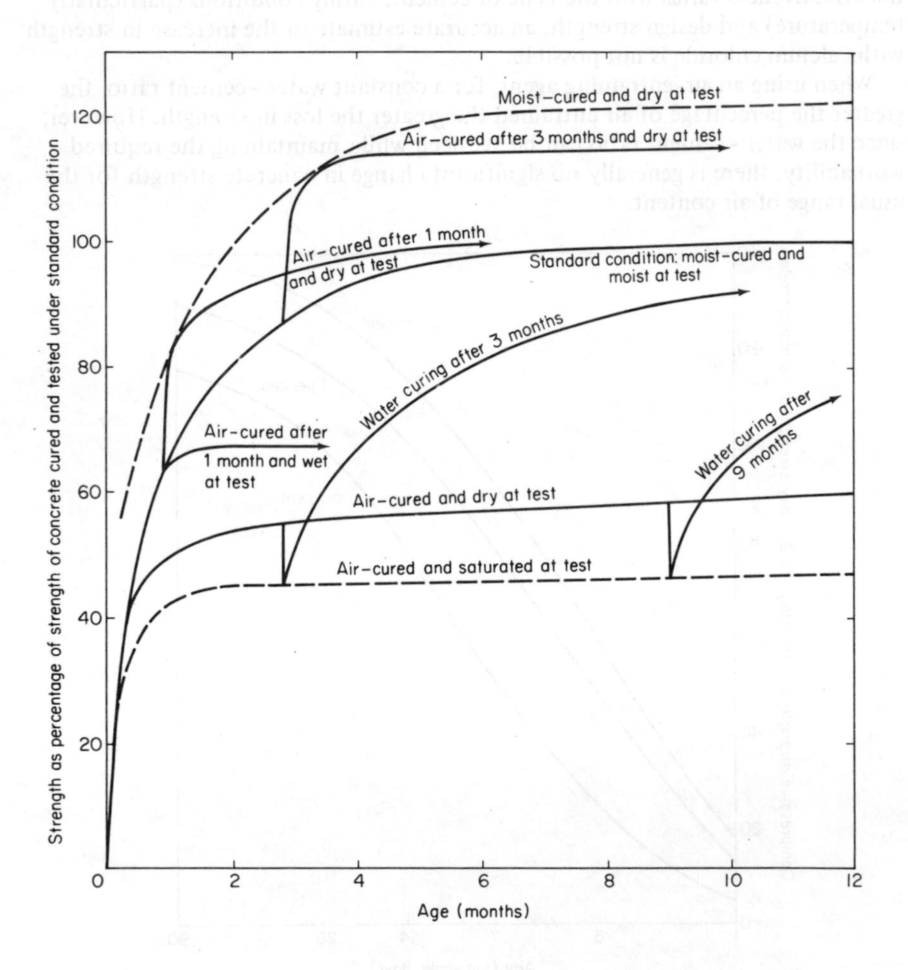

Figure 14.6 *Effect of curing and condition of concrete when tested on concrete strength, based on Gilkey (1937)*

Influence of Curing

Curing of concrete is a prerequisite for the hydration of the cement content. For a given concrete, the amount and rate of hydration and furthermore the physical make-up of the hydration products are dependent on the time-moisture-temperature history.

Generally speaking, the longer the period during which concrete is kept in water, the greater its final strength. It is normally accepted that a concrete made with ordinary Portland cement and kept in normal curing conditions will develop about 75 per cent of its final strength in the first 28 days. The development of concrete strength under various curing conditions is shown in figure 14.6. It is apparent that concrete left in air achieves the lowest strength values at all ages owing to the evaporation of the free mixing water from the concrete. The gain in strength depends on a number of factors such as relative humidity, wind velocity and the size of structural member or test specimen. Figure 14.6 shows that both the increased hydration due to improvements in initial curing (moist or water curing at normal temperature) and the condition of the concrete at the time of testing have a significant effect on the final apparent strength of concrete. It should also be noted that moist (or water) curing after an initial period in air results in a resumption of the hydration process and that concrete strength is further improved with time, although the optimum strength may not be realised.

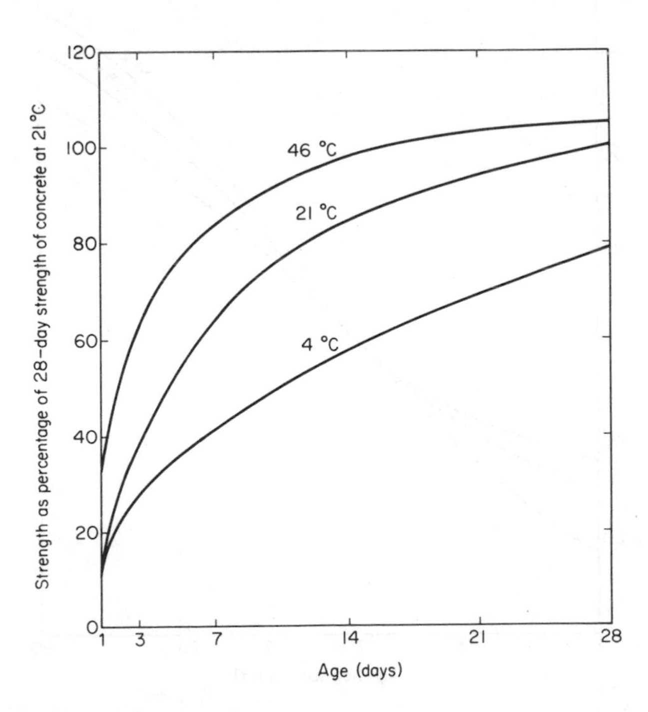

Figure 14.7 *Comparative compressive strength of concrete cast, sealed and maintained at different temperatures, based on Price (1951)*

The temperature at which concrete is cured has an important bearing on the development of its strength with time. The rate of gain in strength of concrete made with ordinary Portland cement increases with increase in concrete temperature at early ages (figure 14.7), although at later ages the concrete made and cured at lower temperatures shows a somewhat higher strength. Figure 14.8 shows how a high temperature during the placing and setting of concrete can adversely affect the development of its strength from early ages. On the other hand, when the initial temperature is lower than the subsequent curing temperature, then higher temperatures during final curing result in significantly higher strengths (figure 14.9). A possible explanation for this behaviour is that a rapid initial hydration appears to form a gel structure (hydration product) of an inferior quality and this adversely affects concrete strength at later ages. Concretes made with other Portland cements would respond to temperature in a somewhat similar manner.

It has been suggested that the strength of concrete can be related to the product of age and curing temperature, commonly known as *maturity*. However, such relationships are dependent on a number of factors such as curing temperature history, particularly the temperature at early ages (see figures 14.8 and 14.9), and are therefore limited in their general applicability for predicting concrete strength.

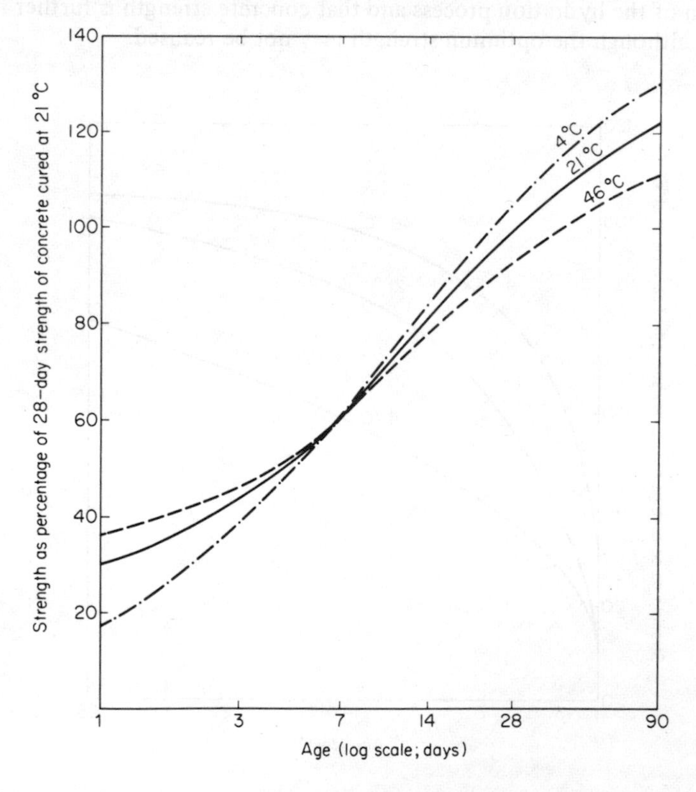

Figure 14.8 *Comparative compressive strength of sealed concrete specimens maintained at different temperatures for 2 hours after casting and subsequently cured at 21 °C, based on Price (1951)*

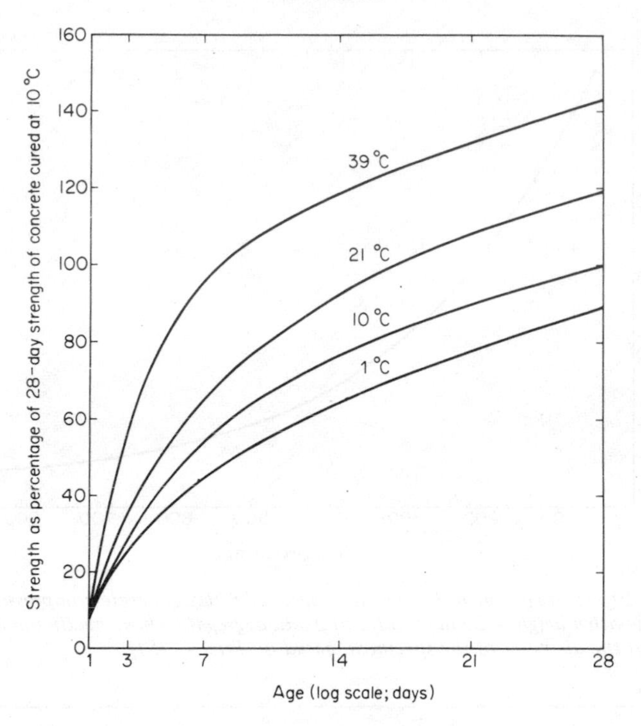

Figure 14.9 *Comparative compressive strength of concrete cast, sealed and maintained at 10 °C for the first 24 hours and subsequently cured at different temperatures, based on Price (1951)*

Influence of Test Conditions

The conditions under which tests to determine concrete strength are carried out can have a considerable influence on the strength obtained and it is important that these effects are understood if test results are to be correctly interpreted.

Specimen shape and size

Three basic shapes used for the determination of compressive strength are the cube, cylinder and square prism. Each shape gives different strength results and furthermore for a given shape the strength also varies with size. Figure 14.10 shows the influence of specimen diameter and it can be seen that as the size decreases, the apparent strength increases. The measured strength of concrete is also affected by the height - diameter ratio (figure 14.11). For height - diameter ratios less than 2 strength begins to increase rapidly owing to the restraint provided by the machine platens. Strength remains sensibly constant for height - diameter ratios between 2 and 3 and thereafter shows a slight reduction. The relative influence of slenderness may be modified by several inherent characteristics of concrete such as strength, air-entrainment, strength of aggregate, degree of moist curing and moisture content at the time of testing.

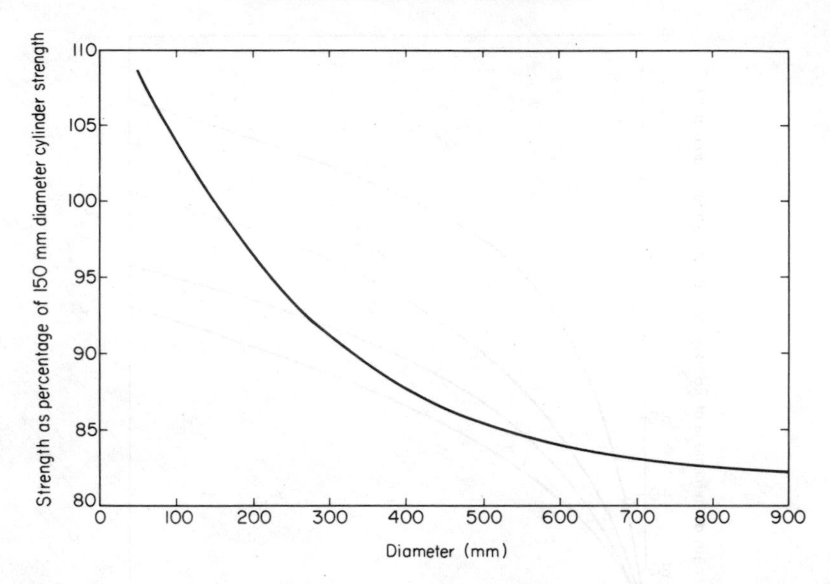

Figure 14.10 *Effect of specimen size on the apparent 28- day concrete compressive strength for specimens with a height - diameter ratio of 2 and aggregate whose maximum diameter is one-quarter of the diameter of the specimen, based on Price (1951)*

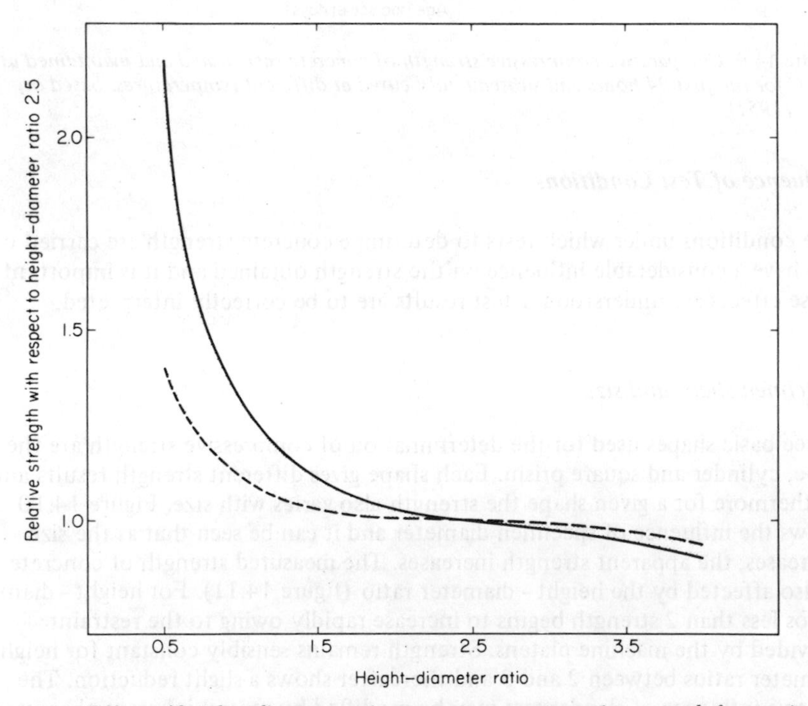

Figure 14.11 *Effect of height - diameter ratio on concrete compressive strength for specimens moist-cured at room temperature and tested wet*

BS 1881: Part 4 specifies the use of concrete cubes for determining compressive strength; 150 mm cubes are widely used in the United Kingdom for quality-control purposes. The same standard permits the use of cored cylindrical specimens from *in situ* concrete. When the height-diameter ratio of such cores is less than 2, the measured strength is required to be corrected using the correction factors given in table 14.2. The equivalent cube strength is then obtained by multiplying the cylinder strength by 1.25. Since the relationship between height-diameter ratio and strength depends on the type of concrete, the use of one set of correction factors of the type given in table 14.2 can only be suitable for a limited range of concrete materials. It should be noted that the correction factors given in table 14.2 are most likely to favour low-strength concretes.

TABLE 14.2

Correction factor for compression tests on cylinders, BS 1881: Part 4

Specimen height-diameter ratio	Strength correction factor
2.00	1.00
1.75	0.98
1.50	0.96
1.25	0.94
1.00	0.92

Specimen moisture content and temperature

This should not be confused with the effect of moisture and temperature during curing. The strength of concrete can be influenced by the absence or presence of moisture and by temperature only when these conditions generate internal stresses which change the magnitude of the external load required to bring about failure. Since the mode of failure in different strength tests is different, it follows that the influence of moisture content and temperature on the apparent strength varies.

In the case of compression tests, air-dry concrete has a significantly higher strength than concrete tested in a saturated condition (figure 14.6). The lower strengths of wet concrete can be attributed mainly to the development of internal pore pressure as the external load is applied.

The flexural strength of saturated concrete is greater however than that of concrete which is only partially dry owing to tensile stresses developed near the surface of the dry concrete by differential shrinkage. The initial drying is therefore critical with the apparent strength reaching a minimum value within the first few days; thereafter it begins to increase gradually as the concrete approaches a completely dry state. A thoroughly dry concrete specimen in which tensile stresses are not being induced has a higher apparent flexural strength than saturated concrete. The indirect tensile or split cylinder strength is lower for saturated concrete than for thoroughly dried concrete.

Since the influence of the moisture content on concrete strength varies with the type of test, standard strength tests (BS 1881: Part 4) should be performed on specimens in a saturated condition.

Method of loading

The compressive strength of concrete increases as the lateral confining pressure increases (figure 14.12).

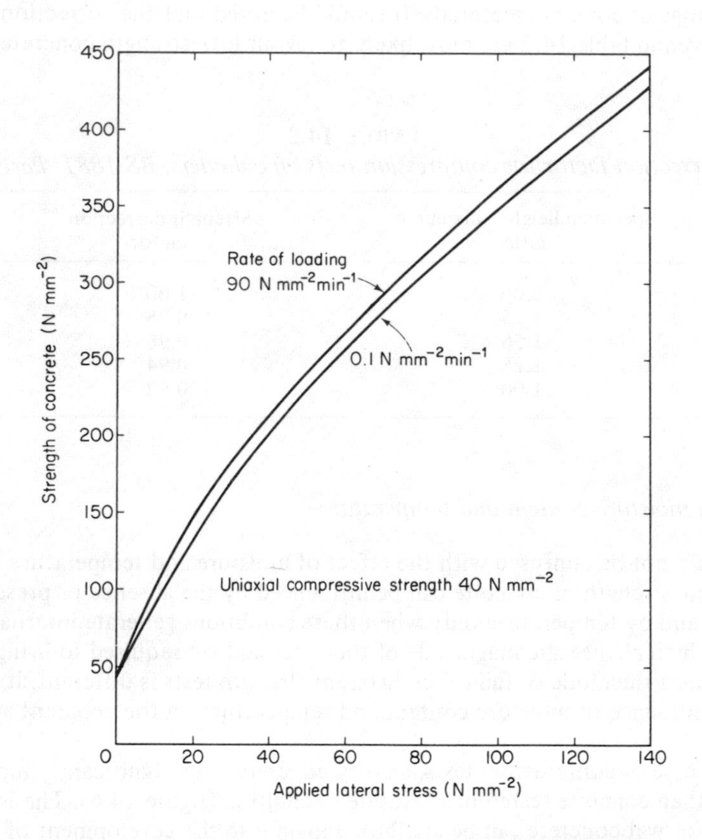

Figure 14.12 *Effect of lateral compression on concrete compressive strength*

The apparent strength of concrete is affected by the rate at which it is loaded. In general, for static loading, the faster the loading rate the higher the indicated strength. However, the relative effects of the rate of loading vary with the nominal strength, age and extent of moist curing. High-strength mature concretes cured in water are most sensitive to loading rate and particularly so for loading rates greater than 600 N mm^{-2} min^{-1}. BS 1881: Part 4 requires concrete in compression tests to be loaded at 15 N mm^{-2} min^{-1}, for which small variations in loading rate will have little effect on strength. The standard rates of loading for flexural and split cylinder tests correspond to rates of increase of tensile stress of 18 N mm^{-2} min^{-1} and 15 N mm^{-2} min^{-1} respectively.

When loads on a structure are predominantly cyclic (repeated loading and unloading) in character, the effects of fatigue should also be considered. This kind of loading produces a reduction in strength. A reduction in strength of as much as 30 per cent of the normal static strength value can take place, although this depends on the stress - strength ratio, the frequency of loading and the type of concrete.

Structural concrete is commonly subjected to sustained loads. It is probable that concrete can withstand higher loads if a constant load is maintained before loading to failure. Improvement in compressive strength can occur for sustained loads up to 85 per cent of the normal static strength although the actual gain in strength depends on the duration and magnitude of load, type of concrete and age. The increase in strength is probably due to consolidation of the concrete under sustained load and the redistribution of stresses within the concrete.

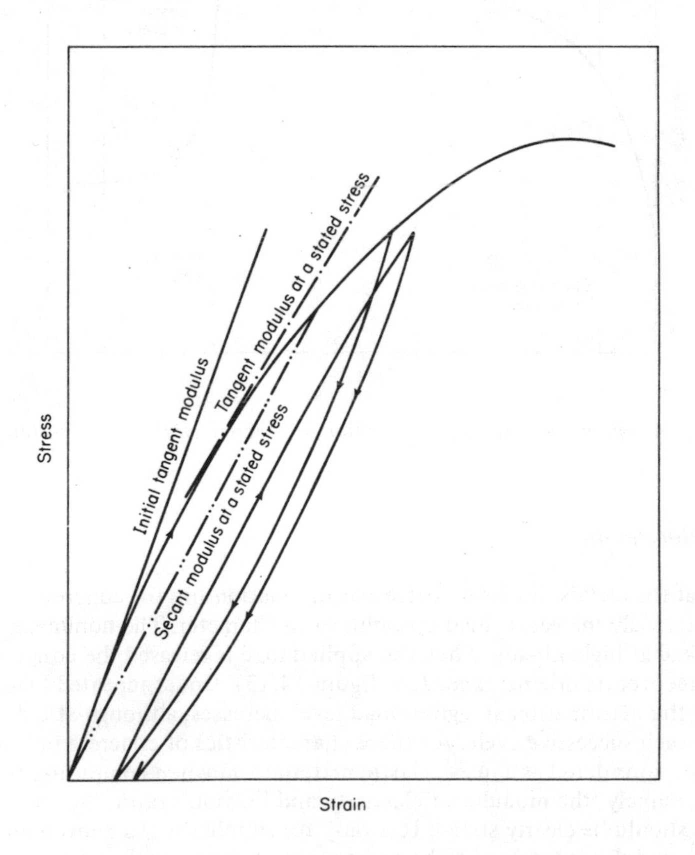

Figure 14.13 *A typical stress - strain curve showing different moduli of elasticity*

14.3 Deformation

Concrete deforms under load, the deformation increasing with the applied load and being commonly known as elastic deformation (figure 14.13). Concrete continues to deform with time, under constant load; this is known as time-dependent deformation or creep (figure 14.14). Deformation due to concrete shrinkage is discussed on p. 173 and is specifically excluded here.

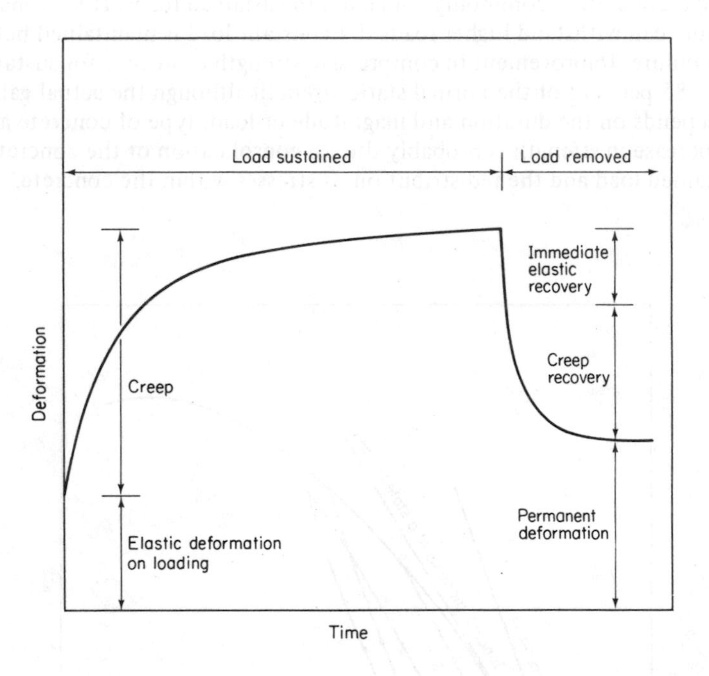

Figure 14.14 *A typical illustration of deformation of concrete subjected to constant load*

Elastic Deformation

Unlike that for metals, the load - deformation relationship for concrete subjected to a continuously increasing load is nonlinear in character. The nonlinearity is most marked at higher loads. When the applied load is released the concrete does not fully recover its original shape (see figure 14.13). Under repeated loading and unloading the deformation at a given load level increases, although at a decreasing rate, with each successive cycle. All these characteristics of concrete indicate that it should be considered as a quasi-elastic material and when computing the elastic constants, namely, the modulus of elasticity and Poisson's ratio, the method employed should be clearly stated. It is only for simplicity and convenience that the elastic modulus is assumed to be constant in both concrete technology and the design of concrete structures.

Modulus of elasticity

This is defined as the ratio of load per unit area (stress) to the elastic deformation per unit length (strain)

$$\text{modulus of elasticity } E = \text{stress/strain} = \sigma/\epsilon$$

The modulus is used when estimating the deformation, deflection or stresses under normal working loads. Since concrete is not a perfectly elastic material, the modulus of elasticity depends on the particular definition adopted (see figure 14.13).

The initial tangent modulus is of little value for structural applications since it has significance only for low stresses during the first load cycle. The tangent modulus is difficult to determine and is applicable only within a narrow band of stress levels. The secant modulus is easily determined and takes into account the total deformation at any one point. The method prescribed in BS 1881: Part 5 requires repeated loading and unloading before the specimen is loaded for determination of its secant modulus of elasticity from a stress - strain curve, which then approaches a straight line for stresses up to about 0.4 of the concrete strength. The modulus of elasticity for most concretes, at 28 days, ranges from 15 to 40 kN mm^{-2}. As a guide, the modulus may be assumed to be 3.8 $\sqrt{}$(concrete strength, N mm^{-2}) kN mm^{-2} for normal weight concrete.

Poisson's ratio

When concrete is subjected to axial compression, it contracts in the axial direction and expands laterally. Poisson's ratio is defined as the ratio of the lateral strain to the associated axial strain and varies from 0.1 to 0.3 for normal working stresses. A value of 0.2 is commonly used.

Factors influencing the Elastic Behaviour of Concrete

Concrete is a multiphase material and its resistance to deformation under load is dependent on the stiffness of its various phases, such as aggregate, cement paste and voids, and the interaction between individual phases. In general the factors which influence the strength of concrete also affect deformation although the extent of their influence may well vary. The modulus of elasticity increases with strength, although the two properties are not directly related because different factors exert varying degrees of influence on strength and modulus of elasticity. Although relationships between the two properties may be derived they are only applicable within the range of variables considered. In short there is no unique relationship between strength and modulus of elasticity.

The influence of stress level and rate of loading on both the axial and lateral strains is shown in figure 14.15 for a concrete which has been cured in water (20 °C ± 1 °C) and air-dried before loading in uniaxial compression. It should be noted that the slower the rate of loading the greater the strains at a given stress level. This is due to the fact that during loading both elastic and creep strains occur, the creep strains increasing with the duration of time for which the load

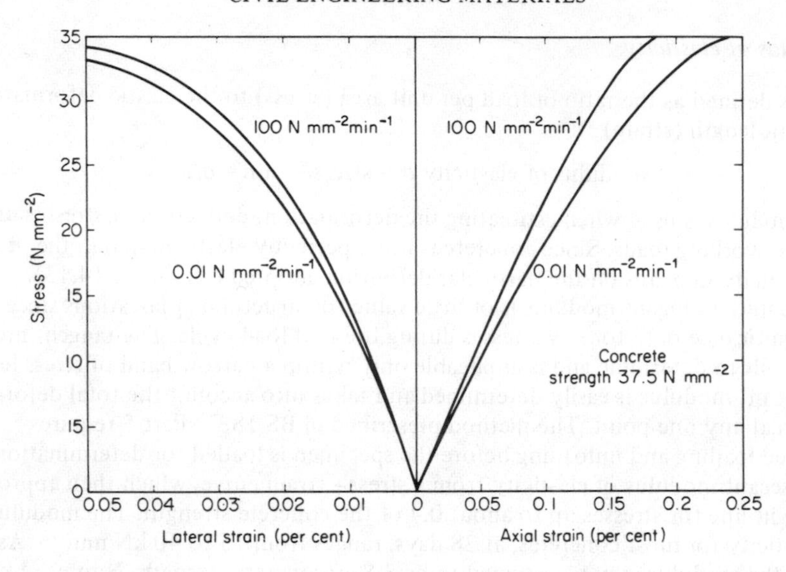

Figure 14.15 *Stress-strain relationships of concrete for different rates of loading, based on Dhir and Sangha (1972)*

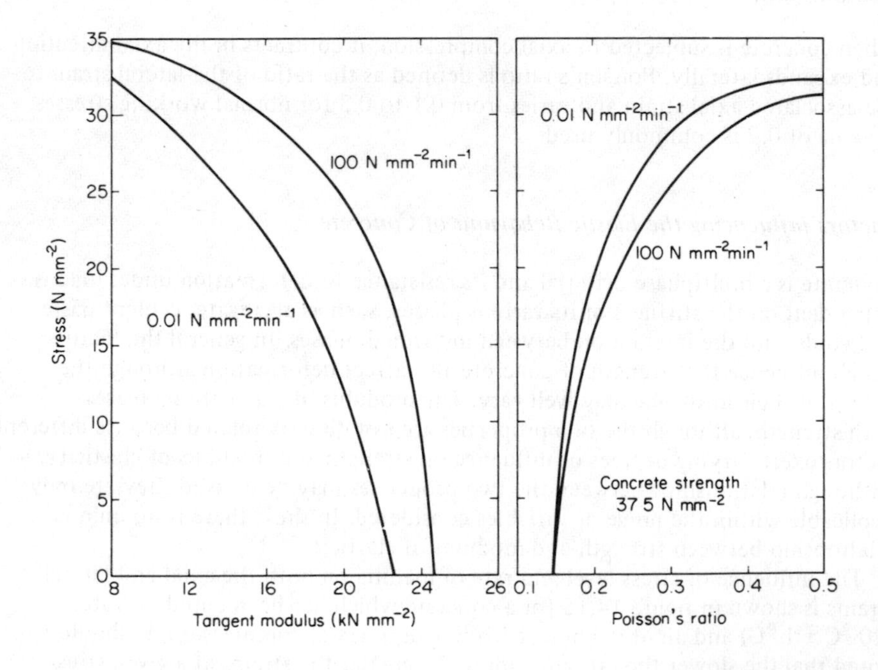

Figure 14.16 *Variation of tangent modulus and Poisson's ratio with stress level and rate of loading, based on Dhir and Sangha (1972)*

acts. The variation of tangent modulus and Poisson's ratio, for the same concrete, with stress level and rate of loading are shown in figure 14.16. The general trends depicted in the figure apply to all concretes although the exact relationship may vary with the type of concrete and the methods employed for measuring the properties.

Creep Deformation

When concrete is subjected to a sustained load it first undergoes an instantaneous deformation (elastic) and thereafter continues to deform with time (figure 14.14). The increase in strain with time is termed creep. It should be noted that after an initially high rate of creep the creep continues but at a continuously decreasing rate, except when the sustained load is large enough to cause failure, in which case the rate of creep increases before failure occurs. The removal of the sustained load results in an immediate reduction in strain and this is followed by a gradual decrease in strain over a period of time. The gradual decrease in strain is called creep recovery. Creep and creep recovery are related phenomena with somewhat similar characteristics. Creep is not wholly reversible and some permanent strain remains after creep recovery is complete.

Since concrete in service is subjected to sustained loads for long periods of time the effect of creep strains, which normally exceed the elastic strains, must be considered in structural design. Failure to include the effects of creep strains may lead to serious underestimates of beam deflections and over-all structural deformation and, in those cases where structural stability is involved, the provision of members with inadequate strength. In prestressed concrete, allowance must be made for the loss of tension in the prestressing tendons resulting from the shortening of a member under the action of creep. Creep strain can also be beneficial in that it can relieve local stress concentrations which might otherwise lead to structural damage. A classic example of this is the reduction of shrinkage stresses in restrained members.

Factors influencing Creep

Both the type of concrete (as described by its ingredients, curing history, strength and age) and the relative magnitude of the applied stress with respect to concrete strength (stress - strength ratio) affect the creep strain. For a given concrete the creep strain is almost directly proportional to the stress - strength ratio for ratios up to about one-third. For the same stress - strength ratio, the creep strain increases as both cement content and water - cement ratio increase (figure 14.17) and decreases as the relative humidity and age at loading increase (figure 14.18).

The influence of the constituent materials of concrete on creep is somewhat complex. The different types of cement influence creep because of the associated different rates of gain in concrete strength. For example, concrete made with rapid-hardening Portland cement shows less creep than concrete made with ordinary Portland cement and loaded at the same age. Normal rock aggregates have a restraining effect on concrete creep and the use of large, high modulus aggregates can be beneficial in this respect.

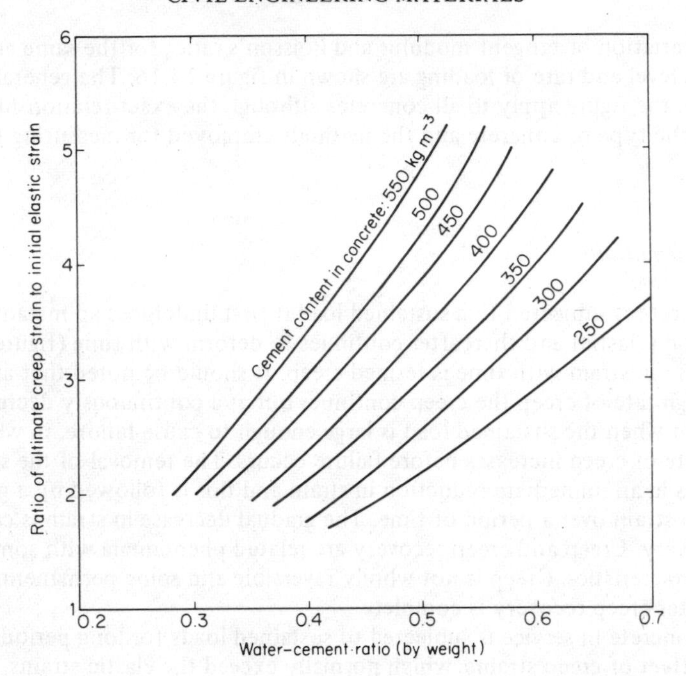

Figure 14.17 *Effect of cement content and water–cement ratio on ultimate creep strain of concrete at 50 per cent relative humidity loaded at an age of 28 days, based on C.E.B. (1964)*

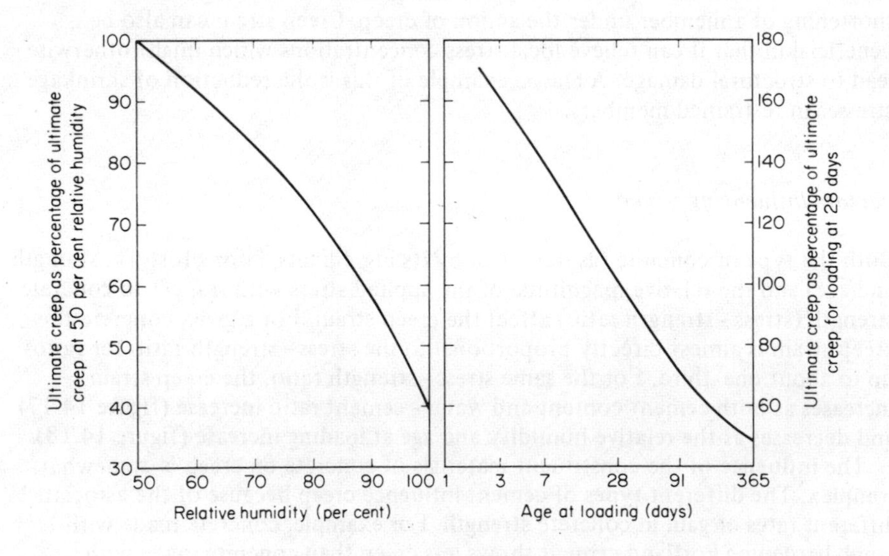

Figure 14.18 *Effect of relative humidity and age at loading on creep, based on Neville (1970)*

The environmental conditions of concrete subjected to a sustained load can also have a marked effect on the magnitude of creep. For example, concrete cured under humid conditions and then loaded in a relatively dry atmosphere undergoes a greater creep strain than if the original conditions had been maintained. A concrete allowed to reach moisture equilibrium before loading, however, is not adversely affected. Creep decreases as the mass of concrete increases owing to to the slower rate at which loss of water can take place within a large mass of concrete.

Several mathematical equations have been proposed for estimating creep strains. Both the rate of creep and its ultimate magnitude are dependent on a multitude of factors. These expressions can only be truly applicable to concretes similar to those for which they were designed. Nevertheless, creep strains are usually estimated for design purposes. It has been noted that, for a given concrete, creep strain depends on the stress - strength ratio. For practical purposes, concrete creep strain may be assumed to be directly proportional to the elastic deformation up to a stress - strength ratio of about two-thirds. On this assumption the ultimate creep strain may be estimated for a given water - cement ratio and cement content using figure 14.17. This creep value may be modified for different environmental moisture conditions and ages on loading using figure 14.18. It may be assumed that 50 per cent of the total estimated creep will take place within one month and 75 per cent within the first six months after loading.

14.4 Durability

Besides its ability to sustain loads, concrete is also required to be durable. The durability of concrete can be defined as its resistance to deterioration resulting from external and internal causes (figure 14.19). The external causes include the effects of environmental and service conditions to which concrete is subjected, such as weathering, chemical actions and wear. The internal causes are the effects of interaction between the constituent materials, such as alkali - aggregate reaction, volume changes, absorption and permeability.

In order to produce a durable concrete care should be taken to select suitable constituent materials. It is also important that the mix contains adequate quantities of materials in proportions suitable for producing a homogeneous and fully compacted concrete mass.

Weathering

Deterioration of concrete by weathering is usually brought about by the disruptive action of alternate freezing and thawing of free water within the concrete and expansion and contraction of the concrete, under restraint, resulting from variations in temperature and alternate wetting and drying.

Damage to concrete from freezing and thawing arises from the expansion of porewater during freezing; in a condition of restraint, if repeated a sufficient number of times, this results in the development of hydraulic pressure capable of

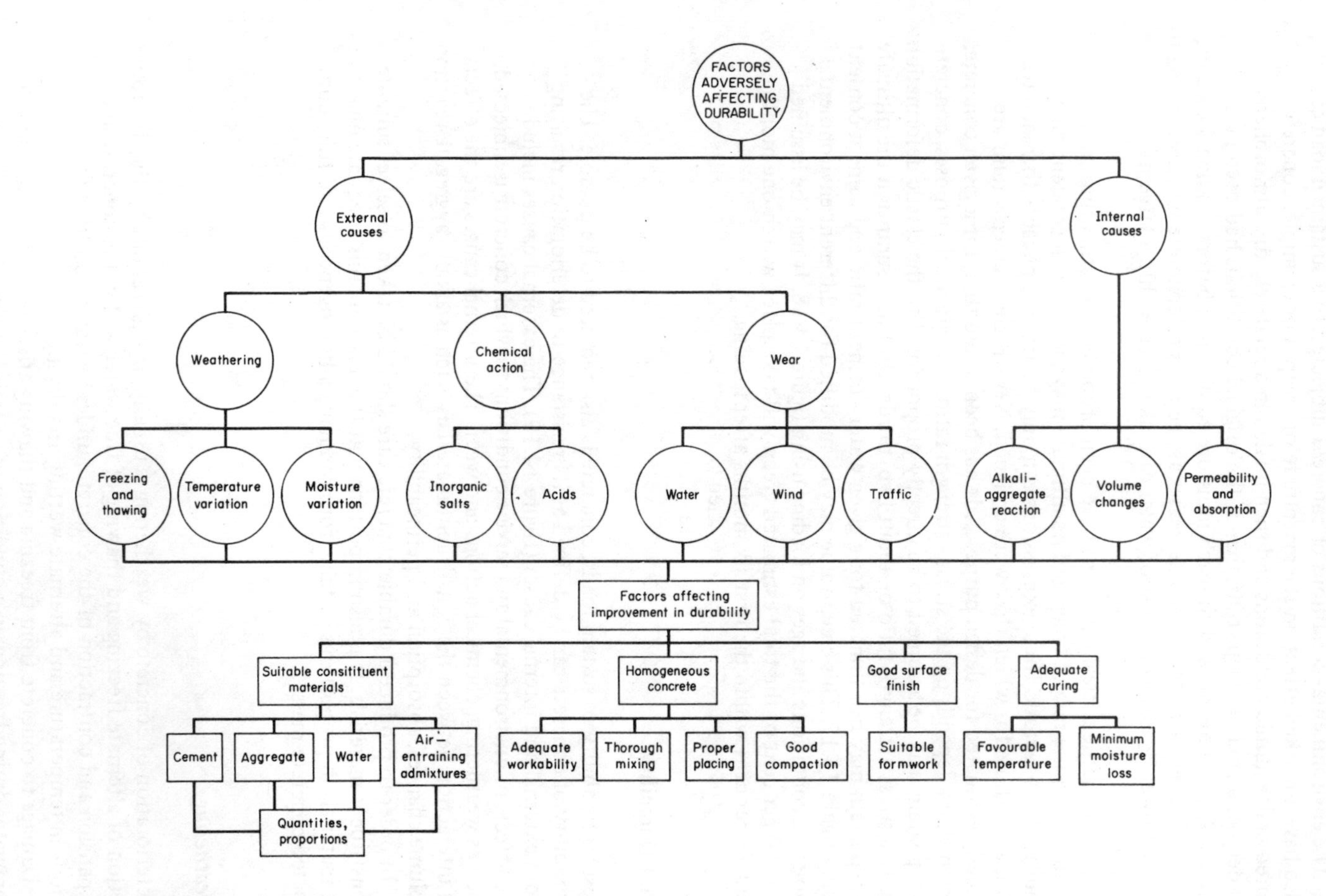

Figure 14.19 *Factors affecting durability of concrete*

disrupting concrete. Road kerbs and slabs, dams and reservoirs are very susceptible to frost action.

The resistance of concrete to freezing and thawing can be improved by increasing its impermeability. This can be achieved by using a mix with the lowest possible water–cement ratio compatible with sufficient workability for placing and compacting into a homogeneous mass. Durability can be further improved by using air-entrainment. The use of air-entrained concrete is particularly useful for roads where salts are used for de-icing. It is also important that, where-ever possible, provision be made for adequate drainage of exposed concrete surfaces. A dry concrete is not affected by freezing. The type of cement used has no effect although during the very early stages of hydration the use of a cement with a high rate of hydration can be beneficial. Damage to a structure resulting from the expansion and contraction of concrete should be minimised by providing joints which permit such movement without restraint.

Chemical Attack

In general, concrete has a low resistance to chemical attack. There are several chemical agents which react with concrete but two forms of attack are most common, namely, leaching and sulphate attack. Chemical agents essentially react with certain compounds of the hardened cement paste and the resistance of concrete to chemical attack therefore depends largely on the type of cement used. The resistance to chemical attack improves with increased impermeability.

Leaching

Calcium hydroxide, Ca $(OH)_2$, in hardened cement paste dissolves readily in water, particularly in the presence of carbon dioxide, CO_2. Thus, if concrete in service absorbs or permits the passage of water through it the calcium hydroxide in the hardened cement is removed, or leached out. Leaching can seriously impair the durability of concrete. Hydraulic structures in which water may pass through cracks, areas of segregated or porous concrete or along poor construction joints suffer from this kind of attack. Concrete may also absorb rain or groundwater and the presence of carbon dioxide in such waters enhances the process of leaching.

A homogeneous and dense concrete with low permeability significantly reduces the effectiveness of the leaching action. Care should be taken in selecting and proportioning the constituent materials and in curing the concrete to ensure that shrinkage cracking is minimised.

Sulphate attack

Most sulphate solutions react with the calcium hydroxide, $Ca(OH)_2$, and calcium aluminate, C_3A, of hydrated cement to form calcium sulphate and calcium sulphoaluminate compounds. Of these, alkaline calcium and magnesium sulphates are most active and occur widely in soils (particularly clays), groundwater and seawater. Although these compounds, unlike calcium hydroxide, do not readily dissolve in water their volume is greater than the volume of the compounds of cement paste from which they are formed. This increase in volume within the hardened concrete contributes towards the breakdown of its structure.

The intensity and rate of sulphate attack depend on a number of factors such as type of sulphate (magnesium sulphate is the most vigorous), its concentration and the continuity of its supply to concrete. The concentration of sulphates in solution is expressed in parts of SO_3 per million (p.p.m.), by weight. Permeability and the presence of cracks also affect the severity of the attack. The type of cement is a very important factor and the resistance of various cements to sulphate attack increases in the following order: ordinary and rapid-hardening Portland cement, Portland blastfurnace cement and low-heat Portland cement, sulphate-resisting Portland cement and supersulphated cement and finally high-alumina cement. Calcium chloride reduces the resistance of concrete to sulphate attack.

To protect concrete from sulphate attack every effort should be made to produce an impermeable concrete. Requirements given in CP 110: Part 1 regarding the type of cement, water-cement ratio and minimum cement content for concrete exposed to various degrees of sulphate attack are reproduced in table 14.3. The number of construction joints should be minimised since these can be particularly prone to attack. For concrete structures housed in sulphate-bearing soils protective coatings such as bitumens, tars and epoxy resins may also be applied on exterior surfaces although some of these coatings may be eroded away by groundwater flow.

Wear

The main causes of wear of concrete are the cavitation effects of fast-moving water, abrasive material in water, wind blasting and attrition and impact of traffic. Certain conditions of hydraulic flow result in the formation of cavities between the flowing water and the concrete surface. These cavities are usually filled with water vapour charged with extraordinarily high energy and repeated contact with the concrete surface results in the formation of pits and holes, known as cavitation erosion. Since even a good-quality concrete will not be able to resist this kind of deterioration the best remedy is therefore the elimination of cavitation by producing smooth hydraulic flow. Where necessary, the critical areas may be lined with materials having greater resistance to cavitation erosion.

In general, the resistance of concrete to erosion and abrasion increases with increase in strength. The use of a hard and tough aggregate tends to improve concrete resistance to wear.

Alkali-Aggregate Reaction

Certain aggregates can react chemically with the alkaline content of cement to form alkaline silica gel. When this happens these aggregates expand or swell

TABLE 14.3

Requirements for concrete exposed to sulphate attack, based on CP 110: Part 1

Concentration of sulphate expressed as SO_3			Minimum cement content (kg m^{-3})			
			Maximum size of aggregate (mm)			
In soil (per cent)	In groundwater (p.p.m.)	Type of cement	37.5	19.0	9.5	Maximum water – cement ratio
Below 0.2	Below 300	Ordinary Portland or Portland blastfurnace	240	280	330	0.55
0.2 to 0.5	300 to 1200	Ordinary Portland or Portland blastfurnace	290	330	380	0.50
		Sulphate resisting	240	280	330	0.55
		Supersulphated	270	310	360	0.50
0.5 to 1.0	1200 to 2500	Sulphate resisting or supersulphated	290	330	380	0.50
1.0 to 2.0	2500 to 5000	Sulphate resisting or supersulphated	330	370	420	0.45
Over 2.0	Over 5000	Sulphate resisting or supersulphated with adequate protective coating	330	370	420	0.45

resulting in cracking and disintegration of concrete. Such aggregates contain silica in its reactive form which occurs in such rocks as cherts, siliceous limestones and certain volcanic rocks. The minimum alkaline content of cement required to produce enough alkaline silica gel (expansive reaction) to damage concrete is 0.6 per cent of the soda equivalent. Moisture is necessary for the alkali - aggregate reaction and increased temperatures accelerate this reaction.

The simplest preventive measure for this type of deterioration is not to use alkaline reactive aggregate with cements having a high alkaline content. Adverse effects of the alkali - aggregate reaction can be minimised by adding a fine pozzolanic material to the concrete mix. Pozzolanic materials combine with the alkaline content of cement while concrete is in its plastic state, thus effectively lowering the alkaline content.

Volume Changes

Principal factors responsible for volume changes are the chemical combination of water and cement and the subsequent drying of concrete, variations in temperature and alternate wetting and drying. When a change in volume is resisted by internal or external forces this can produce cracking: the greater the imposed restraint, the more severe the cracking. The presence of cracks in concrete reduces its resistance to the action of leaching, corrosion of reinforcement, attack by sulphates and other chemicals, alkali - aggregate reaction and freezing and thawing, all of which may lead to disruption of concrete. Severe cracking can lead to complete disintegration of the concrete surface particularly when this is accompanied by alternate expansion and contraction.

Volume changes can be minimised by using suitable constituent materials and mix proportions having due regard to the size of structure. Adequate moist curing is also essential to minimise the effects of any volume changes.

Permeability and Absorption

Permeability refers to the ease with which water can pass through the concrete. This should not be confused with the absorption property of concrete and the two are not necessarily related. Absorption may be defined as the ability of concrete to draw water into its voids. Low permeability is an important requirement for hydraulic structures and in some cases watertightness of concrete may be considered to be more significant than strength although, other conditions being equal, concrete of low permeability will also be strong and durable. A concrete which readily absorbs water is susceptible to deterioration.

Concrete is inherently a porous material. This arises from the use of water in excess of that required for the purpose of hydration in order to make the mix sufficiently workable and the difficulty of completely removing all the air from the concrete during compaction. If the voids are interconnected concrete becomes pervious although with normal care concrete is sufficiently impermeable for most purposes. Concrete of low permeability can be obtained by suitable selection of its constituent materials and their proportions followed by careful placing, compaction and curing. In general for a fully compacted concrete, the permeability decreases with decreasing water - cement ratio. Permeability is affected by both the fineness and the chemical composition of cement. Coarse cements tend to produce pastes with relatively high porosity. Aggregates of low porosity are preferable when concrete with a low permeability is required. Segregation of the constituent materials during placing can adversely affect the impermeability of concrete.

14.5 Shrinkage

Shrinkage of concrete is caused by the settlement of solids and the loss of free water from the plastic concrete (plastic shrinkage), by the chemical combination of cement with water (autogenous shrinkage) and by the drying of concrete (drying shrinkage). Where movement of the concrete is restrained, shrinkage will produce tensile stresses within the concrete which may cause cracking. Most concrete structures experience a gradual drying out and the effects of drying shrinkage should be minimised by the provision of movement joints and careful attention to detail at the design stage.

Plastic Shrinkage

Shrinkage which takes place before concrete has set is known as plastic shrinkage. This occurs as a result of the loss of free water and the settlement of solids in the mix. Since evaporation usually accounts for a large proportion of the water losses plastic shrinkage is most common in slab construction and is characterised by the appearance of surface cracks which can extend quite deeply into the concrete. Preventive measures are usually based on methods of reducing water loss. This can be achieved in practice by covering concrete with wet hessian or polythene sheets or by spraying it with a membrane curing compound.

Autogenous Shrinkage

In a set concrete, as hydration proceeds, a net decrease in volume occurs since the hydrated cement gel has a smaller volume than the sum of the cement and water constituents. As hydration continues in an environment where the water content is constant, such as inside a large mass of concrete, this decrease in volume of the cement paste results in shrinkage of the concrete. This is known as autogenous shrinkage because, as the name implies, it is self-produced by the hydration of cement. However, when concrete is cured under water, the water taken up by cement during hydration is replaced from outside and furthermore the gel particles absorb more water, thus producing a net increase in volume of the cement paste and an expansion of the concrete. On the other hand if concrete is kept in a dry atmosphere water is drawn out of the hydrated gel and additional shrinkage, known as drying shrinkage, occurs.

Several factors influence the rate and magnitude of autogenous shrinkage. These include the chemical composition of cement, the initial water content, temperature and time. The autogenous shrinkage can be up to 100×10^{-6} of which 75 per cent occurs within the first three months.

Drying Shrinkage

When a hardened concrete, cured in water, is allowed to dry it first loses water from its voids and capillary pores and only starts to shrink during further drying when water is drawn out of its cement gel. This is known as drying shrinkage and in some concretes it can be greater than 1500×10^{-6}, but a value in excess of 800×10^{-6} is usually considered to be undesirable for most structural applications. After an initial high rate of drying shrinkage concrete continues to shrink for a long period of time but at a continuously decreasing rate (see figure 14.20). For practical purposes, it may be assumed that for small sections 50 per cent of the total shrinkage occurs in the first year.

When concrete which has been allowed to dry out is subjected to a moist environment, it swells. However, the magnitude of this expansion is not sufficient to recover all the initial shrinkage even after prolonged immersion in water. Concrete subjected to cyclic drying and wetting approaches the same shrinkage level as that caused by complete drying (figure 14.20). A test procedure for determining shrinkage is described in BS 1881: Part 5.

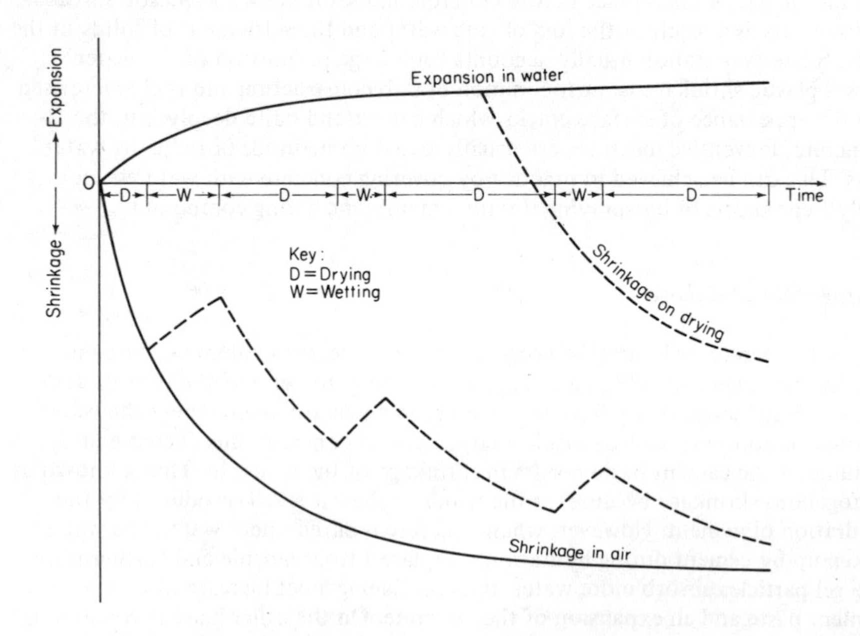

Figure 14.20 *Drying shrinkage and expansion characteristics of concrete*

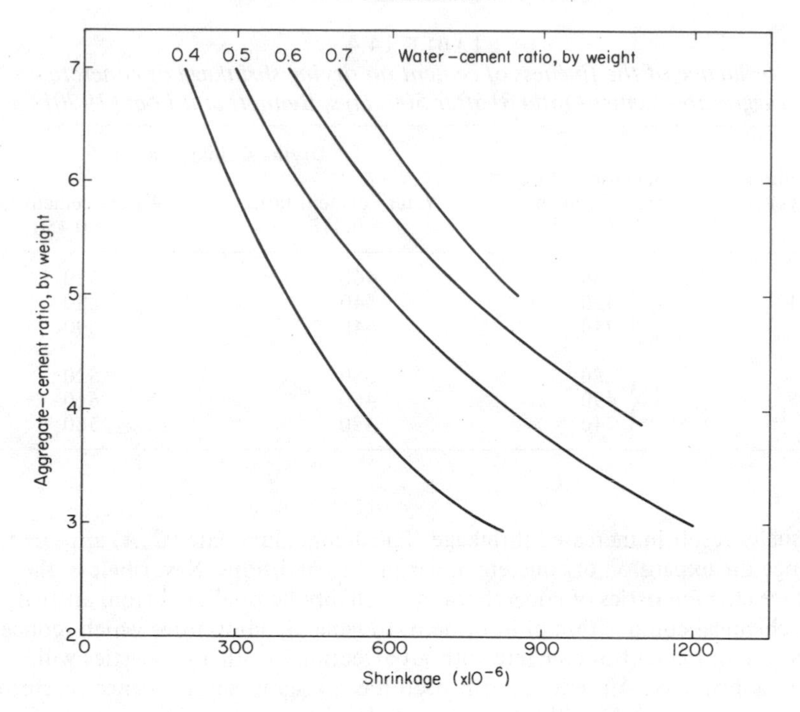

Figure 14.21 *Effect of water-cement and aggregate-cement ratios on drying shrinkage of concrete at 20 °C and 50 per cent relative humidity, based on Lea (1970)*

Factors affecting Drying Shrinkage

Several factors influence the over-all drying shrinkage of concrete. These include the type, content and proportion of the constituent materials of concrete, the size and shape of the concrete structure, the amount and distribution of reinforcement and the relative humidity of the environment.

In general, drying shrinkage is directly proportional to the water-cement ratio and inversely proportional to the aggregate-cement ratio (see figure 14.21). Because of the interaction of the effects of aggregate-cement and water-cement ratios, it is possible to have a rich mix with a low water-cement ratio giving higher shrinkage than a leaner mix with a higher water-cement ratio. For a given water-cement ratio shrinkage increases with increasing cement content.

Since the aggregate exerts a restraining influence on shrinkage the maximum aggregate content compatible with other required properties is desirable. When the aggregate itself is susceptible to large moisture movement, this can aggravate shrinkage (or swelling) of the concrete and may result in excessive cracking and large deflections of beams and slabs. The composition and fineness of cement can also affect its shrinkage characteristics. In general, shrinkage increases as the specific surface area of cement increases (table 14.4) although this effect is slight and is usually overshadowed by the effects of water-cement ratio and aggregate-cement ratio. Increases in dicalcium silicate (C_2S) content and ignition

TABLE 14.4
Influence of the fineness of cement on drying shrinkage of concrete (aggregate - cement ratio 3) after 500 days, Bennett and Loat (1970)

Moist curing (days)	Specific surface area of cement (m^2 kg^{-1})	Drying shrinkage $\times 10^{-6}$	
		Water - cement ratio = 0.375	Water - cement ratio = 0.450
1	280	460	520
	490	540	680
	740	540	690
28	280	380	520
	490	460	610
	740	420	580

loss usually result in increased shrinkage. Tricalcium aluminate (C_3A) appears to influence the expansion of concrete under moist conditions. Nevertheless, the shrinkage characteristics of concrete cannot reliably be predicted from an analysis of the chemical composition of its cement. In general, admixtures which reduce the water requirement of concrete without affecting its other properties will reduce its shrinkage. Air-entrainment itself has no significant influence on shrinkage. Calcium chloride may considerably increase shrinkage.

The size and shape of a specimen affects the rate of moisture movement in concrete and this in turn influences the rate of volume change. Since drying begins from the surface, it follows that the greater the surface area per unit mass, the greater the rate of shrinkage. For a given shape, the initial rate of shrinkage is greater for small specimens although there will be little difference in the ultimate drying shrinkage, if this stage is ever reached for very large masses of concrete.

The shrinkage of reinforced concrete is less than that of plain concrete owing to the restraint developed by the reinforcement. This restraint induces tensile stresses in the concrete which may be large enough to cause cracking.

The relative humidity and temperature of the environment have a significant effect on both the rate and magnitude of shrinkage in as much as they affect the movement of water in concrete. The duration of initial moist curing has little effect on ultimate shrinkage although it affects the initial rate of shrinkage.

14.6 Evaluation of the Quality of Concrete From Nondestructive Testing

The quality of concrete is usually taken to mean its strength and durability although other properties such as resistance to deformation and shrinkage can be significant in determining structural behaviour. In general most of the properties of concrete improve with increasing strength and for this reason the quality of concrete is often judged by its strength. Nondestructive testing, as the name implies, requires that the material under test is not damaged during testing.

although factors such as 4 c's are very important

Direct measurement of the strength of concrete involves destructive stresses and thus cannot be used for determining the quality of concrete in structures. Furthermore, the compressive strength test, as described in BS 1881: Part 4 can only indicate the potential strength of the mix. The actual concrete strength within a concrete unit or structure depends on the conditions of placing, compaction and curing. Although test samples can be cored from the structure and tested for evaluating the quality of concrete, this operation is too expensive for general use. Moreover, only a limited number of cores can be taken without damaging the structure.

It is for these reasons that attempts have been made, in the last three decades, to determine some suitable nondestructive test for determining the quality of concrete. Several tests have now been developed but those which have been most widely accepted include vibrational methods for estimating strength, durability and uniformity and for detecting flaws, and hardness methods for estimating strength. Although the tests are simple to perform they have certain limitations. Nevertheless, when applied rationally the techniques often provide information which cannot otherwise be obtained by direct methods. The vibrational methods can also be beneficially employed in laboratory investigations where progressive changes in the quality of concrete due to environmental effects are to be evaluated.

Resonance Method

This method is used to determine the dynamic modulus of elasticity. A concrete specimen of a well-defined shape, such as a beam similar to the one used in the flexural test, is subjected to vibration in either the longitudinal, flexural or torsional mode. Basically the experimental technique is similar in each case and the only difference is the way in which the beam is supported and excited. For the longitudinal mode of vibration, as described in BS 1881: Part 5, the beam is clamped at its centre and the exciter and pick-up units are brought into contact with the ends of the beam, without exerting restraint, so that the ends are free to vibrate in the longitudinal direction (see figure 14.22). The exciter unit is driven by a variable-frequency oscillator and forces the specimen into longitudinal vibration. These vibrations are received by the pick-up unit and after amplification their amplitude is indicated on a meter.

During the test the frequency of the oscillator is varied so that resonance is obtained at the fundamental frequency, indicated by maximum deflection on the meter. The dynamic modulus of elasticity E_d is given by

$$E_d = Fn^2\rho = 4n^2 l^2 \rho\, 10^{-12}\ \text{MN m}^{-2}$$

where n is the natural frequency of the fundamental mode of longitudinal vibration of the specimen (Hz), l the length of the specimen (mm) and ρ its density (kg m^{-3}). The value of the constant F depends on the shape of the specimen and the mode of vibration.

In general, changes in the quality of concrete are related to changes in the dynamic modulus of elasticity but these relationships are not unique and are affected by several factors such as the constituent materials, particularly the type

Figure 14.22 *A typical arrangement for measuring the longitudinal resonance of a concrete beam*

of aggregate, mix proportions and curing conditions. However, for a given concrete the variation in dynamic modulus of elasticity can give a good indication of variations in strength and static modulus of elasticity, and this test is particularly useful for assessing the progressive change in strength and durability as affected by various factors such as the action of alternate freezing and thawing and sulphate solutions. The main disadvantage of the resonance method is that it cannot be used for assessing the properties of concrete in an actual structure since it requires specimens of shapes for which relationships between frequency and the dynamic modulus of elasticity are known.

Ultrasonic Pulse Method

In this method the velocity of an ultrasonic pulse passing through the concrete is determined (BS 4408: Part 5). This technique is now widely used for assessing the quality of concrete in structures and several models of the apparatus are commercially available. The pulse, produced by an electro-acoustical transducer placed in contact with the concrete under test, passes through the concrete and is picked up and converted into an electrical signal by a second electro-acoustical transducer (figure 14.23). The time taken by the pulse to travel through the concrete is measured by an electrical timing unit, to the nearest 0.1 μs, and the

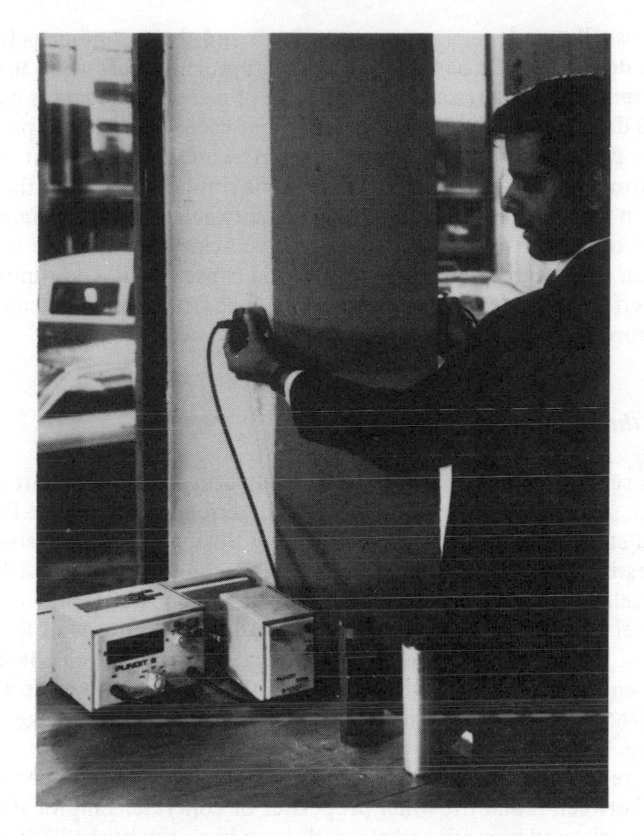

Figure 14.23 *Ultrasonic pulse apparatus*

pulse velocity v is calculated using the relationship

$$v = \frac{L}{t} \text{ m s}^{-1}$$

where L is the length of the path (m) and t is the time taken (s). The accuracy of the velocity thus obtained depends on the length of the path although after a certain length the sharpness of the signal decreases and there is no further gain in the accuracy of the measurement. Concrete thicknesses between 0.1 and 15 m can be tested quite satisfactorily.

The main advantages of the ultrasonic pulse method are that it can be employed for assessing the quality of concrete in structures and there are no restrictions concerning the shape of the concrete mass although access to the structure from both sides is desirable. As for the dynamic modulus of elasticity, there is no unique relationship between the pulse velocity and strength as it is influenced by the concrete constituents and curing conditions. The effect of the coarse aggregate is of particular significance since its influence on the pulse velocity is more marked than its influence on strength. Thus the evaluation of the quality of concrete in structures is usually made on a comparative basis and the technique is frequently employed to detect inferior parts within a structure. However, when the mix proportions remain constant and only one type of coarse aggregate is used then it is possible to determine a specific relationship between strength and pulse velocity for *in situ* concrete. Since the pulse velocity is affected by moisture it is important to have moisture conditions in the test specimens similar to those in the *in situ* concrete when establishing strength - pulse-velocity relationships. The pulse is transmitted most effectively by solid media and cracks or cavities are indicated by a reduction in the pulse velocity. The method is useful for a continuous assessment of the effects of deteriorating agencies such as frost and chemicals for both *in situ* and laboratory concrete.

Hardness Method

The Schmidt rebound hammer is the most widely accepted instrument for measuring the surface hardness of concrete, as described in BS 4408 : Part 4. A fixed mass of steel is charged with kinetic energy through a spring system by gradually pressing the plunger against the surface to be tested (figure 14.24). The steel mass is released and impinges on the plunger, which remains in contact with concrete. After this impact the mass rebounds and the magnitude of the rebound is a measure of the hardness of the surface, indicated by a rider on a linear scale graduated in empirical rebound numbers. When the hammer is used to test a non-vertical surface the rebound reading is corrected because of the change in the impact energy.

Hardness (rebound number) is a relative property and there can be no physical relationship between it and the other properties of concrete. Empirical relationships between rebound number and strength have been established and in general the higher the rebound number, the greater the strength. As for the dynamic modulus of elasticity and ultrasonic pulse velocity, there is no unique relationship between rebound number and strength. For this reason it is advisable to deter-

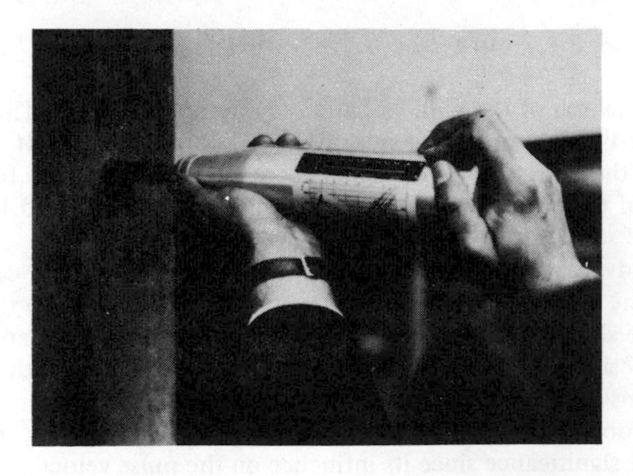

Figure 14.24 *Schmidt hammer*

mine the strength - hardness relationship for each concrete instead of relying on *secondhand* values. When the hammer is used for assessing the strength of *in situ* concrete the test procedure and environmental conditions should be similar to those employed during calibration. Since for a given concrete the rebound number can vary because of differences in hardness between aggregate and matrix and the possible variation in aggregate mineralogy, it is necessary that several readings are taken and the average value used. Control over the preparation of the test surface is important for proper use of the hammer. Provided the limitations of the method are borne in mind and the hammer is used intelligently, it can be a useful tool for assessing the strength of concrete in structures. It is an inexpensive mechanical device and is easy to use. The technique can also be applied to assess the uniformity of concrete, for example, to locate the possible existence of an area of unsatisfactory concrete in a wall.

Other Methods

These include the use of gamma rays for detecting voids in concrete and for locating reinforcement, electrical methods for measuring moisture content and electromagnetic methods for measuring the depth below the concrete surface of steel reinforcing bars. Some of these are described in BS 4408: Part 1 and Part 3.

15

Concrete Mix Design and Quality Control

Concrete mix design can be defined as the procedure by which, for any given set of conditions, the proportions of the constituent materials are chosen so as to produce a concrete with all the required properties for the minimum cost. In this context the cost of any concrete includes, in addition to that of the materials themselves, the cost of the mix design, of batching, mixing and placing the concrete and of site supervision.

Quality control refers in the first instance to the supervision exercised on site to ensure both that the materials used in the production of the concrete are of the required quality and that the stated mix proportions are adhered to as closely as is possible with the available site facilities. It also refers to the control testing of the quality of the hardened concrete to ensure that it conforms with the design requirements and, where it deviates significantly from this, the taking of the necessary corrective action.

Mix design, quality control and the over-all cost of the finished concrete product are all interdependent and this will become increasingly apparent as the factors affecting mix design are discussed.

In this chapter the basic requirements for any concrete and the way in which these may be incorporated into concrete mix design are described. The essentials of quality control associated with the production of concrete on site have been considered earlier. Some of the statistical methods of quality control used for the interpretation of control tests on the hardened concrete are described here.

15.1 Required Concrete Properties

The basic requirements for concrete are conveniently considered at two stages in its life.

In its hardened state (in the completed structure) the concrete should have adequate durability, the required strength and also the desired surface finish.

In its plastic state, or the stage during which it is to be handled, placed and compacted in its final form, it should be sufficiently workable for the required properties in its hardened state to be achieved with the facilities available on site. This means that

(1) the concrete should be sufficiently fluid for it to be able to flow into and fill all parts of the formwork, or mould, into which it is placed;

(2) it should do so without any segregation, or separation, of the constituent materials while being handled from the mixer or during placing;

(3) it must be possible to fully compact the concrete when placed in position; and

(4) it must be possible to obtain the required surface finish.

If concrete does not have the required workability in its plastic state, it will not be possible to produce concrete with the required properties in its hardened state. The dependence of both durability and strength on the degree of compaction has been noted earlier. Segregation results in variations in the mix proportions throughout the bulk of the concrete and this inevitably means that in some parts the coarser aggregate particles will predominate. This precludes the possibility of full compaction since there is insufficient mortar to fill the voids between the coarser particles in these zones. This results in what is descriptively known as honeycombing on the surface of the hardened concrete with reduced durability and strength as well as unacceptable surface finish.

The means by which each of the required concrete properties may be achieved are now considered.

Durability

Adequate durability of exposed concrete can frequently be obtained by ensuring full compaction, an adequate cement content and a low water - cement ratio, all of which contribute to producing a dense, impermeable concrete. The choice of aggregate is also important particularly for concrete wearing surfaces and where improved fire resistance is required. Aggregate having high shrinkage properties should be used with caution in exposed concrete. For particularly aggressive environments additional precautions may be necessary, for example, the use of sulphate-resisting cement where the concrete will be exposed to sulphate attack. For airfields and roads where salt is likely to be used for de-icing purposes, the use of air-entrained concrete is recommended.

Durability is not a readily measured property of the hardened concrete. However, for a correctly designed concrete mix any increase in the water - cement ratio on site with the associated reduction in durability, will be accompanied by a reduction in concrete strength. The latter can be determined quite easily using control specimens and for this reason the emphasis in control testing is on the determination of concrete strength.

Strength

Although durability is sometimes the overriding criterion, the strength of the concrete is frequently an important design consideration particularly in structural applications where the load-carrying capacity of a structural member may be closely related to the concrete strength. This will usually be the compressive strength although occasionally the flexural or indirect tensile strength may be more relevant.

The strength requirement is generally specified in terms of a *characteristic strength* (CP 110: Part 1) coupled with a requirement that the probability of the strength falling below this shall not exceed a certain value. Typically this may be 5 per cent or a 1 in 20 chance of a strength falling below the specified characteristic strength, this generally being the 28-day strength.

An understanding of the factors affecting concrete strength on site, and of the probable variations in strength, is essential if such specifications are to have any real meaning at the mix design stage. A histogram showing the frequency of cube compressive strengths for a road contract is shown in figure 15.1. What might appear to be large variations in strength can be seen, with actual cube strengths ranging from 8.5 to 34.0 N mm^{-2}. Some of the differences between individual cube strengths may be attributed to testing errors including sampling, preparation and curing of the test cubes, the testing machine itself and to the actual test procedure. Differences in strength can also occur owing to variations in the quality of the cement but the principal factor affecting the strength is the quantity of water, or more specifically the water - cement ratio, in the concrete mix.

If the proportions of aggregate and cement and also the quality of the aggregate are maintained constant, the water - cement ratio can be controlled very effectively at the mixer by adding just sufficient water to give the required workability. Once a suitable mix has been obtained the workability can be assessed quite satisfactorily

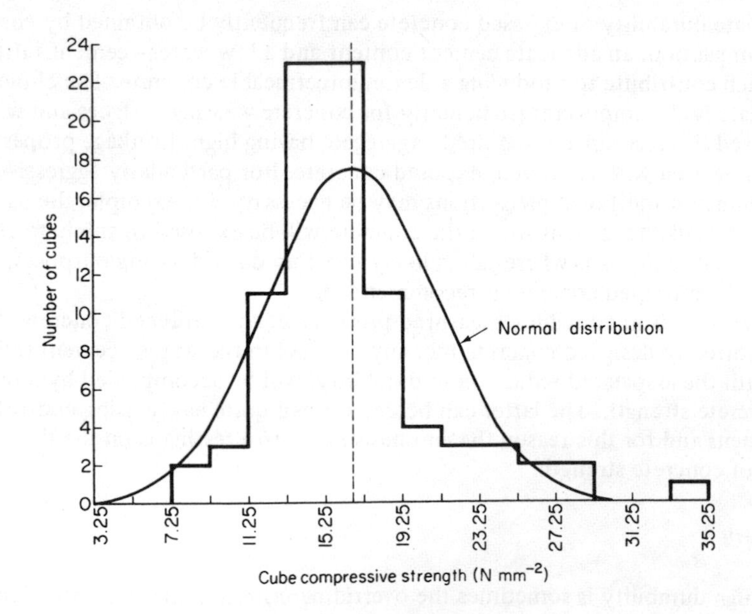

Figure 15.1 *Histogram for cube compressive strength*

by an experienced mixer operator, with periodic control tests of the workability. However, human error will inevitably result in some variation in the water - cement ratio either side of the desired value. Any variation in mix proportions or significant changes in the aggregate grading will affect the quantity of water needed to maintain the required workability and this too will result in variations in the water - cement ratio and hence in concrete strength.

All these factors tend to give water - cement ratios which are as likely to be greater as they are to be less than the target value. The actual water - cement ratios tend therefore to have a *normal* or gaussian distribution about the mean, or target, value. The relationship between water - cement ratio and concrete strength is non-linear. Nevertheless over a limited range the relationship will be approximately linear and it might be expected that concrete strengths will also tend to have a normal distribution. This can be seen in figure 15.1 in which a typical bell-shaped normal distribution curve is shown.

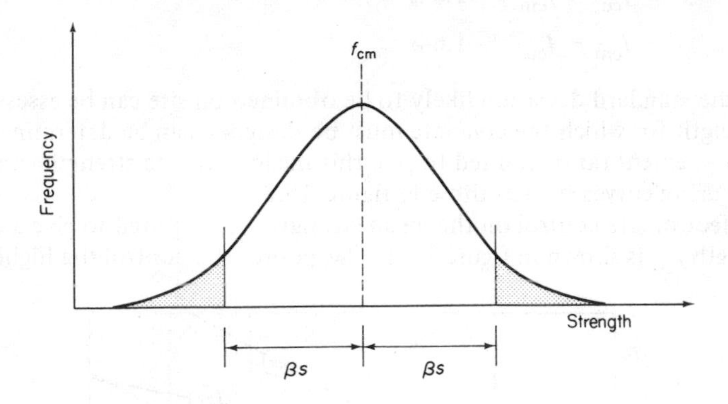

Figure 15.2 *Normal distribution curve*

Design Mean Strength

The assumption of a normal distribution of concrete strengths forms the basis of mix design and statistical quality control procedures for satisfying the strength requirement.

For a normal distribution, the probability of a strength lying outside specified limits either side of the mean strength can be determined. These limits (figure 15.2) are usually expressed in terms of the standard deviation s, defined by

$$s = \left[\frac{\Sigma\,(f_c - f_{cm})^2}{n - 1}\right]^{\frac{1}{2}} = \left[\frac{\Sigma(f_c)^2 - (\Sigma\,f_c)^2/n}{n - 1}\right]^{\frac{1}{2}} \text{ N mm}^{-2}$$

where f_c is an observed strength, f_{cm} is the best estimate of the mean strength, equal to $(\Sigma\,f_c)/n$, and n is the number of observations. The probabilities of a strength lying outside the range $(f_{cm} \pm \beta s)$ for different values of β are given in table 15.1, in which the probabilities of strengths falling below the lower limit $(f_{cm} - \beta s)$ are also given.

TABLE 15.1
Probability values

Probability of an observed strength lying outside the range ($f_{cm} \pm \beta s$)	β	Probability of an observed strength being less than ($f_{cm} - \beta s$)
1 in 50	2.33	1 in 100
1 in 20	1.96	1 in 40
1 in 10	1.64	1 in 20

If the specified characteristic strength f_{cu} is the strength below which not more than 1 in 20 of the population of strengths shall fall, it follows that

$$f_{cu} = f_{cm} - 1.64s$$

or

$$f_{cm} = f_{cu} + 1.64s$$

Hence if the standard deviation likely to be obtained on site can be assessed, the mean strength for which the concrete must be designed can be determined. The free water - cement ratio required to give this mean concrete strength can then be estimated using curves such as those in figure 15.3

The effect of site control on the mean strength f_{cm} required to give a characteristic strength f_{cu} is shown in figure 15.4. The poorer the control the higher the

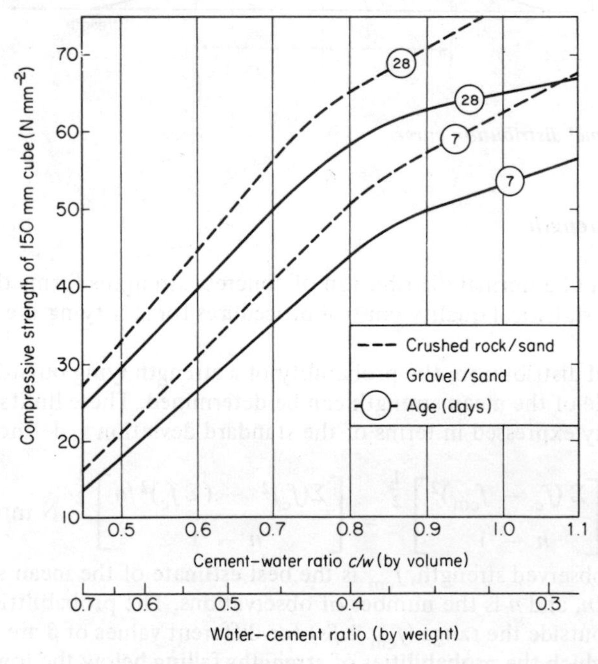

Figure 15.3 *Cube compressive strength relationships for concrete made with ordinary Portland cement*

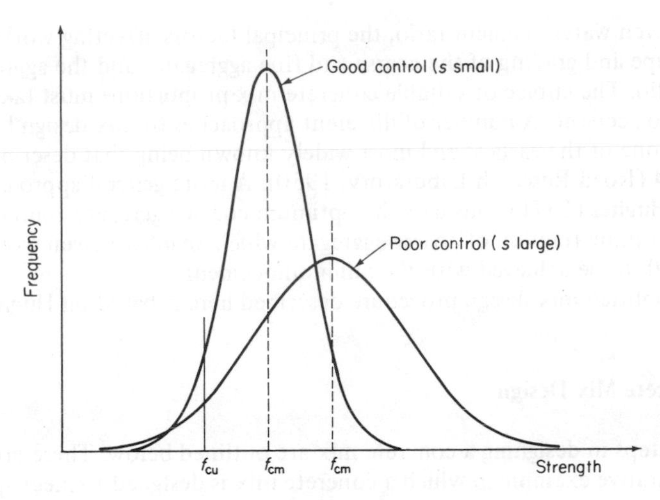

Figure 15.4 *Effect of control on required mean strength*

mean strength required and this will mean a lower water - cement ratio. This in turn will require a smaller aggregate - cement ratio, or larger cement content per unit volume, for a given workability and a consequent increase in the total cost of materials.

Workability

Suitable workabilities for different types of work are given in table 15.2, workability in this context referring to those qualities measured by the slump, compacting factor and V - B tests. In this table the bracketed values are the least reliable although the associated tests can generally be used for control purposes once the actual value corresponding to a suitable workability has been determined.

TABLE 15.2
Workability requirements

Type of work	V - B (s)	Compacting factor	Slump (mm)	Workability
Heavily reinforced sections with vibration. Simply reinforced sections without vibration	(3)	0.92	(25 - 100)	Medium
Simply reinforced sections with vibration. Mass concrete without vibration	6	(0.86)	(10 - 50)	Low
Mass concrete and large sections with vibration. Road slabs vibrated using power-operated machines	12	(0.80)	–	Very low

For a given water - cement ratio, the principal factors affecting workability are the shape and grading of the coarse and fine aggregates and the aggregate - cement ratio. The choice of suitable concrete mix proportions must take all these factors into account. A number of different approaches to mix design have been proposed, one of the earliest and most widely known being that described in Road Note No. 4 (Road Research Laboratory, 1950). A more general approach proposed by Hughes (1971) considers the optimum coarse aggregate content, this being the volume fraction of coarse aggregate which enables a given workability and strength to be achieved with the minimum cement.

The simplified mix design procedure described here is based on Hughes' approach.

15.2 Concrete Mix Design

The basic steps in designing a concrete mix are outlined below. These are followed by an illustrative example in which a concrete mix is designed to meet specific requirements.

(1) *Durability.* For the degree of exposure determine the type of cement, the minimum cement content and maximum water - cement ratio, being guided by relevant specifications.

(2) *Strength.* Where a maximum water - cement ratio has been obtained in step 1, determine the associated concrete strength using relationships such as those shown in figure 15.3. Compare this with the specified characteristic strength and accept the larger value as the required characteristic strength f_{cu}.

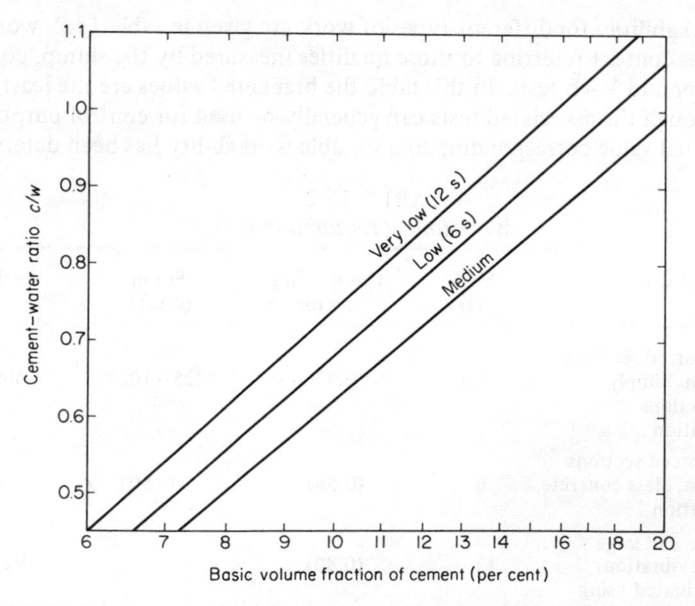

Figure 15.5 *Basic volume fraction of cement for different workabilities, based on Hughes (1971)*

(3) Determine the required mean concrete strength from

$$f_{cm} = f_{cu} + \beta s$$

using an appropriate value of β. The standard deviation s should be based on recent test results for conditions similar to those which will occur on site. Where no such data are available, conservative values should be adopted. For example, when $\beta = 1.64$, values of $\beta s = 0.67 f_{cu}$ for $f_{cu} \leqslant 22.5$ N mm^{-2} and $\beta s = 15.0$ N mm^{-2} for $f_{cu} \geqslant 22.5$ N mm^{-2} might be used.

(4) Determine the cement - water ratio c/w (by volume) associated with the mean strength f_{cm} using, for example, figure 15.3.

(5) *Workability*. Determine the *basic* volume fraction of cement required to give the desired workability (see table 15.2), for the cement - water ratio c/w obtained in step 4, using figure 15.5.

(6) Using table 15.3 and the gradings of the available coarse and fine aggregates determine the corresponding grading moduli, G_a and G_b respectively, and the equivalent mean diameter D_b of the fine aggregate.

(7) Using figure 15.6 obtain the corrections to the *basic* volume fraction of cement required when (a) the coarse aggregate is crushed rock; (b) the fine aggregate is crushed rock; (c) the coarse aggregate grading modulus G_a differs from 0.46; or (d) the fine aggregate grading modulus G_b is greater than 16 or less than 9. Hence determine the corrected volume fraction of cement c. Compare this with the corresponding durability requirement, step 1, and accept the higher value as the required volume fraction of cement c. Where the durability requirement is the overriding factor increase the cement - water ratio c/w accordingly so that the workability remains unchanged, using figure 15.5.

(8) Using the volume fraction of cement c and the volume ratio c/w determine the volume fraction of water w.

(9) Evaluate the fine-to-coarse size ratio $G_a D_b$. The optimum coarse aggregate content depends on the relative sizes of the coarse and fine aggregate particles, more space between the coarse aggregate particles being required to accommodate the fine aggregate particles as the relative size of the latter increases. It has been shown (Hughes, 1968) that a satisfactory measure of the ratio of fine-to-coarse aggregate size is given by $G_a D_b$. Using table 15.4, an estimate of the

TABLE 15.3
Aggregate grading parameters

BS 410 test sieve (mm)	37.50	20.0	10.0	5.00	2.36	1.18	0.60	0.30	0.15	0.075
Grading modulus (× 0.90 mm^{-1})		$\frac{1}{4}$	$\frac{1}{2}$	1	2	4	8	16	32	63
Equiv. mean diameter (× 0.90 mm)				8	4	2	1	$\frac{1}{2}$	$\frac{1}{4}$	$\frac{1}{8}$

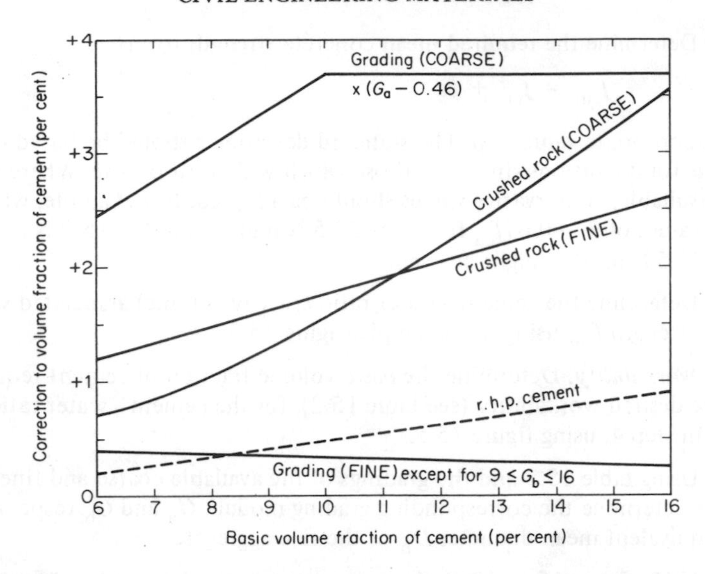

Figure 15.6 *Corrections to basic volume fraction of cement*

solids fraction of loose coarse aggregate a_b is obtained. Hence determine the recommended volume fraction of coarse aggregate a, using figure 15.7.

(10) The sum of the volume fractions a, c, w and the volume fraction of fine aggregate b must be unity. The volume fraction of fine aggregate b is equal therefore to $1 - (a + c + w)$.

(11) Determine batch quantities per cubic metre using the specific gravities of the constituent materials, the values for the aggregates being those for saturated surface-dry aggregates.

(12) Trial mixes in the laboratory are generally recommended so that any necessary preliminary adjustments to the mix proportions can be made before full-scale trial mixes under site conditions are carried out.

Trial Mixes

The necessity for trial mixes arises, for example, since the relationships between concrete strength and water-cement ratio (see figure 15.5) depend to some

TABLE 15.4

Solids fraction of loose coarse aggregate

	Solids fraction a_b	
Aggregate size	20 – 10 mm	20 – 5 mm
Irregular gravel	0.56	0.58
Crushed rock	0.46	0.48

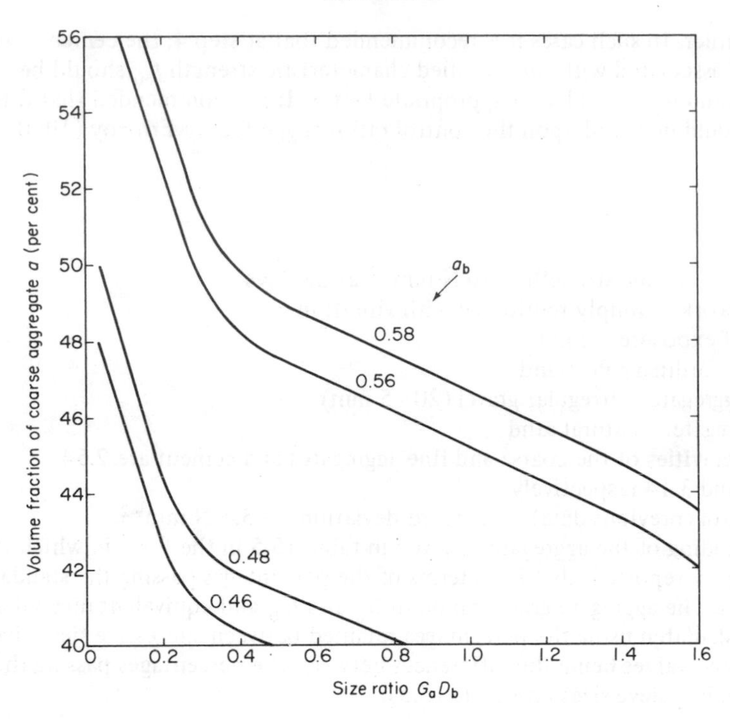

Figure 15.7 *Recommended volume fraction of coarse aggregate, based on Hughes (1971)*

extent on the local aggregates and the source of cement. Experience in a particular locality of the quality and characteristics of both cement and aggregates can be of considerable value at the mix design stage.

Segregation of the water and cement fines from the concrete can occur with lean concretes having a high workability. The cohesiveness of such mixes can be improved, and hence the tendency for wet segregation to occur reduced, by reducing the water content and/or increasing the proportion of fine to coarse aggregate. Either of these causes some reduction in workability and some reduction in the volume fraction of the combined fine and coarse aggregates is generally necessary if the workability is to be maintained. Alternatively the use of an air-entraining agent can be beneficial in such cases, only minor adjustments being necessary to allow for the volume fraction of the entrained air. Typically this will include some reduction in the volume fractions of both water and fine aggregate.

High-strength Concrete

For very high-strength concretes a skew distribution of concrete strength might be expected, as opposed to a normal distribution, as the upper limit of the attainable strength for the materials used is approached. The actual water-cement ratio might still be expected, however, to follow a normal distribution as discussed earlier. In such cases it is recommended that at step 4, the cement-water ratio c/w associated with the specified characteristic strength f_{cu} should be obtained and increased by an appropriate factor. It is recommended that this factor should be based upon the control ratios suggested by Erntroy (1960).

Example

Characteristic cube strength 30 N mm^{-2} at 28 days
Type of work simply reinforced with vibration
Degree of exposure mild
Cement ordinary Portland
Coarse aggregate irregular gravel (20 - 5 mm)
Fine aggregate natural sand
Specific gravities of the coarse and fine aggregates and cement are 2.54 2.60 and 3.14 respectively
Site control (previous data) standard deviation s = 5.5 N mm^{-2}

The grading of the aggregates is given in table 15.5 in the form in which this is frequently reported, that is, in terms of the percentages passing the standard sieve sizes. The aggregate grading moduli G_a and G_b and equivalent mean diameter D_b are calculated using the percentages retained between successive sieve sizes, the required values being the differences between the percentages passing the corresponding sieve sizes (see table 15.5).

TABLE 15.5
Aggregate grading

BS 410 test sieve (mm)	37.5	20.0	10.0	5.00	2.36	1.18	0.60	0.30	0.15	0.075
					per cent passing					
Irregular gravel	100	98	46	4	0					
Natural sand			100	96	78	60	36	8	3	
					per cent retained					
Irregular gravel		2	52	42	4					
Natural sand				4	18	18	24	28	5	3

Mix design

(1) Durability (CP 110: Part 1) minimum cement content = 250 kg m^{-3}
(2) –
(3) Strength $f_{cm} = f_{cu} + \beta s = 30 + 1.64\,(5.5) = 39.2$ N mm^{-2}
(4) From figure 15.3 c/w by volume = 0.595
(5) Workability, table 15.2 low workability appropriate
From figure 15.5 basic volume fraction of cement = 8.5 per cent
(6) Using table 15.3

Gravel (G_a)	Sand (G_b)	Sand (D_b)
$2 \times \frac{1}{4} = \frac{1}{2}$	$4 \times 1 = 4$	$4 \times 8 = 32$
$52 \times \frac{1}{2} = 26$	$18 \times 2 = 36$	$18 \times 4 = 72$
$42 \times 1 = 42$	$18 \times 4 = 72$	$18 \times 2 = 36$
$4 \times 2 = 8$	$24 \times 8 = 192$	$24 \times 1 = 24$
76.5	$28 \times 16 = 448$	$28 \times \frac{1}{2} = 14$
	$5 \times 32 = 160$	$5 \times \frac{1}{4} = 1\frac{1}{4}$
	$3 \times 63 = 189$	$3 \times \frac{1}{8} = \frac{3}{8}$
	1101	179.7

$G_a = 0.90\,(76.5) \times 10^{-2} = 0.689$ mm^{-1}; $G_b = 0.90\,(1101) \times 10^{-2} = 9.91$ mm^{-1}; $D_b = 0.90\,(179.7) \times 10^{-2} = 1.62$ mm

(7) From figure 15.6 (a) –
(b) –
(c) 3.2 (0.689 – 0.46) = + 0.74 per cent
Corrected volume fraction $c = 8.5 + 0.74 = 9.3$ per cent
Durability requirement = 250 kg m^{-3}

$$\equiv \frac{250 \times 100}{(\text{s.g. cement})\ 1000} = 8.0 \text{ per cent} \quad \text{O.K.}$$

Therefore volume fraction of cement $c = 0.093$
(8) From step 4, $c/w = 0.595$
Therefore volume fraction of water $w = 0.093/0.595 = 0.156$
(9) Using step 6, $G_a D_b = 0.689 \times 1.62 = 1.12$
From table 15.4 $a_b = 0.58$
From figure 15.7 volume fraction of coarse aggregate $a = 0.460$
(10) $a + c + w = 0.709$
Therefore volume fraction of fine aggregate $b = 0.291$
(11) Batch quantities

water	0.156 × 1000	= 156 kg m^{-3}
cement	0.093 × 1000 (3.14)	= 292 kg m^{-3}
sand	0.291 × 1000 (2.60)	= 757 kg m^{-3}
gravel	0.460 × 1000 (2.54)	= 1168 kg m^{-3}
		2373 kg m^{-3}

Adjustment to Mix Proportions

The foregoing example illustrates the basic mix design procedure. Some of the possible corrections to the mix proportions thus obtained are now considered.

Let trial mixes for the concrete designed in the example show that the desired mean strength f_{cm} is being achieved but the workability is too low, the V – B time being 8 s and the compacting factor 0.83. Referring to figure 15.5, to increase the workability by the required amount, namely, about one-half of the interval between successive degrees of workability, the volume fraction of cement should be increased by about 0.4 per cent. Hence c is increased from 0.093 to 0.097, w is increased accordingly from 0.156 to 0.163 to maintain the same c/w ratio, a remains unchanged at 0.460 and b, the sand content, is reduced to 0.280 to maintain the relationship $a + b + c + w = 1$. The corresponding batch quantities are obtained as before.

Alternatively let the workability be satisfactory but the estimated mean strength f_{cm} be only 35 N mm^{-2}. From figure 15.3, to increase the strength by 4 N mm^{-2} requires an increase of 0.045 in the c/w ratio, giving a final value of $c/w = 0.640$. From figure 15.5, the corresponding basic volume fraction of cement is 9.1 per cent, that is, an increase of 0.6 per cent over the value previously obtained. The corrected volume fractions are therefore $c = 0.099$, $w = 0.155$, $a = 0.460$ (as before) and $b = 0.286$. The change in w is very small and may frequently be neglected in which case w and a will be unchanged and the increase in c, namely, 0.006, will be reflected by a corresponding decrease in b. The required corrections to the batch quantities will be an increase in cement content of 19 kg m^{-3} and a reduction in sand content of 16 kg m^{-3}.

Where both workability and strength need minor adjustments, similar considerations treating each correction in turn enable the required adjustments to mix proportions to be obtained.

15.3 Statistical Quality Control

It is necessary to check that the desired quality of concrete is being obtained on site and for this purpose control testing is required. Tests on the hardened concrete are usually performed on 150 mm cubes prepared from samples of the concrete used on site. The importance of correct sampling and preparation, curing and testing of the control cubes cannot be overemphasised. It is usual for at least two control specimens to be made from each sample, one of these being tested at 28 days and the other at 7 days or earlier in some cases, for example, when accelerated curing techniques are used. The early test results enable estimates to be made of the probable 28-day concrete strengths once a relationship between the strengths at the two ages has been established. Any necessary remedial action can then be taken at a much earlier stage than would otherwise be possible.

The control testing may indicate that the site concrete has the desired mean strength but a greater standard deviation than had been assumed at the design stage. In these circumstances, adjustment of the mix proportions is required so as to increase the mean strength. If, on the other hand, the standard deviation is less than had been assumed some economy can be achieved by reducing the mean strength.

It is possible also that a change in mean strength, or standard deviation, can occur unintentionally during the course of construction for some reason or other. It is important that any such changes are detected as soon as possible so that the necessary action can be taken and the reasons for the changes determined at the earliest opportunity. This requires a continuous assessment of the strengths of the control specimens. It is here that statistical methods of quality control are invaluable, generally incorporating visual aids in the form of control charts.

Shewart Control Charts

The standard control chart for concrete strength comprises a central horizontal line corresponding to the mean concrete strength f_{cm} with pairs of lines, either side of and equidistant from this central line, corresponding to the limits $f_{cm} \pm \beta s$.

TABLE 15.6
*Control chart data**

Cube reference number	Estimated 28 day strength f_c	$(f_c - f_{cm})$	Cumulative sum of (3) CUSUM M	Range of adjacent results in (2)	(5) minus design mean range	Cumulative sum of (6) CUSUM SD	Mean of group of previous four results
(1)	(2)	(3)	(4)	(5)	(6)	(7)	(8)
1	37.5	−0.5	− 0.5				
2	42.0	+4.0	+ 3.5	4.5	− 1.0	− 1.0	
3	42.5	+4.5	+ 8.0	0.5	− 5.0	− 6.0	
4	29.0	−9.0	− 1.0	13.5	+ 8.0	+ 2.0	37.8
5	42.0	+4.0	+ 3.0	13.0	+ 7.5	+ 9.5	38.9
6	41.0	+3.0	+ 6.0	1.0	− 4.5	+ 5.0	38.7
7	47.5	+9.5	+15.5	6.5	+ 1.0	+ 6.0	39.9
8	31.5	−6.5	+ 9.0	16.0	+10.5	+16.5	40.5
9	40.0	+2.0	+11.0	8.5	+ 3.0	+19.5	40.0
10	38.0	0	+11.0	2.0	− 3.5	+16.0	39.3
11	39.0	+1.0	+12.0	1.0	− 4.5	+11.5	37.2
12	35.5	−2.5	+ 9.5	3.5	− 2.0	+ 9.5	38.2
13	35.5	−2.5	+ 7.0	0	− 5.5	+ 4.0	37.1
14	34.0	−4.0	+ 3.0	1.5	− 4.0	0	36.1
15	45.5	+7.5	+10.5	11.5	+ 6.0	+ 6.0	37.7
16	36.5	−1.5	+ 9.0	9.0	+ 3.5	+ 9.5	37.9
17	41.5	+3.5	+12.5	5.0	− 0.5	+ 9.0	39.4
18	37.5	−0.5	+12.0	4.0	− 1.5	+ 7.5	40.3
19	42.0	+4.0	+16.0	4.5	− 1.0	+ 6.5	39.4
20	37.0	−1.0	+15.0	5.0	− 0.5	+ 6.0	39.5
21	40.5	+2.5	+17.5	3.5	− 2.0	+ 4.0	39.3
22	35.5	−2.5	+15.0	5.0	− 0.5	+ 3.5	38.8
23	34.0	−4.0	+11.0	1.5	− 4.0	− 0.5	36.8
24	35.5	−2.5	+ 8.5	1.5	− 4.0	− 4.5	36.4
25	42.5	+4.5	+13.0	7.0	+ 1.5	− 3.0	36.9
26	36.0	−2.0	+11.0	6.5	+ 1.0	− 2.0	37.0
27	31.0	−7.0	+ 4.0	5.0	− 0.5	− 2.5	36.3
28	38.5	+0.5	+ 4.5	7.5	+ 2.0	− 0.5	37.0

* Continued on p. 196

TABLE 15.6 (Continued)

Cube reference number	Estimated 28 day strength f_c	$(f_c - f_{cm})$	Cumulative sum of (3) CUSUM M	Range of adjacent results in (2)	(5) minus design mean range	Cumulative sum of (6) CUSUM SD	Mean of group of previous four results
(1)	(2)	(3)	(4)	(5)	(6)	(7)	(8)
29	44.5	+ 6.5	+11.0	6.0	+0.5	0	37.5
30	37.5	− 0.5	+10.5	7.0	+1.5	+ 1.5	37.9
31	34.0	− 4.0	+ 6.5	3.5	−2.0	− 0.5	38.6
32	42.0	+ 4.0	+10.5	8.0	+2.5	+ 2.0	39.5
33	37.5	− 0.5	+10.0	4.5	−1.0	+ 1.0	37.7
34	43.5	+ 5.5	+15.5	6.0	+0.5	+ 1.5	39.2
35	42.0	+ 4.0	+19.5	1.5	−4.0	− 2.5	41.2
36	39.5	+ 1.5	+21.0	2.5	−3.0	− 5.5	40.6
37	27.0	−11.0	+10.0	12.5	+7.0	+ 1.5	38.0
38	39.0	+ 1.0	+11.0	12.0	+6.5	+ 8.0	36.9
39	39.5	+ 1.5	+12.5	0.5	−5.0	+ 3.0	36.3
40	47.5	+ 9.5	+22.0	8.0	+2.5	+ 5.5	38.2
41	42.0	+ 4.0	+26.0	5.5	0	+ 5.5	41.9
42	38.0	0	+26.0	4.0	−1.5	+ 4.0	41.7
43	32.5	− 5.5	+20.5	5.5	0	+ 4.0	40.0
44	36.0	− 2.0	+18.5	3.5	−2.0	+ 2.0	37.1
45	30.0	− 8.0	+10.5	6.0	+0.5	+ 2.5	34.1
46	41.0	+ 3.0	+13.5	11.0	+5.5	+ 8.0	34.9
47	28.5	− 9.5	+ 4.0	12.5	+7.0	+15.0	33.9
48	42.5	+ 4.5	+ 8.5	14.0	+8.5	+23.5	35.5
49	36.5	− 1.5	+ 7.0	6.0	+0.5	+24.0	37.1
50	37.5	− 0.5	+ 6.5	1.0	−4.5	+19.5	36.2
51	37.0	− 1.0	+ 5.5	0.5	−5.0	+14.5	38.4
52	37.0	− 1.0	+ 4.5	0	−5.5	+ 9.0	37.0
53	35.0	− 3.0	+ 1.5	2.0	−3.5	+ 5.5	36.6
54	39.5	+ 1.5	+ 3.0	4.5	−1.0	+ 4.5	37.1
55	31.5	− 6.5	− 3.5	8.0	+2.5	+ 7.0	35.8
56	33.5	− 4.5	− 8.0	2.0	−3.5	+ 3.5	34.9
57	42.5	+ 4.5	− 3.5	9.0	+3.5	+ 7.0	36.8
58	33.0	− 5.0	− 8.5	9.5	+4.0	+11.0	35.1
59	36.0	− 2.0	−10.5	3.0	−2.5	+ 8.5	36.3
60	34.0	− 4.0	−14.5	2.0	−3.5	+ 5.0	36.4

Characteristic strength f_{cu} = 30.0 N mm^{-2}; design standard deviation s = 5.0 N mm^{-2}; design mean strength f_{cm} = 38.0 N mm^{-2}; design mean range = 1.13s = 5.5 N mm^{-2}.

With only one pair the lower limit will be the characteristic strength f_{cu}.

A convenient visual presentation of the variation of the test results from the mean strength is obtained by plotting, on this control chart, the individual test results as these become available. For $\beta = 1.64$, if more than three results in any forty consecutive results fall below the lower limit the degree of control is immediately suspect and the need for some remedial action is indicated.

Typical test results are given in table 15.6 together with the specified characteristic strength and design data. The associated control chart is shown in figure 15.8.

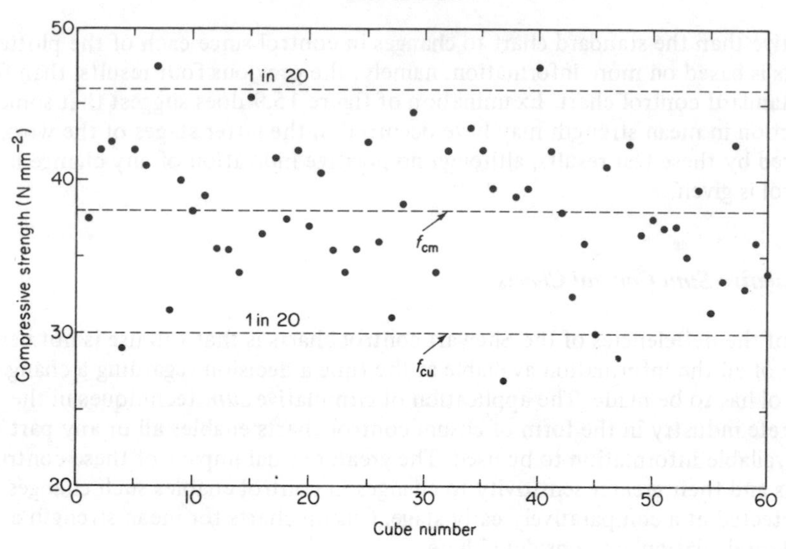

Figure 15.8 *Control chart (individual results)*

In this case it would appear that adequate control is being exercised. Unfortunately this type of control chart is relatively insensitive to changes in control and a reduction in mean strength during construction may not become apparent until possibly thirty or forty results are available following such a reduction. Furthermore no real indication is given of the actual mean strength at any time, although of course when forty or more results are available average values may be calculated.

Another similar control chart is also in current use where, instead of individual results, the mean strengths of groups of consecutive results are plotted. In this case the upper and lower limits, corresponding to $f_{cm} \pm \beta s$ in the standard control chart, are $f_{cm} \pm (\beta s/\sqrt{N})$ where N is the number of results in each group. In table 15.6, column 8, the mean strengths of groups of four results are given and the associated control chart is shown in figure 15.9. This modified chart is more

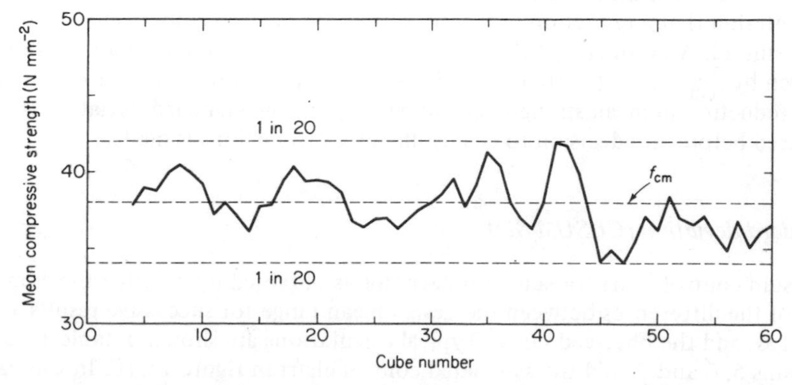

Figure 15.9 *Control chart (groups of four consecutive results)*

sensitive than the standard chart to changes in control since each of the plotted points is based on more information, namely, the previous four results, than for the standard control chart. Examination of figure 15.9 does suggest that some reduction in mean strength may have occurred in the latter stages of the work covered by these test results, although no positive indication of any change in control is given.

Cumulative Sum Control Charts

One of the deficiencies of the Shewart control charts is that full use is not being made of all the information available at the time a decision regarding a change in control has to be made. The application of *cu*mulative *sum* techniques in the concrete industry in the form of cusum control charts enables all or any part of the available information to be used. The greater visual impact of these control charts and their greater sensitivity to changes in control enables such changes to be detected at a comparatively early stage. Cusum charts for mean strength and standard deviation are considered here.

Mean strength (CUSUM M)

To construct a cusum control chart for mean strength, the cumulative sum of the algebraic differences between the assumed or design mean strength and each test result is plotted as successive results become available. The mean slope over any portion of the cusum plot is then directly proportional to the difference between the actual and assumed mean strengths. An estimate of the actual mean strength $\overline{f}_{cm}$ may be obtained from

$$\overline{f}_{cm} = f_{cm} + \frac{\text{(change in CUSUM M over } n \text{ results)}}{n}$$

Typical calculations are shown in table 15.6, columns 3 and 4, and the associated control chart is shown in figure 15.10. It is clear from this form of presentation that a reduction in the mean strength has probably occurred following result 42. An estimate of the actual mean strength between results 42 and 60 is given by $\overline{f}_{cm} = 38 + (-40.5/18) = 36$ N mm^{-2}. In practice, the reason for any such reduction in mean strength should be sought. The standard deviation is required before any decision to change the mix proportions is made.

Standard deviation (CUSUM SD)

A cusum control chart for standard deviation is obtained by plotting the cumulative sum of the differences between the design mean range for successive results, equal to 1.13*s*, and the observed range. Typical calculations are shown in table 15.6, columns 5, 6 and 7, and the associated control chart in figure 15.10. In this case, an estimate of the actual standard deviation $\overline{s}$ may be obtained from

$$\bar{s} = s + \frac{(\text{change in CUSUM SD over } n \text{ results})}{1.13\,n}$$

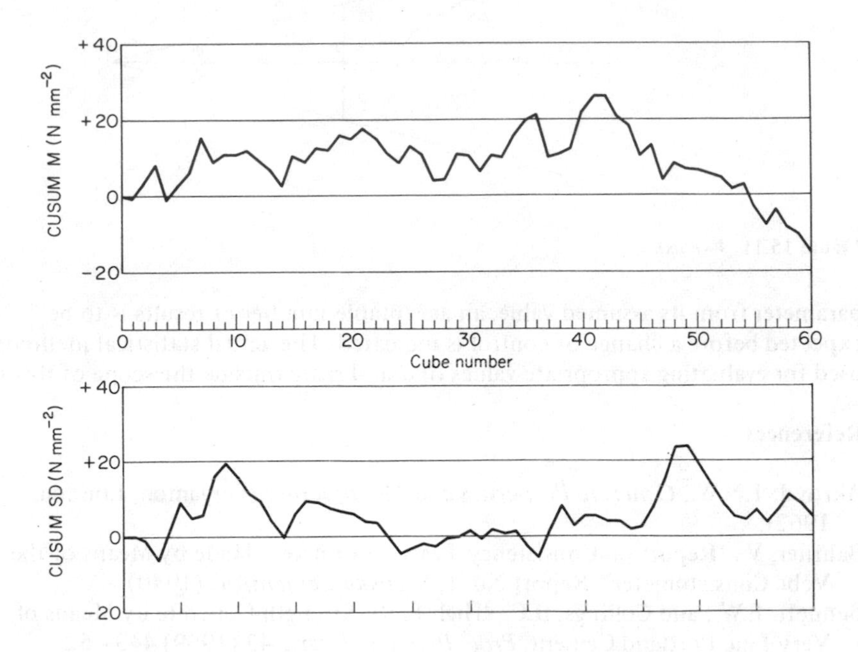

Figure 15.10 *Control charts for mean strength (CUSUM M) and standard deviation (CUSUM SD)*

In figure 15.10 it is seen that, despite apparent local variations, the standard deviation remains close to the assumed value. In the circumstances, the fact that the actual mean strength, after result 42, is 2 N mm^{-2} lower than assumed indicates that some adjustment to the mix proportions must be considered if the characteristic strength requirements are to be satisfied.

Local variations in the slope of the cusum plot, due to random variations of the test results, may be expected even when there is no real change in the associated control parameter. Such local variations can be seen in figure 15.10. In general the smaller the number of results used the less reliable will be the estimated value of the parameter in question. To standardise the decision-making process and reduce the probability of making either unnecessary or delayed adjustments to concrete mix proportions, some form of transparent V-mask, (figure 15.11) is used in practice. On this mask, lines are marked on each side of and at an angle ϕ to the horizontal. A point P, distance d from the apex of the V, is placed on each point as it is plotted on the cusum chart. A change in control from that assumed is indicated whenever the cusum plot intersects one of the lines forming the V (see figure 15.11). The magnitudes of d and ϕ are so chosen that, for a given standard deviation (allowance being made for the random variations of results referred to above) and different departures of the control

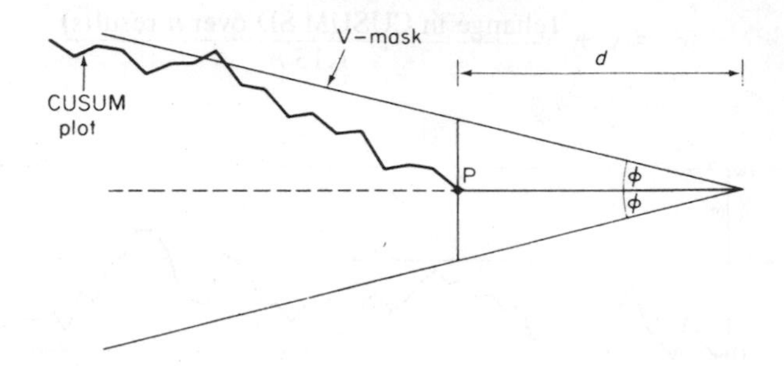

Figure 15.11 *V-mask*

parameter from its assumed value, an acceptable number of results is to be expected before a change of control is indicated. The actual statistical methods used for evaluating appropriate values of d and ϕ are outside the scope of this book.

References

Akroyd, T.N.W., *Concrete Properties and Manufacture* (Pergamon, London, 1962).

Bährner, V., 'Report on Consistency Tests on Concrete Made by Means of the Vebe Consistometer', Report No. 1, *Svenska Cementför.* (1940).

Bennett, E.W., and Collings, B.C., 'High Early Strength Concrete by Means of Very Fine Portland Cement, *Proc. Inst. civ. Engrs,* 43 (1969) 443 - 52.

Bennett, E.W., and Loat, D.R., 'Shrinkage and Creep of Concrete as Affected by the Fineness of Portland Cement, *Mag. Concr. Res.,* 22 (1970) pp. 69 - 78.

B.R.E. Digest 111, *Lightweight Aggregate Concretes – 3: Structural Applications* (H.M.S.O., London, 1969).

B.R.E. Digest 123, *Lightweight Aggregate Concretes* (H.M.S.O., London 1970).

BS 12 : 1978 Specification for ordinary and rapid-hardening Portland cement

BS 146 : 1973 Portland-blastfurnace cement

BS 410 : 1976 Specification for test sieves

BS 812 : Part 1 : 1975 Sampling, size, shape and classification

BS 812 : Part 2 : 1975 Physical properties

BS 812 : Part 3 : 1975 Mechanical properties

BS 877 : 1973 Foamed or expanded blastfurnace slag lightweight aggregate for concrete

BS 882 : Part 2 : 1973 Aggregates from natural sources for concrete (including granolithic)

BS 915 : Part 2 : 1972 High alumina cement

BS 1014 : 1975 Pigments for Portland cement and Portland cement products

BS 1047 : Part 2: 1974 Specification for air-cooled blast furnace slag coarse aggregate for concrete

BS 1165 : 1966 Clinker aggregate for concrete

BS 1370 : 1974 Low heat Portland cement

BS 1881 : Part 1 : 1970 Methods of sampling fresh concrete

BS 1881 : Part 2 : 1970 Methods of testing fresh concrete

BS 1881 : Part 3 : 1970 Methods of making and curing test specimens
BS 1881 : Part 4 : 1970 Methods of testing concrete for strength
BS 1881 : Part 5 : 1973 Methods of testing concrete for other than strength
BS 1926 : 1962 Ready-mixed concrete
BS 3148 : 1959 Tests for water for making concrete
BS 3587 : 1963 Calcium chloride (technical)
BS 3681 : 1973 Methods for the sampling and testing of lightweight aggregates for concrete
BS 3797 : 1976 Lightweight aggregates for concrete
BS 3892 : 1965 Pulverised-fuel ash for use in concrete
BS 4027 : Part 2 : 1972 Sulphate-resisting Portland cement
BS 4246 : 1974 Low heat Portland - blastfurnace cement
BS 4248 : 1974 Supersulphated cement
BS 4408 : Part 1 : 1969 Electromagnetic cover measuring devices
BS 4408 : Part 3 : 1970 Gamma radiography of concrete
BS 4408 : Part 4 : 1971 Surface hardness methods
BS 4408 : Part 5 : 1974 Measurement of the velocity of ultrasonic pulses in concrete

C.E.B., *Recommendations for an International Code of Practice for Reinforced Concrete* (Cement and Concrete Association, London, 1964).

Cornelius, D.F., 'Air-entrained Concretes: a Survey of Factors Affecting Air Content and a Study of Concrete Workability', *Road Research Laboratory Report LR* 363 (Ministry of Transport, London, 1970).

CP 110 : Part 1 : 1972 The structural use of concrete

Cusens, A.R., 'The Measurement of the Workability of Dry Concrete Mixes', *Mag. Concr. Res.*, 8 (1956) pp. 23 - 30.

Dewar, J.D., 'The Workability of Ready-mixed Concrete', *RILEM Seminar, Fresh Concrete: Important Properties and their Measurement* (University of Leeds, 1973).

Dhir, R.K., and Sangha, C.M., 'A Study of the Relationships Between Time, Strength, Deformation and Fracture of Concrete', *Mag. Concr. Res.*, 24 (1972) pp. 197 - 208.

Erntroy, H.C., *The Variation of Works Test Cubes*, Research Report 10 (Cement and Concrete Association, London, 1960).

Gilkey, H.J., 'Moist Curing of Concrete', *Engng News Rec.* 119 (1937) pp. 630 - 33.

Glanville, W.H., Collins, A.R., and Matthews, D.D., 'The Grading of Aggregates and Workability of Concrete', *Road Research Tech. Paper No. 5* (H.M.S.O., London, 1947).

Goetz, H., 'The Mode of Action of Concrete Admixtures', *Proceedings of the Symposium on the Science of Admixtures* (Concrete Society, London, 1969) pp. 1 - 5.

Hughes, B.P. 'Rational Concrete Mix Design', *Proc. Inst. civ. Engrs*, 17 (1960) pp. 315 - 32.

Hughes, B.P., 'A Laboratory Test for Determining the Angularity of Aggregate', *Mag. Concr. Res.*, 18 (1966) pp. 147 - 52.

Hughes, B.P., 'The Rational Design of High Quality Concrete Mixes', *Concrete*, 2 (1968) pp. 212 - 22.

Hughes, B.P., 'The Economic Utilization of Concrete Materials', *Proceedings of the Symposium on Advances in Concrete* (Concrete Society, London, 1971).

Lea, F.M., *The Chemistry of Cement and Concrete* (Arnold, London, 1970).

McIntosh, J.D., *Concrete Mix Design* (Cement and Concrete Association, London 1964).

Neville, A.M., *Properties of Concrete* (Pitman, London, 1968).

Neville, A.M., *Creep of Concrete: Plain, Reinforced and Prestressed* (North-Holland, Amsterdam, (1970).

Orchard, D.F., *Concrete Technology*, vols 1 and 2 (Applied Science, Barking, 1968).

Price, W.H., 'Factors Influencing Concrete Strength', *J. Am. Concr. Inst.*, 47 (1951) pp. 417 - 32.

Road Research Laboratory, *Design of Concrete Mixes*, Road Note No. 4 (H.M.S.O., London, 1950).

Short, A., and Kinniburgh, W., *Lightweight Concrete* (Applied Science, Barking, 1968).

Taylor, W.T., *Concrete Technology and Practice* (Angus and Robertson, Melbourne, 1965).

Further Reading

Council on Tall Buildings, Group CL, *Tall Building Criteria and Loading*, vol. CL of Monograph on Planning and Design of Tall Buildings, ch. 6 Quality Criteria (American Society of Civil Engineers, New York, 1980).

W.H. Mosley and J. H. Bungey, *Reinforced Concrete Design* (Macmillan, London and Basingstoke, 1976).

High Alumina Cement Concrete in Buildings, Current Paper CP 34/75 (Building Research Establishment, Watford, 1975).

P.L. Owens, *Basic Mix Method – Selection of Proportions for Medium Strength Concretes*, Ref. 11.005 (Cement and Concrete Association, London, 1973).

Polymers in Concrete, SP-40 (American Concrete Institute, Detroit, 1973).

IV
BITUMINOUS MATERIALS

Introduction

Engineers through the ages have made use of the excellent durability and adhesive properties of bituminous materials. Bitumen mastics were used in Mesopotamia in the waterproofing of reservoirs about 3000 B.C., and some parts of these and other works, many 2000 years old and more, are still in existence.

The materials used by these engineers of ancient times were naturally occurring bitumens. Such materials are still used today, often for similar purposes. However, the vast bulk of today's bituminous materials are the products of industrial refining processes.

In any discussion of bituminous materials it is difficult to avoid confusion since the same terms have often been used to define different materials. This is particularly confusing when comparing British and American terminology. Throughout this part of the book the terms used have the following meanings (BS 892).

Binder: a material used for the purpose of holding solid particles together.

Bitumen: a viscous liquid or a solid, consisting essentially of hydrocarbons and their derivatives, which is soluble in carbon disulphide, is substantially non-volatile and softens gradually when heated. It is black or brown in colour and possesses waterproofing and adhesive properties. It is obtained by refinery processes from petroleum, and is also found as a natural deposit or as a component of naturally occurring asphalt, in which it is associated with mineral matter. (Note: some petroleum technologists disagree with the phrase 'consisting essentially of hydrocarbons and their derivatives', maintaining that a large proportion of bitumens consist of reduced organic sulphur compounds.)

Asphaltic cement: bitumen, a mixture of lake asphalt and bitumen, or lake asphalt and flux oils or pitch or bitumen, with cementing qualities suitable for the manufacture of asphalt pavements.

Asphalt: a general term for certain mixtures of asphaltic cement and mineral matter.

Tar: a viscous liquid, black in colour, with adhesive properties, obtained by the destructive distillation of coal, wood, shale, etc. Where no specific source is stated it is implied that the tar is obtained from coal.

16

Bituminous Binders

All bituminous materials are for the most part used in mixtures with mineral or other aggregates but before considering these mixtures it is important to understand the properties of the various bituminous binders. These binders all have certain valuable properties in common. They are water resistant, stand up well to ordinary weathering processes, and have good adhesive properties. Moreover, they are relatively cheap and are available in large quantities.

16.1 Elementary Chemical Properties

All binders, whether tars or bitumens, are exceedingly complex materials chemically. It has been estimated that there are more than 10 000 compounds in tar, only a few hundred of which have been separated and identified with any certainty. The chemistry of bitumens is at least as complex.

Different tars and bitumens have been characterised by separating them into fractions according to their solubility in a series of solvents of increasing dispersing power. For tars, solvents such as n-hexane, benzene and pyridine have been used together with fractional distillation. Broadly speaking this allows the tar to be analysed into fractions of increasing molecular weight. Bitumens have been treated in a similar way, and divided into three fractions:

(1) carbenes: the fraction insoluble in carbon tetrachloride;

(2) asphaltenes: the fraction insoluble in light aliphatic hydrocarbon solvent (for example, petroleum ether);

(3) maltenes: fraction soluble in light aliphatic hydrocarbon solvent.

In any discussion of the chemical properties of tars and bitumens it is important to consider the source of the material under consideration.

Tars

Tars can be obtained from the destructive distillation of coal, wood, shale, etc. According to BS 76, where no specific source is mentioned it is implied that the tar is obtained from coal. Knowing that a tar is derived from coal does not necessarily define its chemical properties very specifically. At the present time in the United Kingdom (1975) the great bulk of all tars are coke-oven tars. It may be that this pattern could change. If, owing to the approaching world fuel crisis, coal gas becomes a popular fuel once again, then vertical or horizontal retort tars may predominate.

The fractional distillation of coal yields many oils, some of which are very valuable. These oils vaporise at different temperatures and for economic reasons coal tars for engineering purposes are obtained by distilling off all the lighter oils leaving a hard semi-solid residual material – pitch. Pitch is in itself too hard and viscous a material for some purposes and softer materials, for example, road tars, are obtained by oiling back – fluxing back – the pitch. The choice of the oil to be added back to the pitch is decided on economic and other grounds but in general it is desirable from an engineering viewpoint to use the oil with the lowest volatility.

Many researchers have found that the chemical properties of a tar depend not only on the oil used to flux back the pitch, but also on the source of the pitch itself, whether from vertical retorts, horizontal retorts or coke ovens. Work carried out at the Road Research Laboratory (Lee and Dickinson, 1954) appears to indicate that horizontal retort tars are the most durable.

The whole subject of durability of tars and pitches is exceedingly complex. The main problem from an engineering point of view concerns the 'ageing' or hardening of the material when exposed to the atmosphere. There is some evidence to suggest that coal tars and pitches derived from high temperature processes (that is, coke ovens and horizontal retorts) weather primarily through oxidation, whereas those obtained from lower temperature processes (vertical retorts or as by-products in the manufacture of smokeless fuels) harden largely through evaporation of the lighter oils.

In whatever way the binders are affected by weathering processes, the weathering is a surface phenomenon only and material below the surface is unaffected unless exposed owing to cracking of the weathered surface layer or removal of this brittle harder layer by mechanical processes.

Although organic in nature tars, bitumens and other similar binders do not act as food material for plants, mosses, etc. In fact they appear to contain chemicals which inhibit such growth.(A useful function of tarmacadam or bitumen macadam surfacing of garden paths is the elimination of weeds.)

As with most oily substances bitumens and tars will burn. A knowledge of the temperature at which a particular binder will ignite – the flash point – is often useful in deciding the temperature to which it can safely be heated in a mixing plant. It is also a piece of information often demanded by Customs and transport authorities. The ignition temperature of a binder is mainly a function of the ignition temperature of the light oils vaporising from its surface. When these have been consumed the flame goes out. At a higher temperature – the fire point – the binder will continue to burn. This higher temperature may be of more concern in building construction. For most binders, both flash point and fire point temperatures are much lower than the temperatures likely to be generated when modern buildings are 'on fire'.

Bitumens

All bitumens are obtained by the fractional distillation of petroleum (crude oil). In some cases this process occurs naturally, producing rock or lake asphalt, but the bulk of the world's bitumen is produced by refining crude oil. Essentially this process consists of heating the crude petroleum to a high enough temperature for the lighter oils to vaporise and be drawn off and condensed. These oils include diesel oil, kerosine, naphtha, etc. The remaining liquid material is bitumen. By varying the temperature and rates of flow through the plant an almost unlimited number of bitumens of various degrees of hardness can be obtained, depending on the efficiency with which the various light oils have been removed. In most plants, however, only two basic grades of bitumen are produced, one at the soft and the other at the hard end of the range. Intermediate grades are then produced by blending these in varying proportions (BS 3690).

Chemically, bitumens are similar to tars and are also highly resistant to most natural weathering agencies. They tend to oxidise when exposed to air, heat and light, but the process is somewhat slower than in tars and pitches.

Again, as in tars, oxidation is largely a surface phenomenon and therefore is more important when the bitumen is applied in thin films.

It is noteworthy that naturally occurring bitumens, such as Trinidad Lake Asphalt (TLA), oxidise more quickly when exposed to the weather than bitumen obtained from the distillation of petroleum.

16.2 Physical Properties of Binders

The very term binder used to describe bituminous materials suggests an ability to act as adhesive and 'bind' other materials together.

Adhesion

As with all adhesives when tars or bitumens are used it is important that the materials to be bound together by these binders should be clean, dry and free from dust.

If, for example, in coating stones for road mixes, the binder is too thick or viscous then it will not 'wet' or coat the stone efficiently. If the stone is dusty or dirty then the binder may not reach the stone itself. If there is water present then in many cases it is difficult for a bituminous binder to displace the water and adhere to the stone. It is much easier for the water to displace the binder.

Silica, a common mineral in roadstone, has a weak negative surface charge and water, a polar liquid, is strongly attracted to these charged surfaces. The chemical constituents of bituminous binders on the other hand have little polar activity and are not so attracted.

If a solid surface is already coated with water and a quantity of binder is placed on the surface, then the system is as shown in figure 16.1. If the angle of contact is θ as shown and the energies of the solid/binder, solid/water and binder/

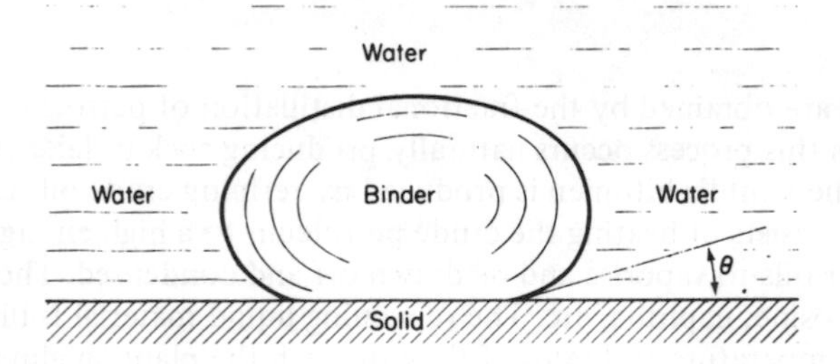

Figure 16.1 *Binder-water-solid system*

water interfaces are γ_{sb}, γ_{sw} and γ_{bw}, then it can be shown that the work W required to displace water from unit area of the solid face is

$$W = \gamma_{sb} + \gamma_{bw} - \gamma_{sw}$$

But for equilibrium to be reached $\gamma_{sb} = \gamma_{sw} + \gamma_{bw} \cos\theta$. It follows that $W = \gamma_{bw} (1 + \cos\theta)$.

For a bituminous binder, θ is always less than 90°, therefore the term in brackets is always greater than 1. An appreciable amount of work is required to make the binder replace the water coating the solid surface. The values of γ_{bw} and θ vary with the type of binder. For most bitumens θ is a small angle, so that the term $(1 + \cos\theta)$ approaches the value 2. For tars, and in particular low temperature tars, θ is a much larger angle so that the term $(1 + \cos\theta)$ is much nearer unity. It follows that, other things being equal, tars and pitches tend to adhere better, particularly in the presence of water, than bitumens.

The values of θ and γ_{bw} can be changed by means of additives. If a cationic surface-active agent is dissolved in the binder then the agent is attracted to the stone surface and the angle θ is increased. At the same time γ_{bw} is reduced and it is easier for the binder to spread out over the surface of the stones.

Stripping of Binders

The stripping of binders by water can be considered in the same way, although initially there is no point of contact between stone and water.

In practice it is found that tars resist such stripping better than bitumens, and also that the harder (high viscosity) binders adhere better in the presence of water than soft (low viscosity) binders.

Viscosity

Bituminous materials are used by engineers and others in a wide variety of consistencies. These range from very hard pitches which shatter when struck to liquids which are free-flowing at ordinary temperatures. Although the harder grades may exhibit some of the characteristics of solids, for example, elasticity, all of them possess the liquid property of viscosity.

When liquids or semi-liquids are subjected to a load, the associated deformation is dependent on the time of application of the load. The longer the load is applied, the greater the deformation which results. The difference in the magnitude of the deformations exhibited by different liquids is due to their different viscosities. Viscosity then is the property of a liquid which retards flow. If the viscosity is high then flow under load is slow, and vice versa. In engineering terms viscosity is defined in terms of shear stress and is the ratio of shear stress to the rate of shear strain.

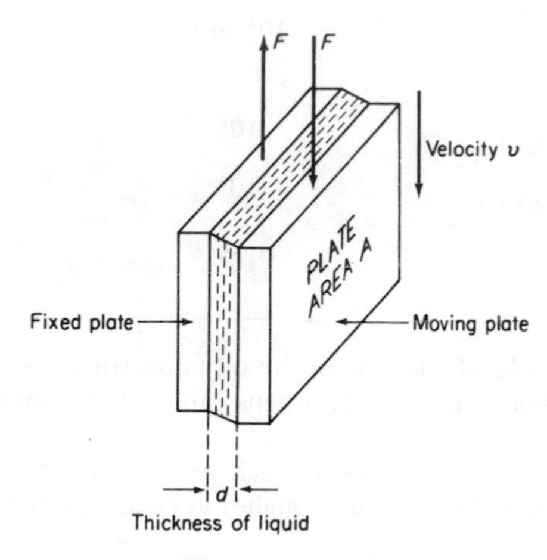

Figure 16.2 *Viscosity parameters*

Referring to figure 16.2, if a shearing force F applied to two plates each of area A and separated by a thickness of liquid d causes one to move with velocity v relative to the other, then the coefficient of viscosity η of the liquid is given by

$$\eta = \frac{F/A}{v/d} = \frac{Fd}{Av}$$

When using SI units the coefficient of viscosity or, more simply, viscosity is measured in N s m^{-2}. In most of the technical literature on bituminous materials however, the unit of viscosity used is the poise. If a tangential shearing force of 1 dyne acting on planes of area 1 cm^2 separated by unit thickness of liquid produces a unit tangential velocity of 1 cm s^{-1} then the viscosity of the liquid is 1 poise. For practical reasons the smaller unit the centipoise (cP) is more convenient, the relationship between the units being

$$1 \text{ cP} = 0.01 \text{ P}$$

and

$$1 \text{ N s m}^{-2} = 1000 \text{ cP}$$

Where the coefficient of viscosity is independent of stress, that is, where doubling the applied stress has the effect of doubling the shear rate (equivalent to a straight line stress-strain graph for solids) the liquid is called a newtonian

TABLE 16.1
Viscosity of different materials

	Viscosity (poise) ($\times 10^{-1}$ N s m^{-2})	State at 20 °C
Water	0.01	Liquid – can be readily pumped
Diesel oil	0.1	
Heavy engine oil (50SAE)	10	
Cutback bitumens and softer tars	10^3 to 10^4	Semi-liquids
300 pen bitumen 50 s tar at 50 °C (50 e.v.t.)	10^5	Soft semi-solids
100 pen bitumen 50 s tar at 60 °C (60 e.v.t.)	10^6	
15 pen bitumen 75 e.v.t. pitch	10^8	Solid

liquid. At higher stress levels and where the oily constituents are not highly aromatic, the behaviour of tars and bitumens can deviate considerably from that of newtonian liquids.

All bituminous binders become very much softer or of lower viscosity as their temperature increases, and harder or of higher viscosity as their temperature falls.

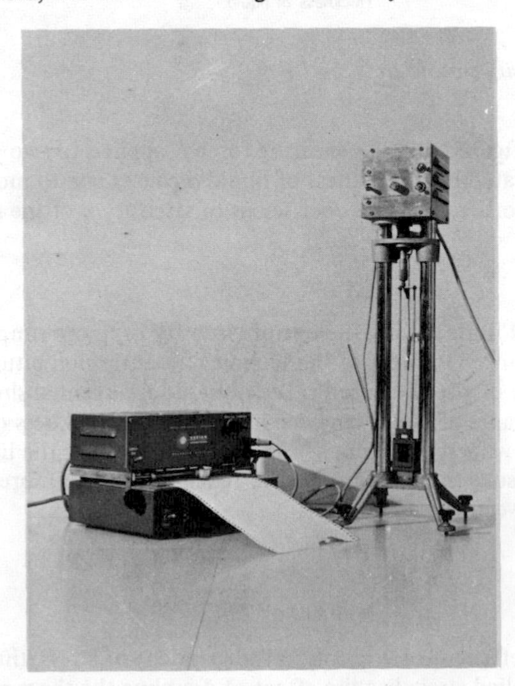

Figure 16.3 *Sliding plate viscometer*

Measuring Viscosity

It follows that in stating the viscosity of a binder, the temperature applicable to this value must be stated. The range of viscosities even of bituminous binders is exceptionally great (see table 16.1) and it follows that no one instrument can be used to measure all viscosities. For non-newtonian binders, viscosity varies not only with temperature but also with rate of change of temperature ($\mathrm{d}T/\mathrm{d}t$). It follows that the tests used for measuring viscosity or, more simply, softness or hardness of binders, are carried out either at constant temperature, and preferably temperatures which have been constant for a reasonable time before the test or, as when determining the softening point, in conditions where temperature is changing at a specific constant rate.

16.3 Viscometers

These are instruments used to measure the viscosity of a liquid. They are of two types, fundamental and empirical. With the fundamental instruments the viscosity is calculated in N s m^{-2}, or in poise when the dimensions of the apparatus and the forces acting on the liquid are known. With empirical instruments viscosity is expressed in terms of a simple unit, usually either seconds or degrees Celsius, when the liquid has been tested in a standard apparatus under specific standard conditions.

Sliding Plate Viscometer

This instrument (see figure 16.3) is based on the fundamental principle of shearing a sample of binder between two parallel pieces of plate glass. This is carried out at constant shearing stress and with the sample and plates immersed in a constant temperature bath. The rate of movement is automatically recorded. The film of binder sheared is usually between 20 and 50 μm thick and since the applied load and the area are also known these together with the recorded velocity of movement of the moving plate allow the viscosity to be calculated directly in poise or N s m^{-2}.

This instrument has many advantages

(1) It requires only small samples and can therefore be used to test samples of material recovered from a road, for example.

(2) It can be used to measure viscosity at temperatures similar to those likely to be experienced in service.

(3) Both stresses and temperatures are easily varied and a wider picture of the properties of the binder can be readily obtained.

(4) The instrument can be used to test binders with a viscosity range from about 10^2 to 10^{11} P.

The sliding plate viscometer is expensive, however, and is primarily a research tool. The accuracy of the instrument also is not as good as that of some of the much cheaper empirical viscometers which test much larger samples.

Capillary Viscometer

This instrument (Institute of Petroleum, 1900) consists basically of a glass capillary tube. A known volume of the binder under test is forced through the capillary by applying either a hydrostatic head or a vacuum. The time of flow is measured in seconds. The viscosity of the binder can then be calculated by applying a known constant, peculiar to the viscometer used. During the test the capillary viscometer is kept immersed in a temperature-controlled bath.

As with the sliding plate viscometer the instrument allows viscosities to be calculated directly in absolute units. Normally different sizes of capillary are used for different viscosity ranges, the total range covered being about 10^2 to 10^5 P.

As can readily be imagined, in order to push a viscous liquid through a fine tube, the test temperature is usually fairly high, about 60 °C, well outside the normal service temperature for most binders. However, the instrument does have the advantage of cheapness and simplicity.

Empirical Tests

The great bulk of all viscosity measurements are made using empirical methods. One of the most common uses efflux viscometers. Since these are all basically similar instruments only one, the standard tar viscometer, is described here.

Figure 16.4 *Standard tar viscometer*

Standard tar viscometer

Under standard conditions of head and temperature (BS 76) the time taken for a given volume of a liquid to flow through a standard orifice is a measure of the viscosity of the liquid. In the standard tar viscometer (see figure 16.4) the viscosity is expressed not in absolute units but as the number of seconds taken for 50 ml of the binder to flow through an orifice 10 mm in diameter and 5 mm in length. The orifice is situated centrally in the base of a brass cup, and is opened and closed by a simple ball valve. The standard head is obtained by always filling the cup to the same level, and the temperature is controlled by means of a water bath which encloses the cup. The volume of tar flowing through the orifice is measured by means of an ordinary measuring cylinder.

In terms of this instrument viscosity is expressed as (say) 50 s at 25 °C. This means that it takes 50 s for 50 ml of the binder to flow through the standard orifice at a test temperature of 25 °C.

For accuracy and convenience the time of efflux must always fall in the range 10 to 140 s. If a particular binder gives a value outside this range then it must be re-tested at a different temperature. For most road tars, test temperatures range from 20 to 50 °C.

The standard tar viscometer is used mainly for testing tars and cutback bitumens.

Penetration test

This is the standard test for ordinary or straight-run bitumens (BS 4691). If a loaded needle is allowed to penetrate a sample of bitumen, the depth of penetration is a measure of the viscosity of the bitumen at that temperature and under those loading conditions.

For normal test conditions in the penetration test, a 1 mm diameter needle is used, ground to a sharp point and loaded with 100 g. The needle just touches the surface of the bitumen at the start and is allowed to fall under gravity for 5 s, the test temperature of the bitumen being 25 °C. The penetration is measured in tenths of a millimetre.

If a bitumen is referred to as 70 pen, it is understood that a penetration of 70 tenths of a millimetre was obtained under the standard conditions of 100 g load, 5 s duration of loading and a test temperature of 25 °C. If any different conditions of loading, time or temperature are used these are always stated.

When using the standard test conditions, penetrations in the range 5 to 500 can be measured, although at the extremes, accuracy begins to suffer. These penetrations are equivalent to viscosities of 5×10^8 P and about 2×10^4 P respectively (5×10^7 and 2×10^3 N s m^{-2}). This is a considerable range of viscosities and much greater than that covered by the standard tar viscometer at any particular test temperature. The standard penetrometer is shown in figure 16.5.

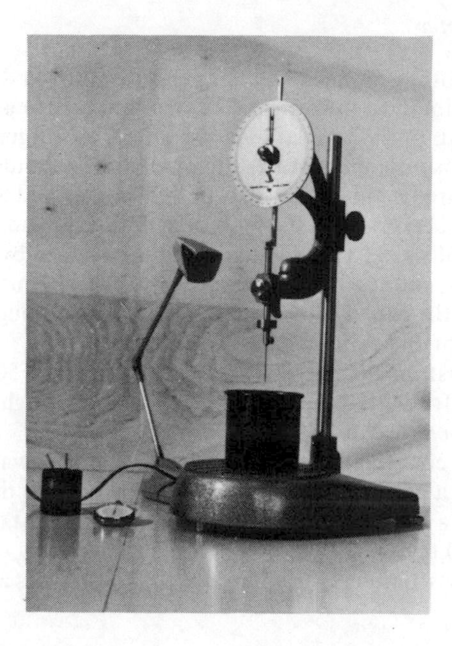

Figure 16.5 *Standard penetrometer*

Equi-viscous Temperature

The range of viscosity covered by any one test temperature with the standard tar viscometer is really quite small, and it is obviously very useful if a single scale can be used to cover all viscosities. Such a scale exists and with it viscosity is expressed as equi-viscous temperature (e.v.t.). This is defined as the temperature in degrees Celsius at which the binder would have a viscosity of 50 s if tested in the standard tar viscometer at that temperature. That is, a binder with an e.v.t. of 25 °C, when tested at a temperature of 25 °C in the standard tar viscometer, takes 50 s for 50 ml of the binder to flow out through the orifice.

Torsion Viscometer

If a metal cylinder is rotated in a fluid a 'drag' or resistance is exerted on the cylinder by the fluid. The higher the viscosity, the greater the resistance to the rotation. Many viscometers make use of this principle. In most of these a steel cylinder is suspended in the liquid by means of a wire and the wire is twisted, imparting a torque to the cylinder; the cylinder then rotates and the speed of the imparted rotation depends on the viscosity of the surrounding liquid. By selecting suitable dimensions for the cylinder, a torsion viscometer, the e.v.t. viscometer (see figure 16.6) has been produced in which a specific ease of rotation of the cylinder occurs when the temperature of the binder tested is its equi-viscous temperature (S.T.P.T.C., 1957).

Figure 16.6 *E.V.T. viscometer*

Since with this instrument the temperature of the binder is slowly increased at a constant rate, the wire can be twisted and the consequent rotation of the cylinder observed at frequent intervals of temperature, until the e.v.t. is reached. The viscosity as measured by e.v.t. can thus be found in one experiment, whereas two or more might be needed when using the standard tar viscometer.

16.4 Other Physical Properties

At low temperatures, and in particular for short durations of loading, many binders show elastic properties, deforming under load and recovering on the removal of load. If loads are applied for an appreciable time, however, viscous flow develops. It has been demonstrated for bitumens that the rate of viscous flow decreases with time (van der Poel, 1954; Huekelom, 1969). This would not be the case for true liquids, and it therefore follows that this apparently viscous flow is made up of two components, a true viscous flow as in liquids, together with a delayed elastic deformation.

In terms of symbols if the ratio of tensile stress to total strain for a binder is called the *stiffness modulus* S_E then

$$\frac{1}{S_E} = \frac{1}{E} + \frac{t}{3\eta} + \frac{1}{D}$$

where E is the modulus of elasticity (instantaneous), t the loading time, η the viscosity and D the modulus of delayed elasticity. The value of E is nearly independent of time and temperature and for most binders is about 27 N mm^{-2}. The value of η depends on temperature, and that of D on both time and temperature. When $t = 0$ the deformation due to E predominates, at $t = \infty$ the term with η is the greatest, and at moderate loading times, the delayed elastic effects are important.

The stiffness modulus S_G which is analogous to the shear modulus G is related to the stiffness modulus S_E by the relation $S_E = 3\ S_G$.

Thixotropy

This is the property whereby a fairly stiff material becomes much more fluid when vibrated or agitated. It is encountered in the gel-type thixotropic paints marketed as 'nondrip'. The property is most marked in blown or oxidised bitumens, although it is exhibited to some extent by many binders.

Softening Point

Most of the tests described so far are generally carried out at temperatures approaching that of a warm summer's day. At this temperature many binders are quite hard, and a question often asked by the engineer is: 'At what temperature will this material soften?'. Hard and soft are of course relative terms; in the case of binders 'soft' is defined in terms of a particular piece of apparatus, and the test is called the ring and ball test (BS 4692).

In this test a sample of the hot binder is poured into a brass ring and allowed to cool, so that a flat disc of binder is enclosed in the ring. The ring is fitted to an opening on a metal carrier and immersed in water at 5 °C. A steel ball 9.53 mm in diameter is placed centrally at the top of the disc of binder using a centering device and the water is heated at a constant rate. As the binder softens it is dragged downwards by the weight of the ball-bearing and the softening point is defined as the temperature of the water at the instant when the binder just touches a plate placed 25 mm below the ring (see figure 16.7).

Temperature Susceptibility

The viscosity of all binders falls quite rapidly with increasing temperatures. For a fairly wide range of temperature a graph of log viscosity against temperature gives approximately straight-line plots (see figure 16.8). From the figure it can be readily seen that the graphs are steeper for tars than for bitumens, and steeper for both than for blown or oxidised bitumens. This indicates that tars are slightly more susceptible to changes in temperature than bitumens; again the consistency of oxidised bitumens alters less with changes in temperature than that of either tars or bitumens.

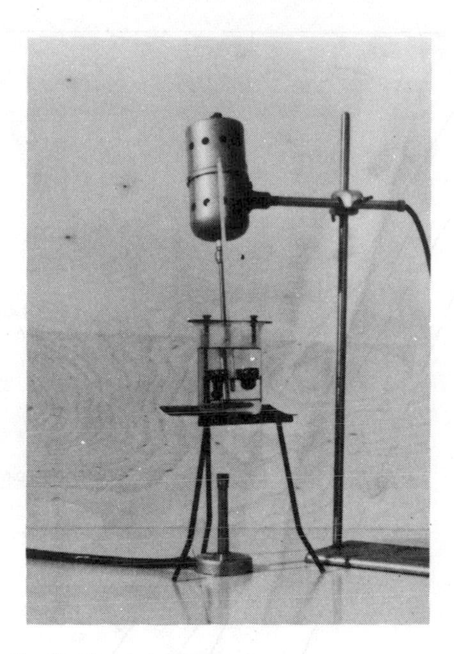

Figure 16.7 *Ring and ball softening point apparatus*

Penetration Index

Observations made in the course of measuring the penetration of bitumens at various temperatures established that at their softening points all bitumens had the same penetration of approximately 800. Taken together with the linear relationship between log penetration and temperature, this led to the development of the penetration index (PI) by Pfeiffer and van Doormaal (1936). Penetration index is defined according to the relationship

$$\frac{d(\log \text{pen})}{dT} = \left(\frac{20 - \text{PI}}{10 + \text{PI}}\right)\frac{1}{50}$$

If the penetration at the normal temperature (usually 25 °C) is measured and the ring and ball softening point determined, this can then be written as

$$\frac{\log 800 - \log \text{pen}}{T_{\text{R+B}} - T} = \left(\frac{20 - \text{PI}}{10 + \text{PI}}\right)\frac{1}{50}$$

where log pen is the logarithm (to base 10) of the measured penetration, $T_{\text{R+B}}$ is the softening point in degrees Celsius and T is the temperature at which the penetration test was carried out, usually 25 °C.

Although the penetration index purports to reflect the temperature susceptibility of a bitumen it is possibly more useful in defining the rheological type of bitumen tested, that is, in indicating by how much it deviates from behaving like a newtonian fluid.

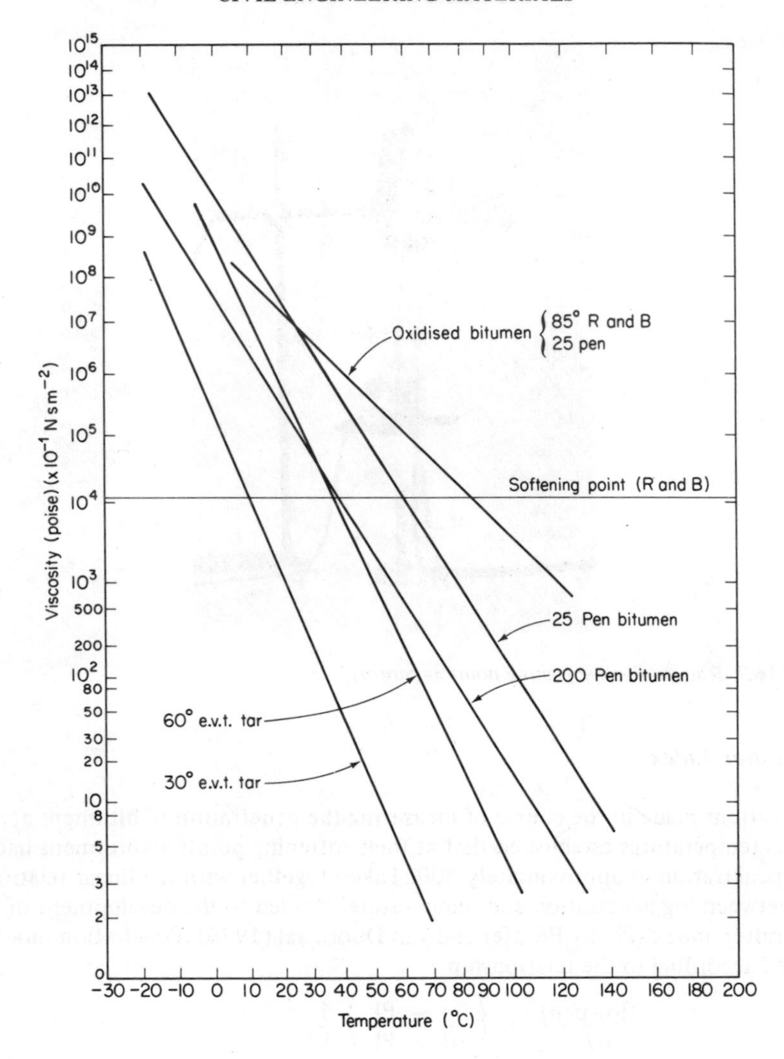

Figure 16.8 *Viscosity – temperature relationships*

PI less than – 2 : substantially newtonian bitumens, brittle at low temperatures, highly susceptible to temperature changes; properties resemble those of pitch.

PI between – 2 and + 2 : normal bitumens, showing some elasticity and a little thixotropy; penetration grades.

PI greater than + 2 : non-newtonian, showing considerable elasticity and thixotropy; less susceptible to temperature changes; this group includes oxidised bitumens.

It should be noted that most bitumens used for road purposes have a PI between –1 and +1.

Handling of Binders

At ordinary temperatures most binders are much too stiff and hard to handle. In order that they can be sprayed, pumped and mixed, or compacted in a stone-binder mixture their viscosity must first be greatly reduced. Suitable ranges of viscosity are

spraying operations	0.05 to 0.1 N s m^{-2} (50 to 100 cP)
pumping	0.5 to 1 N s m^{-2} (500 to 1000 cP)
rolling (mixed materials)	1 to 50 N s m^{-2} (1000 to 50 000 cP)

Reference to figure 16.8 will make it clear that for many binders these operations can only be carried out at fairly high temperatures. In the case of mixed materials, where the binder usually forms less than 10 per cent of the mixture, the aggregate also must be heated if the mixed material is to arrive at the site in a fit condition to be spread and compacted. Obviously this entails very high fuel costs and other solutions must be used in order to reduce these.

16.5 Binder Mixtures

Cutbacks and Emulsions

Cutbacks are mixtures of binders with light volatile oils, the resultant mixture having a much lower viscosity than the original binder, allowing the various handling operations to be carried out at a much lower temperature than would otherwise be the case. In thin layers, for example, in coating a piece of aggregate, the light oil quickly evaporates leaving a coating of the original binder. Kerosene and creosote are commonly used to cut back or flux binders. Emulsions are comparatively stable systems of small globules of binder (which may be tar, bitumen or a blend of both), dispersed in water and kept in permanent suspension by means of an emulsifier. Emulsions are quite stable when stored in bulk, but break down or 'crack' when applied to surfaces in thin layers, the water evaporating to leave a coating of the original binder on the surface. A method for determining the viscosity of cutback bitumen and road oil is given in BS 4693.

The emulsions of bituminous binders and water are generally referred to as anionic or cationic depending on the type of emulsifier used (BS 434). Anionic emulsions are soaps, for example, sodium stearate. In water this salt ionises and the stearate anion $CH_3(CH_2)_{16}COO^-$ is soluble in bitumen. Each bitumen globule in the emulsion is therefore surrounded by an absorbed layer of negatively charged stearate ions. Consequently the globules tend to repel one another and there is no tendency for them to coalesce.

Cationic emulsifiers in effect do the opposite : the positive cation is soluble in bitumen, so that each globule of bitumen is surrounded by a positively charged layer. Cetyl trimethyl-ammonium bromide is an emulsifier of this type, ionising into the negative bromine anion Br^- and the positive cetyl trimethyl-ammonium cation $C_{16}H_{33}(CH_3)_3 N^+$. In theory cationic emulsions should adhere better to some rocks, many of which have a weak negative charge, and in fact this charge does attract the positively charged globules of bitumen, expediting the break-down of the emulsion. The effect although rapid is not very important since the negative charge on the stone is rapidly neutralised, and further breaking of the emulsion will depend on evaporation.

Rubberised Binders

Attempts to modify bituminous binders by adding natural rubber to them have been made since the beginning of the century. It is only in the last thirty or so years, however, that the addition of rubber to binders has become in any way popular. The rubber can be mixed with the binder either as latex, sheet rubber, rubber powder or ground tyre-tread. Of these, unvulcanised rubber powder is probably the most easily handled and disperses more easily in the binder. Latex gives problems at temperatures over 100 °C since it contains much water. Vulcanised powders, for example, from tyres, are more difficult to disperse but are usually cheaper than unvulcanised powders. Sheet rubber demands certain pre-treatment which makes it uneconomic.

The addition of small quantities of rubber was originally made in an attempt to render binders more elastic and less susceptible to brittle fracture. There is no doubt that rubber even in small proportions if effectively dispersed throughout the binder can do both. The binder is invariably bitumen since rubber and tar appear to be largely incompatible, especially at elevated temperatures such as those encountered in mixing plants.

The effect of adding quantities of rubber to bitumens can be summarised as follows.

(1) Even small proportions increase viscosity and softening point and decrease penetration. There is some evidence to suggest that the effects are more marked on softer bitumens.

(2) Rubber in bitumens renders them less susceptible to temperature change.

(3) Elasticity of binders is increased by the addition of rubber.

It should be noted that

(a) The proportion of rubber added to bitumens is generally very small, seldom more than 5 per cent and often half of one per cent or even less.

(b) In a bitumen-rubber mixture the properties obtained will be greatly altered if the mixture is subjected to prolonged heating or high temperature. The rubber under these conditions may be greatly changed and may even act as a fluxing liquid, thereby softening the binder.

Binder Mixtures for Roads

In road work it had long been observed that asphalts made from natural bitumens, in particular, Trinidad Lake Asphalt (TLA) had superior nonskid properties to mixtures using refinery bitumens. TLA, however, was a more expensive material, and in many cases mixtures consisting usually of equal parts of the natural asphalt and refinery bitumen were used. There appears to be no valid reason why a 50/50 mixture was preferred to any other ratio, but this was the common United Kingdom practice. These mixtures were found to have good nonskid properties and this was attributed to the faster oxidation rate of TLA and the gritty nature of the oxidised material. A surface of stone and refinery bitumen alone tended to be more slippery since, when the stone wore smooth, the refinery bitumen still had a smooth unweathered surface giving little grip.

Pitch-Bitumen Blends

It was realised as long ago as 1950 that a mixture of coal tar pitch and bitumen might give a binder which would oxidise more quickly than bitumen alone, and enhance the nonskid properties of road mixes, giving a gritty sandpaper type of surface similar to that given by the TLA - bitumen mixtures. The possible mixtures of pitch and bitumen that can be used are limited by the fact that the mixture is highly unstable if the proportion of pitch exceeds 30 per cent. Most mixtures used consist of about 20 per cent of coal tar pitch (ring and ball softening point not more than 80 °C) and about 80 per cent of bitumen with a penetration of about 100.

17

Bituminous Mixtures

Binders, no matter the type, are seldom used alone. They are invariably mixed with quantities of mineral matter, ranging in size from fine dust or short asbestos fibres to coarse aggregates. Most commonly tar or bitumen binders are used in combination with mineral aggregates and the various mixtures are very similar, whether they are used for highway or hydraulic applications.

All practical mixtures of binders and aggregates lie between two extremes. At one extreme are the mastic asphalts consisting of binder with a fairly small proportion of aggregate 'floating' in the binder (BS 988, BS 1446 and BS 1447). At the other extreme are open-textured macadam mixes, almost analogous to 'no-fines' concrete, consisting largely of single-sized aggregate with a coating of binder as a cementing agent to hold the aggregate particles together (BS 4987).

Most of the mixes commonly specified are not designed by any rational method but are the result of lengthy practical experience. Mixes are specified by 'recipe' rather than 'performance' specifications. Engineers specify the various constituents and their proportions, the methods of mixing, laying and compacting, in the knowledge that a given material treated in a particular way has proved satisfactory in similar applications before.

In selecting mixes for any particular application the desired properties virtually dictate the type of mix adopted. Since long life and low maintenance costs are an essential of many engineering structures it frequently appears that *durability* is the one overriding property a mix must have. All bituminous binders are extremely durable and hence this property leads to the selection of mixes rich in binder with small aggregate particles added to act as extenders. The properties of such mixes are mainly dependent on the binder used. Provided they are well compacted with few voids they weather only on the surface and are virtually impermeable. If it is desired that such a mix also has considerable mechanical strength, then a high viscosity binder, perhaps stiffened by the addition of filler will be used. Such a material can only be mixed, laid and compacted at high temperatures and it therefore follows that it will be an expensive material because of both the high binder content and the high fuel costs.

Where bituminous mixtures are protected from weathering agents other properties such as mechanical strength, flexibility and economy become more important. It may be that a mixture consisting largely of aggregate can give the desired properties, and if the aggregate particles are suitably shaped and graded will give considerable strength. The role of the binder in such a mixture is simply to help the compacted mixture to hold together better and to act as a lubricant between the aggregate particles during the compacting operation. Since the strength of such a mixture depends largely on the mechanical interlocking of the aggregate, the binder used can have a low viscosity and the material can be laid and compacted at comparatively low temperatures.

17.1 Properties

Apart from certain waterproofing treatments, for example, on roofs, and the surface dressing of roads when a layer of stone may be stuck to a tack coat of binder, all bituminous mixtures are mixed in mixing plants of some description before laying or applying. The majority of all mixes contain varying proportions of four materials

Coarse aggregates : crushed stone, gravel or slag, with a grain size exceeding that of sand (material retained on a 2.36 mm sieve).
Fine aggregates : mineral material of sand sizes (material passing a 2.36 mm sieve).
Filler : fine-grained mineral material, often limestone or Portland cement with a grain size smaller than 0.075 mm.
Binder : tar, bitumen, etc.

Of great importance in any mixture is the property of workability.

Workability

The mixed material must be easy to handle and capable of being laid and compacted to the desired density. If a mix is not adequately compacted, it will more readily deform under load and it will be more permeable, allowing air and water to penetrate below the surface and hence reduce its durability. This loss of durability is brought about by oxidation of the binder and by loss of adhesion between the binder and the mineral materials.

If a bituminous mixture is to be laid at the correct workability it is obviously important that it should be laid and compacted at the correct temperature; this could vary from 15 °C for a mixture using a cutback binder to 200 °C for a mastic asphalt using highly viscous binders with a high proportion of filler. For the cold-laid mixes there is no great problem and even if the temperature is a few degrees different from the ideal, the material can still be laid and compacted properly. With materials that have to be laid hot, difficulties arise, particularly if they are to be transported over any great distance between the mixer and the site. Laying temperatures become much more important if proper compaction is to be obtained. In this respect, mixes made with bitumen as a binder are slightly less sensitive to temperature variations than those with a binder of road tar. For materials that have to be laid while very hot, transporting premixed materials over any appreciable distance is out of the question and most of the hot-laid mixes have to be heated in small quantities on the site where they are to be laid.

Strength and Flexibility

Bituminous mixes can be made strong enough to resist deformation under considerable loads, both static and dynamic. At the same time it is often desired that they should also be sufficiently flexible to accommodate both long-term movements of the material or structure supporting them, and repeated short-term loading of the type likely to cause failure by fatigue.

Most standard mixtures have their constituents specified by weight. This can be somewhat misleading when endeavouring to understand how these mixtures behave in practice, owing to the different specific gravities of aggregates and binders. It is useful to look at figure 17.1, which shows the proportions of typical compacted mixes by volume.

As can be seen from the figure a mastic asphalt can consist of one-third or more of binder and have no air entrapped. This implies that all voids between aggregates are completely filled, the aggregate particles are unlikely to be touching one another and the strength and flexibility of the mixture will be largely determined by the mechanical properties of the filler-binder mortar. By using a filler-binder mixture rather than a binder only, stronger and stiffer mixtures are obtained. It is

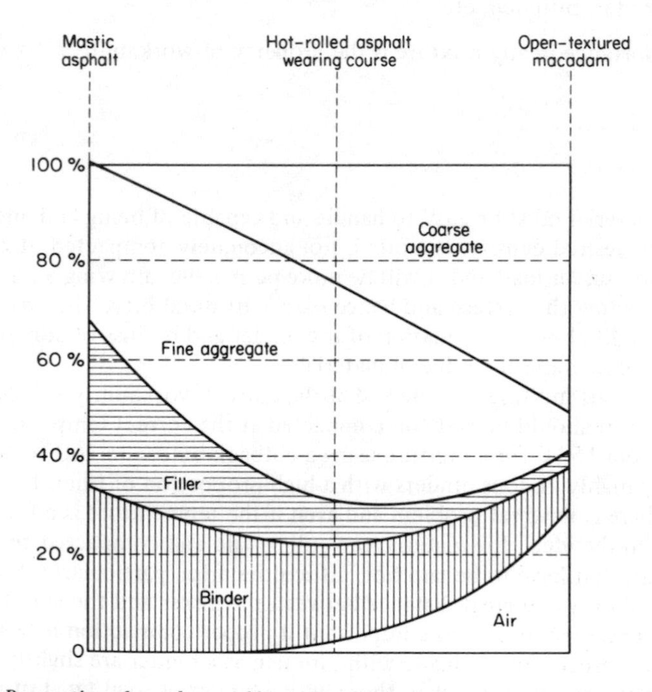

Figure 17.1 *Proportions by volume of bituminous mixtures*

thus possible to make and lay strong, dense, stiff mixtures at relatively low temperatures. The quantity of filler present in the filler - binder mortar is very important, too much and voids are created in the mortar, too little and the mortar develops the strength properties of the binder only. When the proportion of aggregate (coarse and fine) to mortar in the mixture increases, it becomes more difficult to achieve full compaction, air voids are created, and aggregate particles actually touch one another. The strength and flexibility of the mixture becomes dependent also on the aggregate properties including grading, surface texture, particle shape, interlocking of particles, and so on.

Strength of a mixture or, perhaps more accurately, resistance to deformation under load, can be obtained either by using high viscosity binders in the binder-filler mortar or by using crushed stones with rough angular surfaces as the aggregate, or both at the same time. Flexibility is most easily obtained by using mixtures with higher binder contents and binders of fairly low viscosity.

In discussing strength and flexibility one has to specify whether the loads are being applied for a short or long time. For loads applied only for a short time, when bituminous mixtures tend to behave in an elastic manner, the stiffness of the mixture depends only on the stiffness of the binder and the proportion of aggregate in the mixture. If, however, the load is applied over a long period of time, when the fluid properties of the binder become more important, the stiffness of the mixture becomes influenced much more by the aggregate properties such as grading, surface texture, shape and interlocking of particles. The behaviour will also be influenced by the confining conditions of the mixture, that is, how it is contained or supported in the structure.

In any discussion of strength and flexibility of bituminous mixtures, temperature cannot be forgotten. An increase in temperature will have similar effects on a mixture as an increase in time of loading.

Durability

The durability of bituminous mixtures is first of all a function of the durability of the binder used. Exposure of the binder to air, particularly when hot, results in evaporation (of some of the lighter oils), oxidation and polymerisation, all of which may occur together and all of which produce hardening of the binder. It follows that a hot mixture with its large surface area exposed during mixing will tend to undergo changes in the binder. Further hardening will occur during transportation and laying, but once the mixture is laid, compacted and cooled further changes proceed at a very much reduced rate. The viscosity of a binder from the time of mixing to 12 months after placing may increase by up to five times or more.

Durability of mixtures is also influenced by the adhesion between binders and mineral aggregates. Poor adhesion in the first instance may be a result of water or dust films preventing intimate contact between binder and aggregate during mixing. At later times adhesion may fail owing to the penetration of water or other contaminant liquids into the mixture, therefore proper compaction with few voids is always important. An adhesion failure could also be caused by using a binder with an unsuitable viscosity for the temperature or loading conditions experienced by the mixture; with increased temperature the hot material might become too fluid to maintain the bond or too brittle at low temperatures to resist impact loading.

17.2 Design of Bituminous Mixtures

Perhaps the use of the term 'design' in this connection is not truly warranted; more often 'selection' of bituminous mixtures would be a more accurate description. Mixes are usually selected for a particular use by considering the strength, flexibility, durability, imperviousness, stiffness, skid resistance (for road surfaces) and perhaps most important of all the economy required.

Probably in the majority of cases the engineer or architect draws on his knowledge and experience when specifying, and demands a mixture he knows has proved satisfactory in the past. This might be called the recipe method of design.

Recipe specifications define mixtures of materials which have proved successful in the past, and the majority of British Standards for bituminous materials are of this type.

Specifying mixtures by the recipe method has some obvious advantages. The specifier has little difficulty in defining his requirements, the contractor merely has to produce the required mixture, and verifying that the mixture has been prepared to the specification is a reasonably simple matter (BS 598). The method has many disadvantages, however. The performance of the mixture depends on workmanship perhaps more than on composition, and workmanship is more difficult to specify. The material may be used in a situation different from that envisaged by the specifier. If the material is proved not to be made to specification it is difficult to judge how serious this is. Recipe specifications also tend to inhibit the use or even the trial of new materials or new ideas generally.

Design by Performance Tests

In the same way as concrete cubes are tested for compressive strength, it should not be too difficult to devise tests for application to small samples of bituminous mixtures in order to verify that some parameter, for example, strength, has reached a specified limit.

In devising tests, the steps are : selecting a property for measurement, constructing a test to measure that property, and setting satisfactory limits for performance. Unfortunately, in practice, the quality of bituminous mixtures with regard to any particular physical property is seldom the cause of failure to perform as desired. For example, in the waterproofing of flat roofs bituminous membranes of various types have been used for many years. Such roofs frequently leak, but in the majority of cases the failure is a result of damage to the membrane from foot traffic on the roof, inadequate details at parapets or around roof vents or rooflights, poor workmanship, or incorrect specification in the first instance. In view of this, there seems little point in devising quite elaborate laboratory tests to be used on small sample specimens of the bituminous material.

Testing Procedures

The testing procedures most commonly used may be grouped together as follows.

(1) Tests which attempt to simulate conditions in practice. The best known of these is the indentation test (BS 988) for mastic asphalt to be used in flooring, roofing and damp-proof courses. This consists of measuring the depth of penetration of a loaded steel pin 6.35 mm in diameter into a sample at a temperature of 25 °C when a stress of 10 N mm^{-2} is applied for 60 s. The wheel-tracking test devised at the Road Research Laboratory (Broome and Please, 1958) for hot-rolled asphalt is not yet a standard design tool.

(2) Tests which attempt to measure fundamental properties. As might be expected attempts have been made to adapt the tools of soil mechanics for this purpose. Triaxial and shear box tests have been used to measure cohesion and angle of internal friction in mixtures. Unfortunately the present knowledge of stress set up in layered road systems, for instance, is not sufficient to make logical use of the values obtained from the tests.

(3) Arbitrary tests. All have the disadvantage that they require somewhat elaborate apparatus, and samples are usually tested at 60 °C, which is much higher than temperatures likely to be encountered by the material in practice. Nevertheless one of these, the Marshall test, has been adopted as both a design and a control tool for road asphalts in many countries. In Britain the test can be used as a design method for rolled asphalt for roads (BS 594) and for airfields. In the Marshall test asphalt mixes are made at several different binder contents (the test applies only to asphalts) with aggregates graded within specified limits. Cylindrical specimens 63.5 mm high and 101.6 mm in diameter are compacted in a steel mould with a standard compaction hammer. They are tested for compacted density, which allows the percentage of voids to be calculated. The cylindrical samples are then tested at 60 °C by loading them in compression, the load being applied to the curved surface and at right angles to the longitudinal axis of the cylinder (see figure 17.2). The maximum load that a specimen will carry before failure, the *stability*, is recorded, together with the measured deformation up to the moment of maximum load, the *flow*. Using the Marshall test mixes can be specified in terms of minimum stability, maximum flow and a void content.

Performance tests such as the Marshall have certain advantages

(1) They allow the design to be varied to suit different conditions fairly easily.
(2) They allow materials to be used which might otherwise be regarded as below standard.
(3) It is possible to estimate the effect of materials which are not quite up to specification.
(4) The tests can often be used as control tests as well as for design purposes.

There are disadvantages, however : the existing tests can only be applied to certain types of bituminous mix, and they are not generally applicable to all mixtures; the laboratory equipment is rather more elaborate, and more highly skilled personnel

Figure 17.2 *Marshall testing machine*

are required; the reproducibility of the results is no better than for conventional analysis methods; the test temperatures for many of the performance tests are on the high side.

Control of Recipe-specified Mixes

When bituminous mixtures are specified by the recipe method an analysis procedure is normally used to verify that the correct material has been supplied.

Although BS 598 lists five or six different methods of analysis, basically they are all very similar. A sampling procedure is employed to obtain a weighed quantity of the bituminous mixture; the binder is dissolved off by washing with a solvent (usually dichloromethane or trichlorethylene); all the solid material is collected by filtration and weighed, and the difference between this and the original weight gives the weight of soluble binder in the mixture. There are variations, since filtration tends to be a slow business. In the sieving extractor method the coarser mineral material is removed by sieving, the filler particles left in suspension in the contaminated solvent are removed by centrifuging, and the quantity of binder is calculated by distilling off the solvent from a known quantity of the solvent-binder mixture. Thus the total quantity of aggregate and binder is known, and the quantity of filler is found by subtracting from the original weight.

In carrying out such tests it is important to know what binder has been used, since not all are wholly soluble in the solvents used, and allowance must be made for the insoluble proportion.

Binder recovered from such a test will not have the same properties as the original binder added at the mixing plant, and any tests made on the binder should be carried out on samples taken at the mixing plant before the binder is added to the mixture.

18

Uses of Bituminous Materials

The various ways in which bituminous materials have been used have arisen directly from

(1) their relative cheapness and availability in large quantities;
(2) their durability;
(3) their adhesive and waterproofing qualities; and
(4) the ease with which they can be handled at elevated temperatures, but quickly become stiff and resistant to deformation at normal temperatures.

All these properties have considerable bearing on the popularity of bituminous mixtures for road construction.

18.1 Bituminous Road Materials

Bituminous road materials normally make up the whole or part of a layered system of road construction.

In the majority of road pavings the total pavement is made up of a number of elements, as shown in figure 18.1.

The *subgrade* is the soil on which the road is supported. This may be in its natural state or may have had its natural properties altered by installation of drains, compaction by rolling or other means.

The *formation* is the name given to the surface of the subgrade shaped to the required camber or fall.

The *sub-base* is a layer usually of granular material inserted for any of a number of possible reasons: for example, to insulate the pavement from a subgrade likely to contaminate it or to provide a good clean surface for construction plant, or to increase the total depth of pavement to help insulate the subgrade from frost, etc.

The *base* is the main structural element in the pavement and its role is to help spread the concentrated loads from traffic over a large enough area of subgrade to sustain them.

The *surfacing* consists usually of the most expensive and highest quality material contained in the pavement. Its function is to supply the desired riding

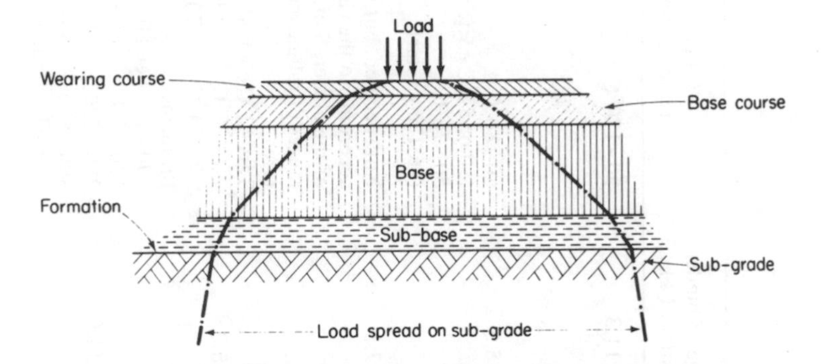

Figure 18.1 *Elements of road pavement*

and nonskid properties to the road; it also serves to protect the less durable base materials from damage and is usually much less pervious to water penetration than the underlying materials. It is often constructed in two layers called *base course* and *wearing course.* In this case these are usually of different mixtures and the wearing course is the more expensive and the higher quality of the two.

Wearing-course Mixtures

It is appropriate to consider these mixtures first since they determine the safety and riding properties of the pavement and to a large extent also the types and dimensions of the underlying layers. Highway authorities are always under pressure from road users to provide them with fast, safe roads which need seldom be closed for repairs. Ideally therefore all surfaces should be stiff, stable ones with good skid resistance, and very durable. Unfortunately to obtain good durability in a bituminous mixture a high binder content is required, whereas it is easier to obtain good stability and high skid resistance with harsh aggregates and low binder contents. Some sort of compromise solution is obviously required; the usual one is to obtain the durability by means of a high binder content, the stiffness and stability by using a high viscosity binder, and the nonskid properties by rolling precoated stone chippings into the top surface. As with most compromise solutions this is an expensive one. High viscosity binders require high mixing and laying temperatures, therefore fuel bills rise. Stiff stable surfacings require fairly rigid supports to prevent them cracking and breaking up, therefore the cost of bases also increases.

To provide the ideal surface for road pavements the total cost of the pavement increases on several counts and the economics of construction and repairs becomes the primary consideration.

It will also be obvious that when expensive pavements are constructed they should be left undisturbed for as long as possible, since repairs and replacements are very costly. It follows that in developing areas where roads are likely to be dug up to install services such as drainage, water supplies, electricity, telephones, etc., inexpensive easily repaired pavements are to be preferred. Expensive surfacings on rigid bases, for example, concrete bases, should only be used where

TABLE 18.1

Typical bituminous wearing-course materials

Material	Relevant specification	Type of aggregate	Type of binder	Range of binder content (%)	Rolling temperature
Hot-rolled asphalt	BS 594	Crushed rock, slag, or gravel with natural or crushed sand and filler	Bitumen or refined lake asphalt or equal proportions of both. Normally 40/60 pen or pitch bitumen blends	5.9 - 11.3	90 - 120°C
Dense tar surfacing	B.R.T.A. (1968)	As above	46 - 54°C e.v.t. tar (grades B46, B50, B54)	5.7 - 8.5	60 - 80°C (50 or 54°C e.v.t. tar) 50 - 70°C (46°C e.v.t. tar)
14 mm nominal size open-textured wearing course macadam	BS 4987 (Tables 29–32)	Crushed rock or slag	100–300 pen bitumen or cutback bitumen 37 - 42°C e.v.t. tar (summer) 34 - 38°C e.v.t. tar (winter) (grades C34, C38, C42 tars)	4.0–7.3	Not specified, but varies according to the viscosity of the binder. Suitable ranges of rolling temperatures Tars 42°C e.v.t. 40 - 60°C
14 mm nominal size dense wearing course macadam	BS 4987 (Tables 41–44)	Crushed rock, slag or gravel	100–300 pen bitumen or cutback bitumen (grades C34 to C42 tars)	5.0–8.0	38°C 30 - 45 34°C 20 - 35 Bitumens 90 - 110 pen 80 - 100°C 190 - 210 65 - 85
Fine cold asphalt	BS 4987 (Tables 53–56)	Crushed rock or slag	100–450 pen bitumen or cutback bitumen (grades C42 and C46 tars)	6.0–11.0	280 - 320 55 - 75 400 - 500 45 - 65 cutback up to 40°C

Material	Relevant specification	Type of aggregate	Type of binder	Range of binder content (%)	Rolling temperature
Compressed natural rock asphalt		Natural rock asphalt containing between 7.5 and 13% of bitumen, ground to a powder and all passing 2.36 mm sieve		7.5 – 12.5	110 – 140°C
Mastic asphalt	BS 1446	Natural rock asphalt ground to powder and all passing 2.36 mm sieve. Crushed rock chippings added later	Bitumen or refined lake asphalt or equal proportions of both, or pitch bitumen blends. 5 – 20 pen	13.0 – 20.0 before chippings added	175 – 220°C
Mastic asphalt	BS 1447	Limestone ground to a powder and all passing 2.36 mm sieve. Crushed rock chippings added later	As above 10 – 25 pen	14.0 – 17.0 before chippings added	175 – 220°C
Asphaltic concrete		Continuous-graded crushed stone or slag	Bitumen or refined lake asphalt or equal proportions of both. 60 – 100 pen	6.0 – 10.0	75 – 90°C

they are likely to be undisturbed for fifteen years or more, that is, for city streets, motorways or other highways where ribbon development alongside the roads is not permitted.

Table 18.1 shows typical bituminous wearing-course mixtures. Of the mixtures tabulated the most rigid and, in most respects, the most expensive materials are the two types of mastic asphalt and compressed natural rock asphalt. All three are expensive not only because of the high laying temperatures but also because all three types are laid by hand. In the case of mastic asphalts the hot material is spread and compacted by hand using wooden floats. The laying is a tiring and slow process since the material is worked by hand for a considerable time. Compressed natural rock asphalt is spread using rakes and compacted using heated metal rammers. Both these materials give a surface which is too slippery in its newly finished state and a nonskid texture is invariably produced by rolling in precoated stone chippings.

The remaining wearing-course materials tabulated are capable of being laid by machine and thus are considerably less expensive than the hand-laid materials. Some of them, however, are still laid at elevated temperatures and contain considerable quantities of high viscosity binders, so they can scarcely be termed cheap. The main attributes of the various wearing courses are now considered in more detail.

Wearing-course Materials

Mastic asphalt

This is the hardest material commonly used; it is exceptionally durable and is highly resistant to distortion under even the heaviest traffic. For this reason it is often used for areas of roadway where repeated braking and acceleration of traffic takes place, for example, at traffic lights and at bus stops (see figure 18.2). Because of its great durability mastic asphalt is sometimes selected for the wearing course of areas where repairs would cause intolerable delays, for example, in vehicle tunnels, on bridge decks, etc. In the case of long-span bridges where dead load is the major loading, a thin surface of mastic asphalt on (say) a steel deck will give long life with little load. The hand trowelling of mastic asphalt produces a smooth slippery surface which must be roughened by rolling in precoated chippings. These are generally of hard clean stone 14 or 20 mm in size coated with about 2 to 3 per cent of bitumen by weight (usually about 70 pen); often 1.5 to 2 per cent of filler is added since this helps to keep the coating intact during handling and transporting. The chippings are spread by hand and rolled into the new surface using a hand or light smooth-wheeled roller. Mastic asphalt is, at low temperatures, a fairly stiff material and requires a rigid support if it is not to crack. For most road work this support is provided by a concrete base. If this base is jointed then movement at the joints under traffic or otherwise may cause the surface to crack. In such a case the mastic asphalt is often laid in two courses, even though the total thickness may be only 50 mm or less.

Figure 18.2 *Deformation due to traffic on a hill at traffic lights.*

Compressed natural rock asphalt

This material has largely been replaced in the United Kingdom by mastic asphalt. It consists of calcareous rock naturally impregnated with bitumen which has been ground to a powder suitable for use, with no artificial binder added. The raw material is expensive, being imported from France, Italy or Switzerland. When laid it gives a very hard impermeable wearing course that is not easily damaged. The asphalt is laid as a hot powder (110 - 140 °C) raked to the desired profile using wooden rakes and compacted by hand ramming using heated heavy iron punners. The thickness of this wearing course is seldom more than 50 mm. It requires roughening and this is done by lightly brushing the surface with a solution of soft bitumen (pen 200) in a solvent such as kerosene or white spirit. Coated chippings are then spread on the surface and rolled in with a light roller. As with mastic asphalt, compressed natural rock asphalt requires a base that is rigid and not affected by high temperatures.

Hot-rolled asphalt

This is a heavy-duty dense road surfacing. It is machine laid, very durable and almost completely impermeable, with few voids. Hot-rolled asphalt is popular with engineers for the surfacing of city streets and heavily trafficked roads. Like mastic asphalt and compressed natural rock asphalt it demands good nonflexible support and in the majority of cases this is provided by a concrete base. Naturally if the road surface has to be opened to install or repair services, the base must be broken through and later replaced. This is a lengthy process and causes considerable delays to traffic. Hot-rolled asphalt is therefore not recommended for roads through developing areas where new services to buildings are in demand.

Hot-rolled asphalt can contain varying amounts of coarse aggregate (stone between 20 and 2.36 mm), but its strength is largely due to the viscous asphaltic mortar. Asphaltic mortar consists of the fine aggregate, filler and an asphaltic cement which may be bitumen, or lake asphalt and bitumen, or a mixture of pitch and bitumen. The pitch – bitumen mixture is thought to give improved weathering and adhesive properties and is usually about 80 per cent of bitumen with 15 to 20 per cent of coal tar pitch with a softening point between 55 and 80 °C.

Hot-rolled asphalt may be laid in one or two courses (if laid on a jointed base two courses are preferred). The lower or base course usually contains much more coarse aggregate (55 to 65 per cent) and consequently requires less binder. The asphalt should not be laid during heavy rain, especially if this is producing ponding on the base. It is compacted by rolling with a roller of at least 60 kN weight and can be opened to traffic as soon as it is cool. If the wearing course contains less than 45 per cent of coarse aggregate the surface is generally roughened by rolling in precoated stone chippings.

As mentioned previously hot-rolled asphalt mixes can be designed by the Marshall method. This consists essentially of making a series of test specimens with the same proportions of aggregates and filler but with binder contents varying, usually in steps of 0.5 per cent by weight. Duplicate specimens for each binder percentage are usually tested in the Marshall tester and a graph is drawn of stability (load at failure) against binder content. Similar optimum binder contents can be found for maximum density of the compacted mix, and flow values (deformation at failure load) are also recorded.

Dense tar surfacing

This material was intended originally to be the tar industry's answer to hot-rolled asphalt. It is a dense durable strong surfacing with fairly high binder contents and a high filler content (B.R.T.A., 1968). Like hot-rolled asphalt it is laid by machine in fairly thin layers. The binder used is usually tar with an e.v.t. of about 50 °C. This is in fact a lower viscosity than that of binder grades used with hot-rolled asphalt, and even with added filler the dense tar surfacing wearing course is liable to deform more readily than the asphalt, particularly in hot weather. Experiments have been tried using higher viscosity tars (60 to 65 ° e.v.t.), but at the necessary mixing temperatures (about 115 °C) large quantities of fumes are given off and volatile constituents are lost in a manner that is difficult to control. With dense tar surfacing the mixture is normally delivered to the laying machines at about 90 °C and rolled at about 70 °C. On a cool day this probably means a time interval of about 10 to 15 minutes and this demands a skilled and experienced workforce to ensure good compaction and a suitable density and finish.

With hot-rolled asphalt the limits are not nearly so finely drawn since the temperature drop between delivery and rolling is normally about 40 °C. Dense tar surfacing however does have some advantages over hot-rolled asphalt. It has good resistance against the softening effect of oil dropped by vehicles on the surface, and has been much used in bus stations, termini and aircraft hard-standings. Although it is without doubt a good surfacing, in general dense tar surfacing has not shown quite the ability to stand up to very heavy traffic exhibited by hot-rolled asphalt.

Asphaltic concrete

This is widely used in the United States for much the same purposes as hot-rolled asphalt in the United Kingdom. Hot-rolled asphalt wearing courses are essentially gap-graded mixtures and there is little mechanical interlock between aggregates whereas the mixtures used in asphaltic concrete are well-graded mixes of coarse aggregate, fine aggregate and filler. The coarse aggregates are usually crushed stone and the stability of the mixture is due more to mechanical interlocking of the aggregate particles and less to the viscous mortar in the mixture. Since asphaltic concrete contains a higher proportion of coarse aggregates, and since the grading is better, it normally requires less binder than hot-rolled asphalt. Wearing surfaces of asphaltic concrete are capable of carrying the very heaviest traffic if well designed and well compacted.

In asphaltic concretes it is essential that the grading of the mineral content is such as will give maximum density. When a suitable grading is produced with the aggregates available, a suitable binder content is arrived at, normally by means of the Marshall test, to give a mix with a specified minimum stability and voids content.

Asphaltic concrete has been used for some time as a runway-surfacing material in the United Kingdom, where its high strength has been found effective in resisting the high pressures from aircraft tyres. Common specifications are

	U.K. runway surfacing	U.S.A. major roads
Minimum stability	8.0 kN	4.5 kN
Maximum flow	4 mm	4 mm
Percentage voids in mix	3 to 4 per cent	3 to 5 per cent

Dense bitumen macadam

This material resembles asphaltic concrete but has a somewhat higher voids content (about 10 per cent).Its stability is largely dependent on mechanical interlock. With high traffic concentrations harder binders (90 - 220 pen) are used, and lower binder contents because, under heavy traffic, the surface tends to compact further while the voids content is reduced and bitumen tends to work up to the surface. Graded aggregates are used with a maximum size of up to 20 mm in a 50 mm thick wearing course. For heavy traffic, slag aggregates are much favoured. Dense bitumen macadams quickly become impervious: they have good load-spreading properties, and are laid at lower temperatures than asphalts. For maximum success with this material the specified grading limits should be attained and the mix must be well compacted.

Tarmacadam and bitumen macadam wearing courses

Specifications for both these materials have been based on the results of full-scale experiments where trial sections of surfacing have been laid on main roads and the performance of the various mixes assessed. Several principles are now fairly well accepted. The main ones are

(1) their performance is greatly influenced by weather;
(2) aggregate shape and gradings are important; and
(3) for heavily trafficked roads lower contents of higher viscosity binders are most successful, particularly if the supporting base is reasonably rigid.

In colder wetter regions, higher binder contents give more durable surfaces, particularly if traffic is not too heavy. In warmer drier areas with heavy traffic the tendency is for the surface to 'fatten up' and become slippery unless lower proportions of harder binders are used. In all these mixes crushed stone or slag aggregates are preferred to gravel, and the resistance of the stone to polishing under traffic is an important factor (BS 4987). These mixtures are invariably machine laid and compacted by rolling with a smooth-wheeled roller of 60 kN weight or more. They may be laid in one or two courses with a total thickness of up to about 130 mm. The upper or wearing course is seldom thicker than 40 mm. The maximum size of aggregates used in a wearing course is usually about half the nominal thickness of the course, and the proportion of fine aggregate (passing 2.36 mm sieve) is often less than 30 per cent. These wearing-course mixtures are used either as the top surface on a new road or to provide a satisfactory surface on an existing road with adequate strength but a poor surface. Tarmacadam and bitumen macadam surfaces are generally fairly pervious when first laid, but become progressively less so under compaction by traffic.

Fine cold asphalt

This is mainly used as a regulating or patching material (see BS 4987). Since low-viscosity bitumens are used as the binder it can be laid warm or even cold and can be stored for quite long periods. With no coarse aggregate it can be laid in thin layers, and to a feather edge if necessary. It is rolled by a roller of at least 50 kN and can be opened to traffic immediately after rolling.

Base-course Mixtures

Most modern wearing-course materials are laid by machine to the required falls, cambers and super elevations. The required riding quality of the surface, unspoiled by load undulations or irregularities, is much easier to obtain if the laying machine is moving on a well-regulated surface. To obtain such a surface it is often necessary first to lay a base course. Base courses are also useful in further dispersing the wheel loads and thus reducing the stresses transmitted to the base.

Base courses are generally about twice as thick as the wearing courses they support. Since they are to some extent protected from weather, and almost wholly protected from the traffic, they can be made of mixtures using lower quality aggregates of a larger maximum size, and lower viscosity binders than are used in wearing courses. Normally, if it suits the construction processes, traffic can be allowed to run on base courses for a short time.

Road Bases

A major advantage of bituminous road surfacings is the speed and economy with which they can be repaired. It follows that rigid concrete bases are not used under these road surfacings except in the case of the high-quality materials such as mastic or hot-rolled asphalt. There are a wide variety of 'flexible' bases used under bituminous surfacings. Almost all consist of graded aggregate compacted in layers by rolling and, in order to ease compaction, they are 'wet' or lubricated by some fluid which later stiffens or evaporates. A widely used group come under the heading of *coated macadam bases.*

The strength of these bases is derived from the mechanical interlocking of 50 or 40 mm graded aggregate coated with tar or bitumen, laid in 80 or 100 mm layers and compacted with an 80 to 100 kN roller. The binder helps compaction during laying and improves the strength of the base. The strength is considerably improved if fairly stiff binders are used, 45 - 55° e.v.t. tar or 40 - 70 pen bitumens. If the binder content is fairly high then each layer of the base can be opened to traffic immediately after rolling.

18.2 Bituminous Materials in Hydraulic Works

The earliest known uses of bituminous materials relate to hydraulic uses. Natural bitumens, in some cases mixed with sand and gravel, were used to waterproof a stone embankment built on the Tigris 3000 years ago. Bituminous materials are still widely used for similar purposes. Basically these are

(1) to prevent water passing through or penetrating into structures; and
(2) to provide a strong layer capable of resisting erosion by water and wave action.

Waterproof Coatings

At its simplest this is really a paint treatment of materials. Tar or bitumen can be painted on to parts liable to corrosion from weather or other sources. In some cases these coatings may be applied in the form of cutbacks or emulsions which can be applied cold (BS 3416). In addition some are pigmented and marketed as bituminous paints. Hot-applied bitumen-based coatings for ferrous products are dealt with in BS 4147. Other materials in addition to metals can be protected, for example the North of Scotland Hydro-electric Board has made a practice of coating concrete water surfaces with bitumen where the water contains peaty acids. In aggressive environments coatings of bituminous materials can often give a high degree of protection. A well-tried method of protecting steel and ironwork in sewers has been to coat the metal using the Angus Smith process. In this the metal is preheated to about 320 °C and then dipped into a solution preheated to 150 °C and consisting of four parts bitumen (coal tar and pitch are also used), three parts prepared oil (often linseed) and one part paranaphthalene.

Thin coatings of bitumen paints or emulsions applied to absorptive materials have the effect of sealing capillaries so that both water and water vapour are prevented from moving through the material. Water and vapour barriers of this type are often used to prevent water from the ground rising through concrete floors. In cases where the soil under suspended floors is a source of moisture the bituminous membrane can be spread on the soil to prevent the ingress of moisture into the underfloor area. Such a membrane may require considerable strength to resist the hydraulic pressure from the static head of a differentially higher water table; the membranes may be fairly thick layers of asphalt, tar or pitch.

Tanking

This is the term used to describe the provision of an impervious layer in the floors and walls of basements. Tanking generally consists of three layers of mastic asphalt forming a continuous membrane up to 30 mm thick on floors and 20 mm thick on walls. Alternatively, two or three layers of bitumen sheeting bonded with hot bitumen may be used, three layers generally being used for sheeting weighing 3.5 kg m^{-2} and two layers for sheeting weighing 5.0 kg m^{-2}.

Unless protected, both types of membrane are liable to be damaged. For example, if laid externally they may be punctured by backfill material. Consequently they are always protected by sandwiching them between the structural floor or wall and a protective coat which may be concrete, brickwork, blockwork or a mortar screed. Where membranes are not fully contained they may be extruded under pressure and, in such circumstances, pressures should be limited to 600 N m^{-2} and 1000 N m^{-2} for asphalt and bitumen sheeting respectively.

The condition of the surfaces to which the membranes are applied is important. If very rough they may puncture bitumen sheeting and if very smooth, particularly on vertical surfaces, asphalt may not adhere. Soft, flaky or friable surfaces will not suit either membrane. Junctions between horizontal and vertical layers must be carefully detailed and are generally reinforced, by asphalt fillets in the case of asphalt membranes and, for bitumen sheeting, by internal and external angles constructed from bitumen sheeting.

Flat Roofs

A common use of continuous bituminous membranes is in the waterproofing of flat roofs. There are three different types of bituminous membrane in common use for this purpose. The most expensive and certainly the most durable is asphalt.

Asphalt

This generally consists of two layers of mastic asphalt. In most cases the total thickness of asphalt used gives a weight of about 0.4 kN m^{-2} and this is sufficient to resist any uplift due to wind, so that no fixing or bonding to the roof deck is required. While mastic asphalt accommodates gradual movement of the supporting structure it is sufficiently stiff to act as a solid under impact loads or at low temperatures. The asphalt has a high coefficient of thermal expansion and is therefore usually laid on an isolating layer of felt. This allows differential movement and bridges discontinuities in the supporting deck.

The majority of treatments consist of several layers. These are a waterproofing membrane on top, a structural slab of precast concrete or other structural material, an insulating layer to prevent excessive heat loss through the roof, and a vapour barrier on the underside. The insulating layer often consists of a lightweight concrete screed, but lightweight concrete is only an effective heat insulator when dry and a vapour barrier is required to prevent it absorbing moisture from within the building. The problem of maintaining the insulating layer in a dry condition is always a difficult one and speed of construction and adequate protection from weather is always important. To help speed construction a lightweight aggregate screed, with the aggregate bonded with bitumen, is sometimes used. This allows very speedy construction but has the disadvantage that any leaks which develop produce unsightly staining of ceilings.

Mastic asphalt roofing membranes are generally strong and provided they are adequately supported by a reasonably rigid deck will resist any damage from traffic, even on roof car-parks. They do become gradually harder and shrink slightly owing to solar exposure, and are therefore often given some solar reflective treatment.

Built-up bitumen felt

In roofs where only foot traffic from occasional maintenance personnel is anticipated built-up bitumen felt roofings are frequently adopted. Roofing felts (BS 747) are of many types but generally consist of a mat of organic or inorganic fibres impregnated and usually coated with bitumen. When such a felt weathers over a long period, surface crazing and pitting allows moisture to penetrate to the fibres and further weathering proceeds unhindered. In the United Kingdom three layers of felt are generally specified, each layer being bonded to the preceding layer by hot bitumen. The total weight of membrane provided by three layers of felt is not as much as provided by mastic asphalt, consequently the membrane is likely to be stripped by wind if not held down. It should therefore be bonded, at least partially, to the underlying roof structure. Where joints exist or shrinkage cracks are likely to develop in the supporting roof structure, a fully bonded felt layer may not have sufficient elasticity and might crack. Accordingly strips of unbonded felt 200 mm or so wide should be laid along the line of such cracks prior to laying the membrane.

Many engineers and architects maintain that only mastic asphalt or three layers of properly bonded felt can give a completely reliable waterproof roofing on flat roofs. However, whereas mastic asphalt should give a life of 50 - 60 years, built-up felt may only give a life of 20 - 25 years. On completely flat roofs ponding generally occurs. This is not serious in the case of mastic asphalt but can have serious effects on the built-up felt roof which should always be laid with sufficient fall, say 1 in 60, to avoid the ponding of rainwater.

Proprietary bituminous roofing systems

In the main these consist of thin roofing membranes built up *in situ*. Typically they are formed by spraying a tack coat of bitumen emulsion or cutback on to the roof. This is followed by further coats reinforced by one or more layers of fibre-glass or other felt mesh. When sufficient thickness is attained the whole is finished by trowelling on a mixture of coarse sand and emulsion, or mineral chippings are spread on a cutback or emulsion layer. This final surface gives some mechanical strength and, if light coloured, some protection from solar heat.

These various systems will give good service if used in the correct situations. They are, however, difficult to construct satisfactorily, except during a period of good weather, and generally should only be preferred on roofs with no ponding problems and where the likelihood of people walking on the roofs is small.

Flat roofs are notorious for developing leaks. In most cases these defects are not due to the material employed but to designers and contractors failing to comply with the published recommendations and good practice generally. Without doubt the greatest source of trouble occurs with edge details, around parapets, curbs, ventilators, roof-lights and so on. If water is trapped beneath a membrane, blistering may occur in hot weather. If vertical surfaces are damp or dirty, the membrane may not adhere and sagging may result. If joints in asphalt are not cut back and the fresh surface painted with bitumen before starting to lay the next panel, the joint will certainly leak. For satisfactory results then, meticulous attention to detail is essential both at the design and the construction stage.

Canals, Dams and Other Hydraulic Structures

In the United Kingdom engineers are blessed with seemingly unlimited supplies of high-quality durable rock and cheap Portland cement. Accordingly they have over many years tended to construct hydraulic works in either concrete or masonry. In the past, where bituminous materials have been used, they have often been used only as jointing compounds between precast concrete units or stone pitching.

Such jointing compounds must be capable of accommodating some movement of the blocks and supporting structure without cracking; they must adhere well and not flow from the joints (which are usually sloping) in summer or become brittle in winter. A typical asphalt joint filling designed to be poured into fairly open joints between concrete blocks is

bitumen 40 - 50 pen	35 to 40 per cent
fine sand	60 to 55 per cent
asbestos fibres 2 - 4 mm long	5 per cent

A primer coat of bitumen primer on the jointing faces is advisable before pouring the joint filler.

Canals and reservoirs

When canals are constructed in pervious soils a waterproofing lining of some kind is essential. The simplest one is a thin continuous membrane of binder. For this purpose the binder has to be durable, tough and capable of following any movements in the supporting soil without tearing. In the United States oxidised bitumens are widely used for this purpose, a commonly used grade having a penetration of 55, a softening point of 85 °C and a corresponding PI of 5.2. Generally these linings are only about 6 mm thick and are protected against damage by covering them with a layer of soil 400 mm or so thick.

Macadams or asphalts similar to those used in road works are also used for hydraulic construction. Obviously macadams are not impervious and their main use is to provide mechanical strength to linings of canals and reservoirs. They provide an erosion-resistant lining. In some cases this may be laid above an impermeable lining in order to protect it, but often permeability is desired to prevent the build-up of hydrostatic pressure or to allow drainage into the canal from the surrounding ground. A typical porous revetment of this type would contain

coarse aggregate (15 - 35 mm)	65 per cent
fine chippings (5 - 15 mm)	25 per cent
filler	5 per cent
bitumen	5 per cent

The grade of bitumen could lie between 70 pen and 200 pen, cutbacks often being used for this work.

When asphaltic concretes are used for hydraulic work they are generally intended to supply a strong erosion-resisting and waterproofing revetment. The mixtures are similar to those used for road work but require better workability (since they generally have to be compacted under more difficult conditions) and should finish with very low permeability. They are usually designed as very dense mixes with high filler and high binder contents, often 10 per cent filler and 7 - 8 per cent binder. The binder is slightly softer than for the equivalent road mix, perhaps 60 to 70 pen.

Rockfill dams

Typically, asphaltic concretes have been used as upstream carpets on rockfill dams. The popularity of rockfill dams has come about largely because of the number of works requiring tunnel construction, aqueducts, spillways and other ancillary works which are the source of much of the rock fill. In a sense therefore the rock is 'free' and the main expenses are those of transport, placing and compaction and some method of rendering the dam watertight. In the traditional method watertightness is achieved by means of a clay core, forming a vertical impervious wall in the centre of the dam. This has the disadvantage that the contractor has to construct what is essentially two dams divided by the vertical core wall; in addition, the contractor's working season is likely to be restricted by the weather with consequent delay in construction of the whole dam. When using clay cores the dam on either side has to be constructed at the same speed as the core.

If a dam is made impervious by means of an upstream carpet of asphaltic concrete then construction of the rockfill dam is not impeded in any way, and the upstream carpet of asphaltic concrete can be constructed after the rockfill embankment is completed. Since asphaltic concrete is reasonably flexible it will accommodate the small settlements liable to occur in a well-compacted rockfill dam. Generally, at least two layers of asphaltic concrete are laid on a well-prepared surface giving a total thickness up to about 250mm. In most cases the upstream carpet has to be laid on a slope that is fairly steep, up to 1 in 1.6 or so, and this presents problems. Rolling must be done up and down the slope and in most cases the rollers are winched up and down. The slope on the upstream face of such a dam is more likely to be fixed by the economics of laying and compacting the carpet than by the stability of the fill.

In many parts of the world there is a general shortage of good crushed stone aggregates, notably in parts of Australia, the United States and Holland. In many cases impermeable linings to canals, reservoirs, dykes, etc., are constructed using bituminous mixtures which consist basically only of sand, filler and a binder, usually bitumen. These mixtures called *sand asphalts* consist of sands with 5 to 10 per cent binder, and in some cases up to 5 per cent filler. They are usually constructed using 55 to 70 pen bitumen and as can be imagined often require to be protected by either asphaltic concrete, macadams or simply a thick layer of soil since the sand asphalt itself is lacking in mechanical strength.

Sand mastic, which contains much more binder than sand asphalt, usually about 18 per cent, is a useful material for placing under water since it contains so much bitumen that it requires no compaction. A typical mixture has about 70 per cent fine sand, 10 per cent filler and 20 per cent bitumen (60/70 pen or harder). In Holland, where it is often desired to prevent the scouring of a sandy sea bottom, a technique has been developed of laying an impermeable sand mastic carpet on the sea-bed from a ship specially designed for the purpose. The ship accommodates the hot-mix plant and, according to reports, can lay sand mastic in depths of water of up to 30 m.

The foregoing represents only a small sample of the many uses of bituminous materials. A good working knowledge of the properties of these materials is clearly most important if they are to be used successfully, with maximum benefit from the natural resources from which they are obtained.

References

Broome, D. C., and Please, A., 'The Use of Mechanical Tests in the Design of Bituminous Road Surfacing Mixtures', *J. appl. Chem., Lond.,* 8 (1958) pp. 121-35.

B.R.T.A., *Dense tar surfacing* (British Tar Industry Association, London 1968).

BS 76 : 1974 Tars for road purposes

BS 434 : 1973 Bitumen road emulsion (anionic and cationic)

BS 594 : 1973 Rolled asphalt (hot process) for roads and other paved areas

BS 598 : 1974 Sampling and examination of bituminous mixtures for roads and other paved areas

BS 747 : 1977 Specification of roofing felts
BS 892 : 1967 Glossary of highway engineering terms
BS 988, 1076, 1097, 1451 : 1973 Mastic asphalt for building (limestone aggregate)
BS 1162, 1410, 1418 : 1973 Mastic asphalt for building (natural rock asphalt aggregate)
BS 1446 : 1973 Mastic asphalt (natural rock asphalt aggregate) for roads and footways
BS 1447 : 1973 Mastic asphalt (limestone aggregate) for roads and footways
BS 3416 : 1975 Black bitumen coating solutions for cold application
BS 3690 : 1970 Bitumens for road purposes
BS 4147 : 1973 Hot applied bitumen based coatings for ferrous products
BS 4691 : 1974 Method for determination of penetration of bituminous materials
BS 4692 : 1971 Method for determination of softening point of bitumen (ring and ball)
BS 4693 : 1971 Method for determination of viscosity of cutback bitumen and road oil
BS 4987 : 1973 Coated macadam for roads and other paved areas
Huekelom, W., 'A Bitumen Test Data Chart for Showing the Effect of Temperature on the Mechanical Behaviour of Asphaltic Bitumens', *J. Inst. Petrol.*, 55 (1969) p. 404.
Institute of Petroleum, *Standard methods for testing petroleum and its products* (Institute of Petroleum, London, 1900).
Lee, A. R., and Dickinson, E. J., 'The Durability of Road Tar', *Road Research Technical Paper No. 31* (Department of Scientific and Industrial Research, London, 1954).
Pfeiffer, J. Ph., and van Doormaal, P. M., 'The Rheological Properties of Asphaltic Bitumens', *J. Inst. Petrol.*, 22 (1936) pp. 414 - 40.
S.T.P.T.C., *Standard Methods for Testing Tar and its Products* (Standardisation of Tar Products Tests Committee, Gomersal, Leeds, 1957)
van der Poel, C., 'A General System Describing the Visco-elastic Properties of Bitumens and its Relation to Routine Test Data', *J. appl. Chem., Lond.*, 4 (1954) pp. 221 - 36.

Further Reading

A. J. Hoiberg (ed.), *Bituminous Materials* (Interscience, Chichester, 1966).
Ministry of Transport, *Specification for Road and Bridge Works* (H.M.S.O. 1969).
Road Research Laboratory, *Bituminous Materials in Road Construction* (H.M.S.O., 1962).

V SOILS

Introduction

Every work of construction in civil engineering is built on soil or rock and in many instances these are also the raw materials of construction. The study of soil and rock materials is an important part of a wider area of study often called *geotechnology*. This is an area common to civil engineering and engineering geology and forms a quantitative branch of engineering geology dealing with the mechanics of behaviour of soils and rocks.

The geotechnical engineer has to find answers to difficult questions, for example

Will a given soil provide permanent support for a proposed building?

Will a given soil compress (or swell) on application of the load from a proposed building, and by what amounts, and at what rates?

What will be the margin of safety against failure or excessive or unequal compression of the soil?

Are natural or constructed soil slopes stable or likely to slide?

What force is exerted on a wall when a given soil is packed against it?

At what rate and to what pattern will water flow through a soil?

Can a given soil be improved in any way by treatment or admixture?

All these require an understanding of soils as materials and of the mechanics of materials. Absolute answers are unobtainable and the engineer has to practise the art as well as the science of geotechnology. Construction work proceeds in the state of knowledge existing at the time the questions are asked. This knowledge, both of precedents and discoveries, is continually growing through research and not least in the area of *soil as a material.* Not all researches lead to an immediate or comprehensive solution to engineering problems but a better understanding of the behaviour of soil as an engineering material cannot but help in the amelioration of these problems. The analysis of behaviour of masses of soil when built upon, or cut to new shapes or used as materials of construction falls properly into texts in soil mechanics, for example, Wilun and Starzewski (1975), Lambe and Whitman (1969), Smith (1968). In this part of the book the emphasis is on the nature of soil itself as a material and within the limitations of space available the author has selected topics of importance to the student engineer; any shortcomings in this selection are those of the author.

There is no clear dividing line between the areas of study of soil as a material and soil mechanics, and both are equally important. There is obviously no justification in using the more sophisticated and rigorous methods of analysis developed in soil mechanics if the measurement of the material characteristics of the soil needed in the analysis is not made in a correspondingly refined and perceptive way.

Soils are naturally occurring, largely inert, materials of a particulate kind derived, as products of weathering, from rocks. The term 'soil' as used in this book therefore excludes the harder, massive or cemented and feebly weathered materials of the planet's surface, called *rocks* by the engineer. Again there is no clear division between soils and rocks, as briefly defined – many aspects of behaviour being common.

19

Formation, Exploration and Sampling of Soils

A detailed study of the origin, formation and distribution of soils would be rewarding but is beyond the scope of this book. The student of civil engineering has the complementary subject of geology in his curriculum and in studying this and coupling it with his own observations and experience he will extend his appreciation of the origins of soils and of natural aggregates for concrete and road construction.

The physical characteristics of soil are important factors in most civil engineering constructions and it is the assessment of these physical characteristics that is the principal concern here. Chemical and biological characteristics of soils are of importance in agriculture, horticulture and land management but they generally have less relevance to the subsoils of concern in civil engineering. In civil engineering usage, the term 'soil' describes the uncemented or weakly cemented material overlying the harder rock on the planet's surface. Soils exist in great variety, and are the accumulated result of many separate factors and processes. Their characteristics depend on the parent rocks from which they are derived; on the weathering of these rocks and the weathering of the soil itself at its various stages of formation; on the means of transport bringing the soil to its present location; on the manner of deposition of the soil; on its history of loading, drainage, wetting and drying and on many other processes.

The parent rocks themselves occur in great variety. The rock may have formed by the cooling and hardening of a molten material, this group being termed *igneous rocks.* The igneous rocks include granites, basalts, and gabbros. On the other hand the rock may have formed from the accumulation of transported and deposited weathered rock fragments, subsequently consolidated to give relatively hard rocks; water has usually been the transporting agent. These are the *sedimentary rocks* – the sandstones, limestones, shales and conglomerates. Some of these igneous and sedimentary rocks have been altered over the ages by heat or pressure or both. The result is the *metamorphic rocks* – very different from the original materials. Metamorphic rocks include schist, gneiss, slate and marble. The processes of metamorphism and weathering over aeons of time have transformed some of the rock material into the infinite variety of soils seen today. The continuing process of pedogenesis, as it is called, is one in which the hard rocks are slowly weathered and broken down and their constituent minerals chemically altered.

If the soil debris of today is derived from weathering of the rocks in place, it is described as a *residual soil*. Soils are commonly formed, however, of transported material – the weathered rock fragments having arrived in their present location after periods of transport by glacier, river or wind. The material may be much altered during transport and after deposition it experiences further weathering. Once settled in place the surface of the soil deposit will in time accumulate a growth of organic material, humus. The great variety of the processes of pedogenesis means that the physical and engineering characteristics of soils are likely to differ within short distances. Mapping of these characteristics over even a limited area of ground and in depth below its surface therefore requires a detailed programme of soil sampling and testing.

19.1 Residual Soils

When rocks are weathered and the products accumulate in place, without transport occurring, the products are described as residual soils. In a sense the humus-bearing top soils and weathered subsoils of any formation are residual soils but the term is usually used for rock products such as laterites. Significant thicknesses of these soils are found in various parts of the world and from their distribution it appears that very warm humid regions are most favourable to the kind of weathering that produces residual soils.

19.2 Transported Soils

Glacially Transported Soils

The superficial geology of much of Scandinavia, Britain and north-west Europe is derived from glacial action. Although the main elements of geographical relief were formed earlier, the land surface was modified during the Ice Age by complex erosion processes followed by grinding of the englacial material and then by the deposition of glacial drift to form many of the scenic features of today. This glacial activity ended some 10 000 years ago but was a major means of transport of soils that are encountered now in these areas by the civil engineer. The rock debris carried in or on the glaciers of the times was ultimately deposited in one of a great variety of forms. A study of these forms provides valuable information to the materials engineer in his search for quantities of suitable natural aggregates and fill materials for construction work.

The nonstratified drift is a direct glacial deposit containing a wide range of sizes of soil particles and is commonly termed *glacial till*. The appearance of a till is shown in figure 19.1. The manner of deposition has produced quite heavily consolidated soils so that in their natural state today these tills are rather

Figure 19.1 *Stiff glacial till as exposed in an excavation*

stiff dense soils and are usually fissured. The shape and orientation of the stone particles in them depends on such factors as the nature of the parent rock, and the distance, manner and direction of transport by glacial actions.

Some of the glacial debris was transported with melt water before final deposition and as a result of this is likely to be sorted into size groups of sands and gravels stratified after the manner of river deposits. The terrain diagram of figure 19.2 illustrates the complex and variable nature of soil formations.

Water-transported Soils

Soils formed of waterborne material tend to vary erratically in physical properties from place to place and depth to depth, a consequence of the continual change occurring in the streams and rivers which formed them. Fast-flowing steep streams in mountain areas transport all but the largest rock fragments downstream and the fragments become progressively broken up, worn and deposited. As the steepness of the streams diminishes the larger particles being transported are left behind. On reaching the plains and estuaries only the finer sizes remain in transport. Seasonal extremes of flood and drought, meandering and braiding of stream channels all lead to wide variations in the erosion and deposition patterns of waterborne deposits and hence in the engineering and other characteristics of the soils forming these deposits.

The mechanical process of abrasion and grinding accounts for most of the reduction in particle sizes from the fragments of the parent rocks. Such action may reduce particle sizes to 0.001 mm (1 μm) and in the case of rock flours derived from glacial grinding of rocks to yet smaller sizes. In appearance these soil particles are mainly blocky and angular although perhaps more or less rounded by abrasion during transport. A specimen of sand is shown in figure 19.3. Larger and smaller particle sizes from the gravels through to the silts may have similar appearance and these soils are described as *granular*, *cohesionless* and *coarse grained*.

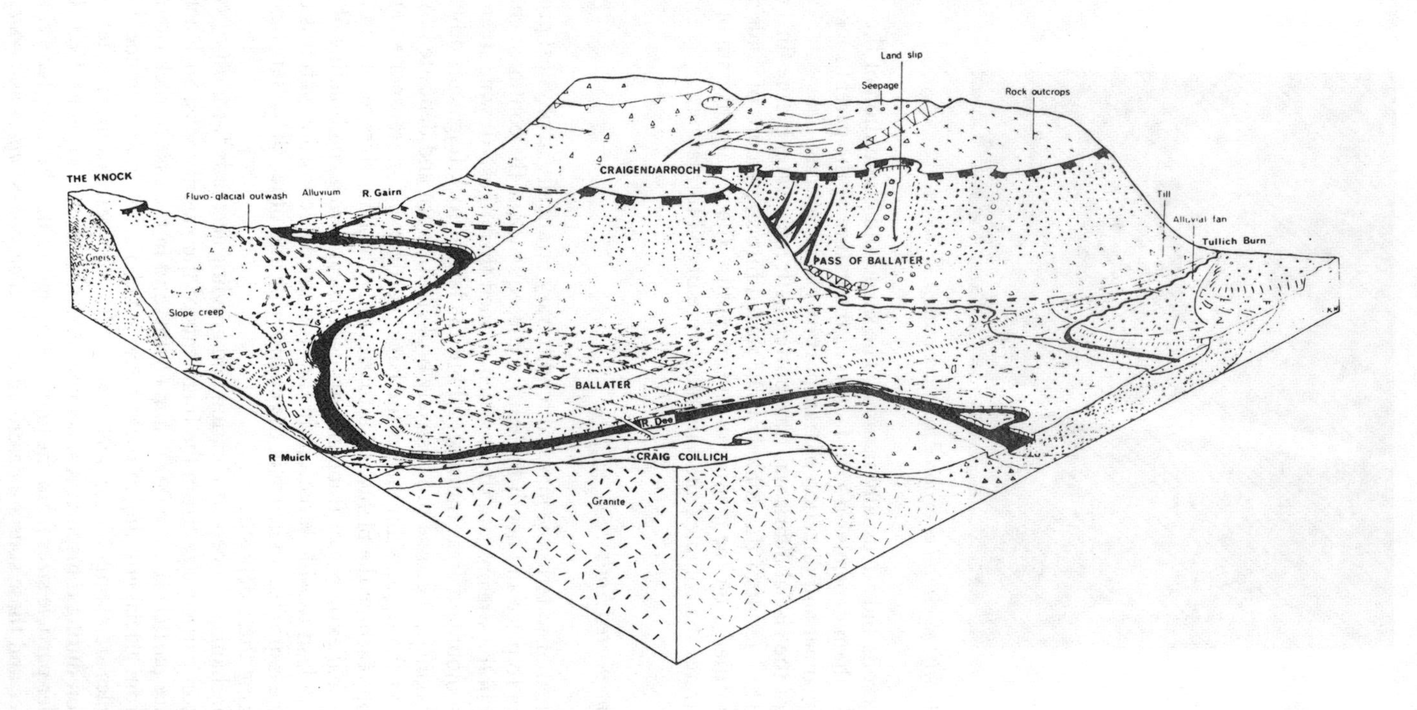

Figure 19.2 *Terrain analysis in perspective – Ballater, Scotland (courtesy of Professor K. Walton, University of Aberdeen)*

Figure 19.3 *Enlarged photograph of a size fraction of a river sand*

Most soil material smaller than about 1 μm is the result of chemical processes of solution and crystallisation and these minute particles form the very significant clay fraction of soils. These clays comprise the *cohesive, fine-grained* fraction. The particles, in addition to being minute in size, are also unlike the coarse-grained soils in appearance, being predominantly plate or rod shaped. The development of electron microscopy with its high magnification has enabled clay particles and to some extent the structural grouping of clay particles to be photographed. In figure 19.4 the appearance of halloysite clay particles is seen and, in figures 19.5*a* and *b* particle arrangements in a mainly illitic silty clay. The smallest soil particles transported in rivers are carried in suspension and on reaching lakes or ponded areas settle there to form soft, mainly clay and often organic deposits termed lacustrine clays. This process is a continuing one with many present-day examples in the course of formation.

The minute particles of clay size are less affected by gravity forces during sedimentation than by the interparticle forces of attraction and repulsion and the latter forces are in turn much influenced by the electrolyte content of the water in which the clays are settling during the formation process. Even so-called fresh water in rivers and lakes contains concentrations of ions and there is a grading of possible depositional environments from fresh through brackish water to marine conditions. These differing environments lead to a complex variety of *micro-fabric forms* in clays and these are a subject of continuing study. In the distant past the British Isles were covered by a succession of seas and one of the marine deposits which has received particular study from engineers is London clay. This is an extensive deposit of the Eocene age some 30 million years ago and today exists as a rather uniform stiff-fissured soil in thicknesses of up to 150 m or so.

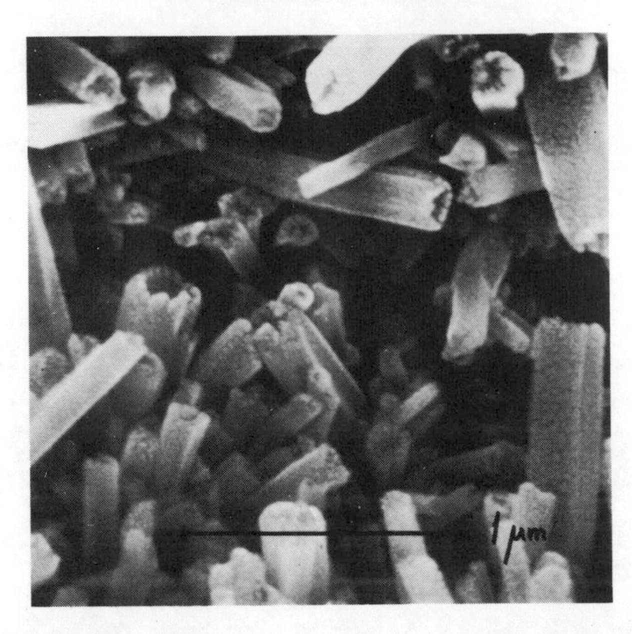

Figure 19.4 *Electron microscope photograph of halloysite from Bedford, Indiana (N.K. Tovey)*

Erosion in late Tertiary and Pleistocene times removed some 200 – 300 m of overlying material, including much of the original clay deposit.

Fine sediments laid down in the late glacial period have certain special features owing to the cyclic nature of their deposition. The lakes were then confined by land formations or by ice damming or both and the sediments reaching the lakes were seasonally released with the melt water from the glaciers. The coarser particle sizes in the sediment settled rapidly to the floor of the lake but the clays settled slowly over the seasonal period, the process being repeated in succeeding seasons. The composite soil as it is encountered today is termed a varved sediment or varved clay or, where the layers are less distinct, a laminated clay. A succession of nearly identical layers or varves is typical of such sediments and, as indicated in figure 19.6, the laminations in a sample are more apparent to the observer if the soil is allowed to dry a little. The engineering properties of such sediments obviously vary with the directions in which the properties are measured. The larger features of clays – the *macro-fabric* – usually have a dominant influence on engineering behaviour. These features include layering and varves, vertical root channels and tension cracks, fissuring and jointing.

In certain parts of the world, notably Scandinavia and South-East Asia, sediments were transported in post-glacial times to the sea and deposited there in salt water. The clay particles tended to flocculate, that is, to form rather uniform groups of platelets in a very loose structure. These marine clays are therefore soft and highly compressible, although some may have been subjected to desiccation, giving a crust of stiffer soil.

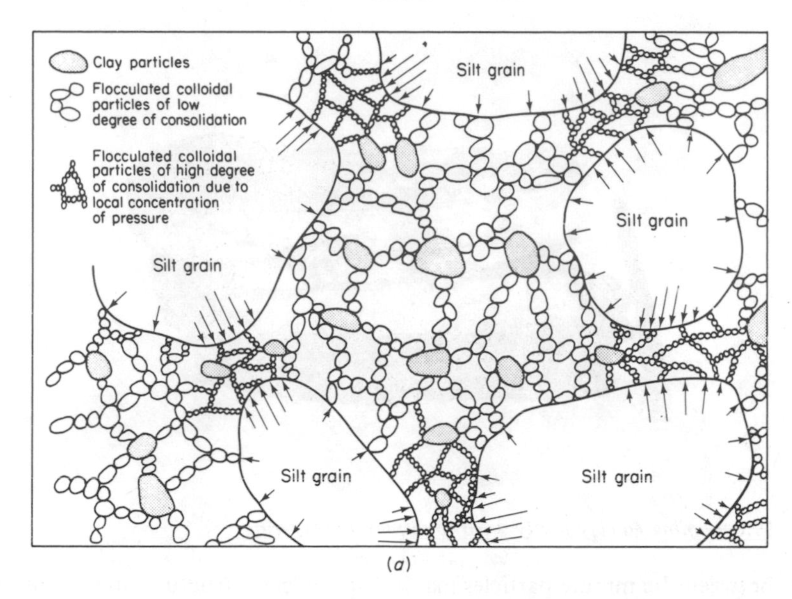

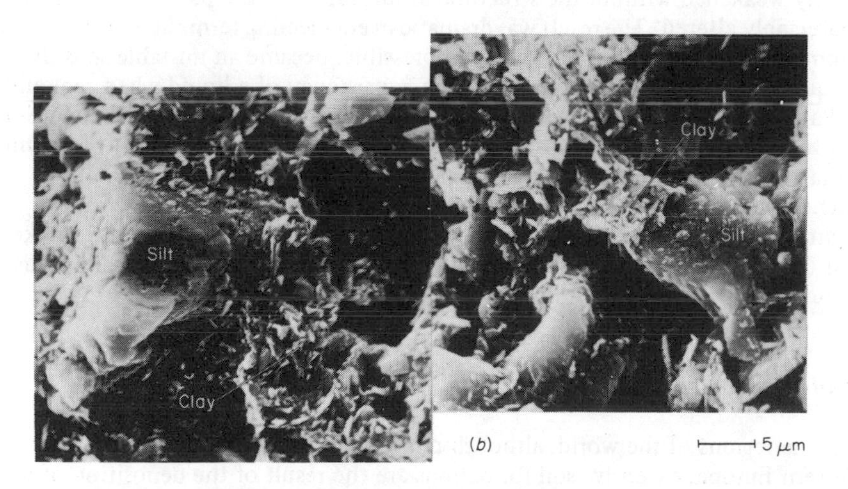

Figure 19.5 *(a) Hypothetical structure of undisturbed marine clay suggested by A. Casagrande in 1932. (b) Electron microscope photograph of a marine clay from the Drammen valley, Norway. This clay is a very sensitive, illitic, marine-deposited, leached silty clay*

Part of the subsequent geological change was the gradual raising of the land surface relative to the sea, a change which is still continuing. This upheaval raised many of the marine-deposited sediments above the present sea level. Artesian or other conditions of groundwater flow then resulted in a slow percolation of fresh water through the clays, gradually exchanging the salt water in the pore spaces in the clays with fresh, the process being called leaching. This leaching process altered the electrolyte content of the porewater and in various ways altered the

Figure 19.6 *Laminated clay partly dried to display the layering*

bonds between the minute particles making up the clay structure; the bonds were greatly weakened without the structural arrangement of the particles being appreciably altered. The result was dramatic in engineering terms in that the leached sediment, originally rather soft and compressible, became an unstable sensitive or 'quick' clay. Such clays today have the consistency of a liquid when remoulded and any disturbance of the natural state leads towards that remoulded condition and to a rapid and usually substantial loss of strength. A second result of leaching is that the already compressible sediment becomes more compressible. These leached marine clays are thus exceptionally difficult materials to deal with in engineering construction works. Some Canadian clays have a history not unlike that described above. The water in which the clays were initially deposited was a more variable electrolyte, fresh, brackish or marine.

Wind-transported Soils

In some regions of the world, although not notably in Scandinavia, Britain or Western Europe, extensive soil formations are the result of the deposition of windborne particles. The principal aeolian or windborne deposit is termed *loess.* This is a rather fine-grained and remarkably uniform soil with a rather loose density. The particles are often cemented and when dry the soils are quite strong. Loess often shows a vertical jointing, attributed to a contraction after deposition, and sloping faces often weather to near vertical faces. The cementing material dissolves on wetting so that the soil is easily eroded by streams and tends to subside or settle quite suddenly when wetted.

Soils Transported by Man

In the consideration of soils in relation to engineering works man himself should be included as an agent of transportation. Soils which have been excavated from one place and transported to and placed in another place are described as *fills*. Fills may be materials other than soils as we have described them – they may comprise randomly dumped rock spoil from a tunnel or other construction works, or colliery waste products or even municipal refuse. Fills are therefore likely to be variable materials in quality and density and if encountered on the site of a proposed construction they should be presumed to be nonuniform and compressible. On the other hand, soil fills may be placed and densified under controlled conditions giving a uniform firm material of predictable and dependable behaviour. The control and compaction of fills is described in chapter 26.

19.3 Exploration and Sampling

To consider soils among the other civil engineering materials it is necessary to distinguish the principal soil groups one from another and, as with the other materials, to establish parameters by which the behaviour of soils can be described quantitatively. The complex formation processes just outlined have, however, produced complex and variable materials. Such engineering characteristics as *permeability, compressibility* and *strength* usually differ in value from point to point in a soil mass and at any one point these characteristics differ with the direction in which they are measured. Soils showing this behaviour are described as *anisotropic.* If the characteristics were practically unvarying in value throughout a soil, that soil would be described as *isotropic.* It is questionable whether any soil is isotropic, although the design engineer will commonly assume isotropic behaviour in soils when he is analysing their projected behaviour in engineering works and will use a simplified sequence of soil strata sufficiently representative of the actual soils to allow design calculations to be made with confidence. This rather sweeping assumption demands an exercise of judgement by the engineer based on knowledge of the actual properties of the soil, measured using the best available methods, together with a good appreciation of the behaviour of soil as a material and an awareness of the susceptibility of his analysis to approximations of soil properties, say, for example, their directional variation. Many case histories exist to support the reliability of this simplified approach when natural strata or fills are sensibly uniform.

It was observed earlier that many soils show a fabric of laminations, varves, fissures, organic matter or root networks. A quantitative description of the engineering characteristics of a particular soil stratum taking account of these features of fabric should apply convincingly to the whole stratum as it exists in nature. The sampling and testing of soils to provide this description is part of an important activity of the civil or geotechnical engineer, namely, site investigation or more particularly *soil exploration.*

Given that all existing information about a site has been assembled, the principal area of new investigation is undertaken by vertical boring into the ground at selected positions to take samples.

Soil Profile

The results of a soil exploration are often presented in the form of a *soil profile*. A zone which is distinguishable from its neighbours above and below is called a soil horizon and the sequence of horizons from the ground surface to the limit of depth of the exploration is called the soil profile. The variation, with depth, of any property of the soil – for example, water content or density – may be displayed by a curve as shown in figure 19.7.

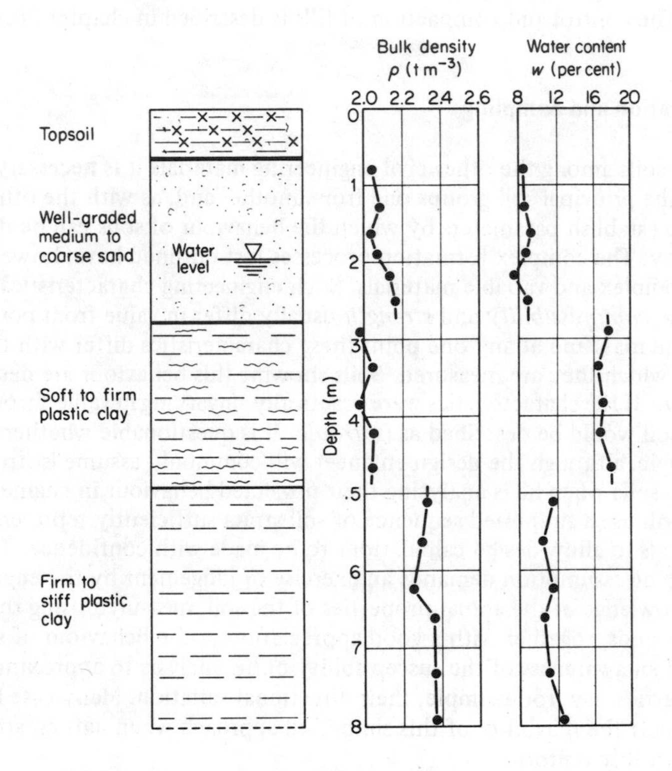

Figure 19.7 *Vertical profile of soil strata showing variations of properties with depth, the properties in this illustration being density and water content*

20

Bulk Properties

As part of a basic description of soil, it is necessary to establish parameters defining some physical properties of the mixtures of solid particles, water and sometimes air or gas which make up a soil. Suppose a sample of natural soil is taken without altering it in any way so that it is entirely typical of the soil stratum from which it came. Let its total volume be denoted by V and its total mass by M. The constituent solid particles, water and air will each have *absolute* volumes and masses which when added together will give V and M respectively. It is convenient to present these constituents diagrammatically in terms of their *absolute* volumes, stacking the absolute volumes one on the other, as in figure 20.1. Thus

$$V = V_s + V_w + V_{air}$$

where V_s is the absolute volume of all solid particles, V_w the volume of porewater and V_{air} the volume of pore air, or gas, in the sample of total volume V. Also

$$V = V_s + V_v$$

where $V_v = V_w + V_{air}$ is the absolute volume of voids or pore spaces in the sample.

The same diagrammatic form is suited to presenting the masses of the constituents of the sample. Thus

$$M = M_s + M_w$$

where M_s is the mass of dry solid particles and M_w the mass of water in the pore spaces in the sample.

The units of volume are usually m^3, l or ml and those of mass Mg, tonne (t), kg or g. The density of the solid particles

$$\rho_s = \frac{M_s}{V_s}$$

and the density of fresh water

$$\rho_w = \frac{M_w}{V_w} = 1000 \text{ kg m}^{-3} = 1 \text{ t m}^{-3} = 1 \text{ g ml}^{-1} \text{ at } 20^\circ \text{ C}$$

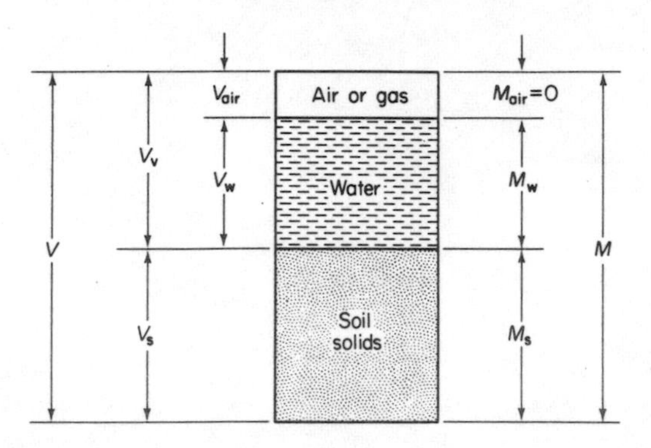

Figure 20.1 *Soil solids, water and gas phases in soil*

20.1 Specific Gravity

The symbol G is used for the specific gravity of the solid particles of soil. Specific gravity is defined as the ratio of the mass of the soil particles to the mass of the same (absolute) volume of water. Thus using the notation of figure 20.1

$$G = \frac{\rho_s}{\rho_w} = \frac{M_s}{V_s\rho_w}$$

and is dimensionless.

The value of G is usually determined as an average for a suitably representative sample of soil, although strictly it may vary among the mineral constituents of the soil and from size fraction to size fraction. Clean sands generally have a specific gravity close to 2.65 and clays a somewhat higher value around 2.72 or more. For many purposes it is sufficient simply to use these values. Low values in natural soils would suggest the presence of organic matter.

Tests for the determination of specific gravity of soil particles are detailed in BS 1377. The average of two determinations is taken and reported to the nearest 0.01. The main difficulty in the test is that of removing air from the soil - water mixture during the procedure for measuring the volume of the soil. If air is not completely removed, a low value of G is obtained. An investigation of the reproducibility of specific gravity results (Sherwood, 1970) has shown that unless the detailed procedures of BS 1377 are closely followed a large proportion of specific gravity results will be outside the desired tolerance of ± 0.05 on the true value required for the proper compaction control of earthworks.

20.2 Water Content

The symbol w is used for the water content of soils and the term is defined as

$$w = \frac{M_w}{M_s} \quad \text{(usually expressed as a percentage)}$$

where w = water content
Mw = Mass of water
Ms = Mass of dry solid particles

The value of w is determined by measuring the loss of water on drying suitably representative samples of soil, again following procedures such as those in BS 1377. Particularly in fine-grained soils, the drying procedure may influence the results obtained. To avoid overheating a drying temperature of 105 - 110 °C is specified. The period of drying varies with the soil type but is usually about 24 hours.

Water content values up to 10 per cent are reported to two significant figures and above 10 per cent to the nearest whole number. Values of natural water content in soils may range from under 5 per cent in gravels and sands to 50 per cent or more in the fine-grained cohesive soils. The presence of organic matter may increase w considerably.

20.3 Void Ratio and Porosity

These are terms describing the proportion of the total volume of a soil which is not occupied by solid constituents. The first term, voids ratio, is the more commonly used. Void ratio e and porosity n are given by

$$e = \frac{V_v}{V_s}$$

and

$$n = \frac{V_v}{V}$$

Hence

$$e = \frac{n}{1 - n}$$

and

$$n = \frac{e}{1 + e}$$

Voids ratio and porosity are derived quantities dependent on a determination of G and bulk density (defined in section 20.6). The values of e and n (usually less than unity) are given to two decimal places.

20.4 Degree of Saturation

This is a measure of the proportion of the available voids volume in the soil which is filled with water. The symbol S_r is used for the degree of saturation, where

$$S_r = \frac{V_w}{V_v} \quad \text{(usually expressed as a percentage)}$$

The determination of S_r is dependent on determinations of G, w and the bulk density of the soil (defined in section 20.6).

Natural soils below the water table (defined in section 23.2) will usually be saturated, that is, $S_r = 100$ per cent. The finer-grained soils may also be saturated for some distance above the water table owing to capillarity and adsorption actions. Fill materials can be compacted to a high degree of saturation but it is not possible by mechanical compaction – rolling, kneading, impact and vibration – to achieve 100 per cent saturation.

20.5 Air Voids Content

The compaction of fill materials is sometimes described in terms of air content. Air voids content is given by

$$\frac{V_v - V_w}{V} \times 100 \text{ per cent}$$

The value will be small in the case of well-compacted fills placed at suitable water contents – usually 5 - 10 per cent.

20.6 Density

The mass of unit total volume of a particulate solid (soil) is described as its density, or bulk density, and is denoted here by the symbol ρ.

The density of a given soil is expressed in a number of different ways; thus, referring again to figure 20.1, natural density ρ, dry density ρ_d, saturated density ρ_{sat} and submerged or buoyant density ρ' are given by

$$\rho = \frac{M}{V} = \frac{G + S_r e}{1 + e} \rho_w$$

$$\rho_d = \frac{M_s}{V} = \frac{G}{1 + e} \rho_w = \frac{\rho}{1 + w}$$

$$\rho_{sat} = \frac{G + e}{1 + e} \rho_w$$

$$\rho' = \rho_{sat} - \rho_w = \frac{G - 1}{1 + e} \rho_w$$

The unit in each case has the form Mg m^{-3}, t m^{-3} or kg m^{-3}. Densities are expressed to the nearest 0.01 t m^{-3}.

In the above equations it is assumed that the porewater is fresh water with a specific gravity $G_1 = 1$. Strictly, if the pore liquid is salt water or some other liquid for which $G \neq 1$, the equations become

$$\rho = \frac{G + G_1 S_r e}{1 + e} \rho_w$$

$$\rho_{sat} = \frac{G + G_1 e}{1 + e} \rho_w$$

$$\rho' = \frac{G - G_1}{1 + e} \rho_w$$

Unit Weight

The weight of unit total volume of a soil is described as its unit weight and is denoted by the symbol γ. It is obtained by multiplying the density ρ by the acceleration due to gravity g.

Thus a soil of density 1.75 t m^{-3} will have a unit weight

$$\gamma = \rho g = 1.75 \times 9.81 = 17.15 \text{ kN m}^{-3}$$

and for fresh water

$$\gamma_w = \rho_w g = 9.81 \text{ kN m}^{-3}$$

Unit weights are required when calculating the forces or pressures exerted by soils.

Relative Density

The natural or compacted density of a given soil will vary from place to place. The drainage characteristics, compressibility and strength and other aspects of engineering behaviour of a soil are related to its density. The variation in density may be examined by determining the variation in natural or dry density values. In cohesionless soils, however, it has been found convenient and useful in practice to refer the actual density of a soil to the range of densities over which the soil can exist or can be placed.

The actual density is then described using the concept of the relative density D_r where

$$D_r = \frac{e_{max} - e}{e_{max} - e_{min}}$$

in which e is the actual void ratio of the cohesionless soil and e_{max} and e_{min} are respectively the greatest and least values of void ratio obtainable with the soil. The term density index I_d is also used for this ratio.

Measures of e_{max} and e_{min} are made by somewhat arbitrary methods which are not strictly standardised. Procedures such as those of Kolbuszewski (1948) are adopted. To obtain e_{max} he suggests taking 1 kg of dry sand, representative of the material, and placing it in a 2000 ml glass graduate which is then stoppered. The graduate is shaken, inverted and inverted again, after which the volume occupied by the sand is read, and e_{max} deduced. The value of e_{min} relevant to normal conditions can be obtained by vibrating a comparable mass of sand until its minimum volume is reached. From this measured volume e_{min} is deduced.

20.7 Derivation of Bulk Properties

These parameters describing the soil bulk are much used and their names are part of the language of geotechnology. They should be studied until the student engineer is wholly familiar with them.

For a sample of a natural or compacted soil the determination of parameters ρ, w and G enables the others to be derived. The following convenient relations between the parameters may also be useful. If $V = 1$ then

$$V_s = \frac{1}{1+e}$$

and

$$V_v = \frac{e}{1+e}$$

allowing easy derivation of the above densities. Also

$$S_r e = wG$$

For a saturated sand therefore $e = 2.65\,w$ approximately.

Some typical values of bulk properties of soils are given in table 20.1 to indicate the magnitudes of the various parameters.

TABLE 20.1
Bulk properties of some natural soils

Description	Porosity n (per cent)	Void ratio e	Water content at $S_r = 1$ w (per cent)	Density ($t\ m^{-3}$) ρ_d	ρ_{sat}
Uniform sand, loose	46	0.85	32	1.44	1.89
Uniform sand, dense	34	0.51	19	1.75	2.08
Mixed grained sand, loose	40	0.67	25	1.59	1.98
Mixed grained sand, dense	30	0.43	16	1.86	2.16
Glacial till very mixed grained	20	0.25	9	2.11	2.32
Soft glacial clay	55	1.2	45		1.76
Stiff glacial clay	37	0.6	22		2.06
Soft slightly organic clay	66	1.9	70		1.57
Soft very organic clay	75	3.0	110		1.43
Soft bentonite	84	5.2	194		1.28

Consistency of fine-grained soils (per cent)

Soil	w	w_l	w_p	I_p	I_l
Soft clay, Chicago	26	32	18	14	57
Stiff clay, London	24	80	28	52	−8
Quick silty clay, Oslo	40	28	20	8	250

21

Coarser- and Finer-grained Soils

21.1 Particle Size Distribution

As already seen in part III (concrete), coarser-grained soils and aggregates are well described by their *particle size distribution* curves. The most common form of the plot is the cumulative diagram displaying the percentage by weight of the material finer than any given size, the latter being presented on a logarithmic scale. Individual soil particles are not of course spherical but for most coarser-grained soils, except those containing quantities of micaceous material, the particles are nearly equidimensional and when the particles are matched against sieves of known aperture sizes a satisfactory estimate of the amounts of material finer than each size is easily obtained. The preparation of samples for seiving (mechanical analysis) and the seive test procedures are described in BS 1377. The lower limit of size analysed by seiving is usually fine sand at 63 μm.

The particle size distribution of the silty fractions of soils are of interest too and these sizes lie below 63 μm. The size distribution over the silt range is indirectly assessed from determinations of the velocities of sedimentation of the particles in water, assuming these to be perfect spheres having the same specific gravity as the soil particles. The silts reach their terminal velocity almost immediately and thereafter fall in water at a constant velocity which, for the silt sizes, is proportional to the square of the particle diameter (Stokes' law). The sedimentation test procedures are also described in BS 1377.

In a soil both procedures may be required to define the particle size distribution curve. An example of such a curve is given in figure 21.1 which also shows the particle size boundaries used in BS 1377 to distinguish silts, sands, gravels and cobbles. Soils with similar size distribution curves would in a general sense be expected to show similar engineering characteristics. Predictions of behaviour are made cautiously because the size distribution alone does not convey the arrangement and density of packing of the particles and other factors which influence the behaviour of the soil.

Although the size distribution plot is in the form of a curve whose shape depends on the chosen method of presentation as well as on the soil itself the plot can be used to describe and classify coarse-grained soils. It has long been known

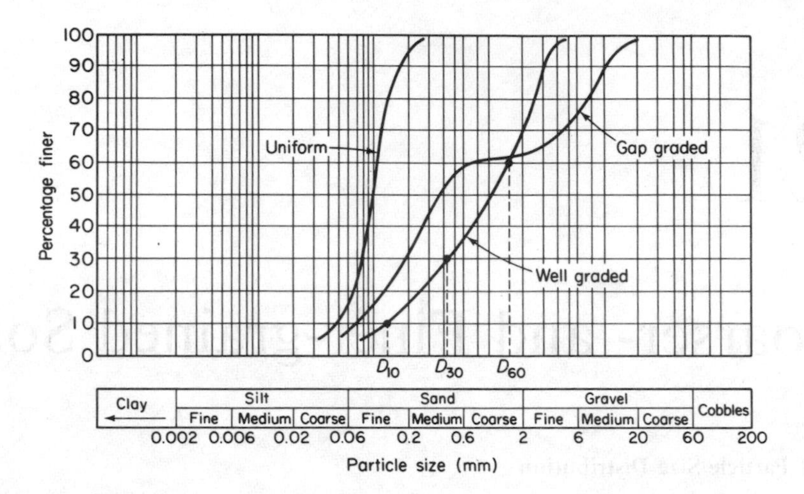

Figure 21.1 *Cumulative particle size distribution curves on the conventional semi-logarithmic plot*

by engineers that the fine grains in a soil exercise the largest influence on its behaviour. Accordingly the term *effective size* or *ten per cent size* (D_{10}) is much used in describing soils. The effective size of a soil is the particle diameter at which 10 per cent by weight of the soil is finer in size. This D_{10} value is conveniently read from the particle size distribution curve as shown in the example in figure 21.1.

The steepness of the curve is given a numerical value in the *uniformity coefficient* C_u where $C_u = D_{60}/D_{10}$ and D_{60} is the particle diameter at which 60 per cent by weight of the soil is finer. If the range of sizes present in a soil is small it is described as a *uniform soil* or *uniformly graded*. Such *poorly graded* soils will have C_u values of 2 or less. A *well-graded* soil is one which can give a dense, rather strong soil. It will have a wide range of particle sizes and the size distribution curve will be smooth and concave upwards, with no deficiencies or excesses of sizes in any size range. A well-graded soil has a C_u value of 5 or more.

The shape of the central portion of the curve is assessed by the *coefficient of curvature* C_c where

$$C_c = \frac{D_{30}^2}{D_{10}D_{60}}$$

For the grain sizes and amounts to be so arranged that a dense packing is possible the distribution curve should be concave upwards giving a C_c value between 1 and 3. Some poorly graded soils have a deficiency of intermediate sizes and are described as *gap graded*. This shows on the size distribution curve as a 'step' and an example is given in figure 21.1.

21.2 Particle Shape and Surface Texture

The coarser-grained soils are commonly equidimensional in particle shape, that is, in any particle the length, width and height are nearly equal. In cases where a more specific description is warranted, reference may be made to BS 812 for definitions and illustrations of form, angularity and texture of particles. Use of a limited range of descriptive terms – flaky, elongated; rounded, angular; smooth, rough and others – is sufficient for this purpose, see table 12.7.

21.3 Atterberg Limits

When a soil contains an appreciable quantity, say 20 per cent or more by weight, of material finer than the 63 μm size a description based on size distribution alone is insufficient. The size distribution of clays and silty clays is of interest but in practical engineering terms too difficult to determine and there are other characteristics which are more simply obtained and of more immediate relevance. The behaviour of clay is related to its mineralogical composition, water content and the micro- and macro-fabric of its particles as outlined in the discussion on sampling and elsewhere. A full study of a clay would require all these and other factors to be accounted for.

The water content is an obviously important and easily measured property of a clay. If the water content of a natural firm clay is allowed to increase, the clay will soften and become more compressible. In a different type of clay the effect will be similar but different in degree. The change in the state of a clay as the water content is changed has been found to be a useful and simple way of distinguishing one clay from another and this is the basis of soil classification applied to clays.

The *Atterberg limits* are empirical and somewhat arbitrary divisions, in terms of water content, between states of a clay. As the water content of a clay element is increased from the dry solid condition it will pass through a semi-solid, friable or crumbly state into a plastic state. In the plastic state the clay can be kneaded without rupture or cracking of the soil. On further wetting the clay has the consistency of a liquid, becoming less viscous as the water content is increased. These states are listed in figure 21.2. The water contents at the boundaries of these states are called the Atterberg limits – they are the *shrinkage limit* w_s, the *plastic limit* w_p and the *liquid limit* w_l. Tests for the liquid limit, the plastic and a

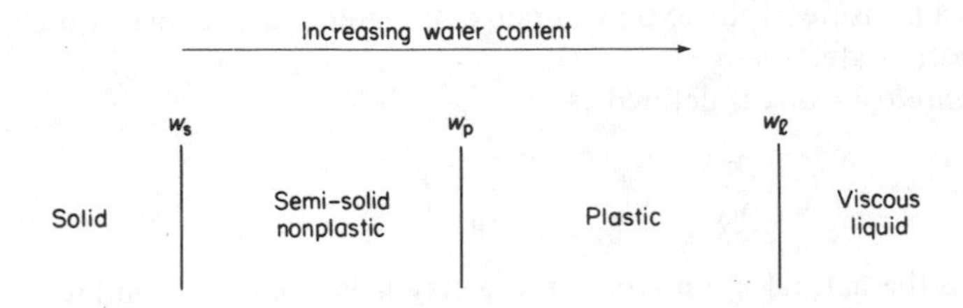

Figure 21.2 *State of clay soil over a range of water contents showing the relative placing of the index limits* w_s, w_p and w_l

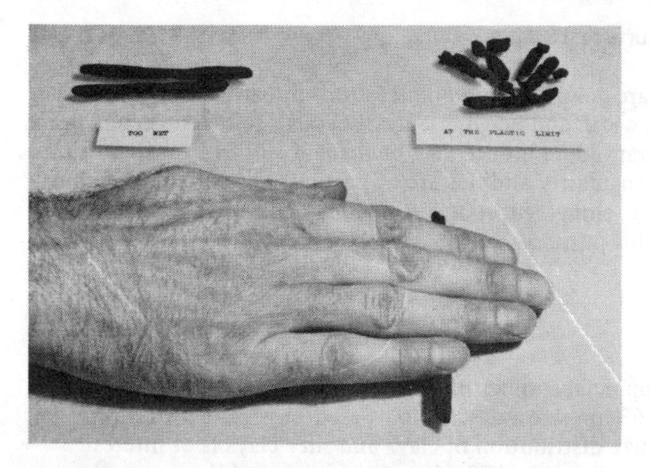

Figure 21.3 *The determination of the plastic limit from BS 1377 : 1975 reproduced by permission of the British Standards Institution, 2 Park Street, London W1A 2BS*

shrinkage limit are made on that fraction of soil finer than 0.42 mm and the procedures are described in BS 1377. These tests are empirical in nature and since the results are to be used in the comparison of soils the tests must follow the standardised procedures.

Essentially the *liquid limit* is determined by relating the water content of a clay to the penetration of a standard cone during a period of 5 ± 1 s when the wet clay is contained in a standard metal cup. The liquid limit w_l is the value of the water content w, expressed to the nearest whole number, corresponding to a cone penetration of 20 mm. The *plastic limit* is determined by rolling threads of the clay on a flat surface. When these threads begin to crumble at just over 3 mm diameter (figure 21.3) the water content is measured as the plastic limit w_p. The *shrinkage limit* is the water content at which an initially dry specimen of the clay is just saturated, without change in total volume. The liquid and plastic limits are the principal limits for classification purposes. The plastic limit is also found useful in deciding on the suitability of soils for use as fill material.

The *plasticity index* is defined as P.I. = L.L − P.L

$$I_p = w_l - w_p$$

PLASTICITY INDEX = LIQUID LIMIT − PLASTIC LIMIT

and gives a measure of the extent or range of water contents over which a soil is in the plastic state.

The *liquidity index* is defined as

$$I_l = \frac{w - w_p}{w_l - w_p} = \frac{w - w_p}{I_p}$$

where w is the natural water content of a clay having limits w_l and w_p. Thus the liquidity index, which is usually expressed as a percentage and which may be negative, gives a measure of the consistency of the natural soil. At a liquidity index of 100 per cent the natural clay is at its liquid limit and it will experience a loss of strength on disturbance or remoulding. A liquidity index of zero signifies that the natural clay is at its plastic limit and will tend to crumble on remoulding. An alternative form, the *consistency index* is sometimes used where

$$\text{consistency index} = 1 - I_l$$

It is worth recalling again the emphasis placed on testing undisturbed soils in the assessment of their future behaviour in engineering works. The liquid and plastic limit tests are made on highly *disturbed* soil-water mixtures and cannot be expected to yield information which depends on the fabric and structure of the soil *in place*. Sherwood (1970) has reported on an investigation of the repeatability of the results of liquid and plastic limit tests when carried out by a single experienced operator, by several different operators and by several different testing laboratories. The liquid limit determinations were made with the Casagrande apparatus, BS 1377. The repeatability was found to be disturbingly poor in view of the important engineering decisions which may rest on the test results. The quality of the information should be good however if a generous number of specimens are tested to give average values for the limits and if the procedures of BS 1377 are meticulously followed.

22

Soil Classification

As emphasised in the introduction, an important aim of the geotechnical engineer is to quantify parameters which describe soil behaviour so that he can communicate these to others or use them in his own design analysis. Individually these parameters may allow a soil to be classified in terms of the degree to which it possesses a particular property – for example, density. A soil may be described as very loose, loose, medium dense, and so on. In a more general way engineers have found it useful to allocate soils to groups, each group having relatively consistent behaviour and distinguishable from soils in other groups. This placing in groups is called *soil classification.*

The purpose of classifying soils is clearly stated in CP 2001 as 'to provide an accepted, concise and reasonably systematic method of designating the various types of materials encountered (in site exploration) in order to enable useful conclusions to be drawn from a knowledge of the type of material. Without the use of such a classification, published information or recommendations on design and construction based on the type of material are likely to be misleading, and it is difficult to apply experience gained to the design of other works'. Several systems have been proposed and as yet no international system is agreed. There are similarities between the Casagrande system and that in CP 2001, which largely adopted the former. The classification of soils for engineering purposes is under review but present practice continues to employ the British system as described in CP 2001.

The general description or classification of soils is based mainly on *particle size distribution* and the *Atterberg limit* tests. In soils of silt sizes and coarser, particle size distribution is the main information required and in clays, where particle size distribution is both difficult to determine and less informative from the point of view of practice, the plasticity properties give the information. Taken together with descriptions of the material in place, these allow soils to be allocated to groups. These groupings provide us with the experience of generations of engineers and give guidance on the prediction of soil behaviour. Used cautiously, such predictions are found of value although without the aid of data from tests to determine specific parameters they may induce a mistaken confidence.

22.1 British System

Table 22.1 adapted from CP 2001 shows the groupings of soils which exhibit similar behaviour. Once a soil is classified into a group it is possible to make some predictions about its behaviour. In some cases this may be sufficient as a basis for design decisions; in others it will give a guide to the programme of tests needed to provide design information.

In the table it is seen that *gravels* and *sands* are identified and classified mainly from their particle size distributions. Their group prefix letters are G and S respectively and according to their grading and fines content carry suffix letters, W, C, U, P and F. The *silts* and *clays* carry group prefix letters M, C and O, the letter M indicating a very fine sand, silt or rock flour, the letter C indicating a clay and O indicating an organic silt or clay. The suffix letters L, I and H are used with these soils to indicate the compressibility of the materials. The *plasticity chart* (figure 22.1) is used to identify the appropriate group symbols for the silts and clays. For any given soil of this range a point in the chart is determinable from the values of w_l and I_p for the soil. The zone in the chart at the point gives the appropriate group symbol. Some examples are given in table 20.1. *Peat* is readily identifiable by visual inspection.

22.2 Field Classification

Where laboratory facilities are not readily available or where an on-the-spot assessment of a soil is required quickly in the laboratory or field, the following simple tests are used instead of those described in BS 1377. Since the results are

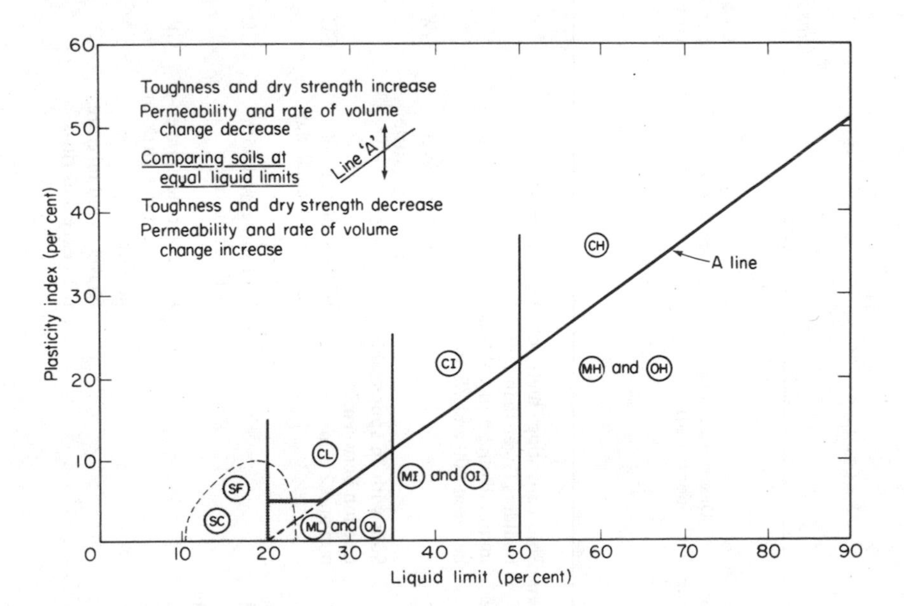

Figure 22.1 *Classification of fine-grained soils : plasticity chart*

TABLE 22.1

*Classification of soils**

	Major divisions	Description and identification	Sub-groups	Casagrande group symbol	Applicable classification tests (carried out on disturbed samples)	Potential frost action	Shrinkage or swelling properties	Drainage characteristics	Bulk dry density at optimum compaction, $t\ m^{-3}$ and void ratio e
COARSE-GRAINED SOILS	Boulders and cobbles	Soils consisting chiefly of boulders larger than 200 mm in diameter or cobbles between 200 and 75 mm	Boulder gravels	–	Mechanical analysis	None to very slight	Almost none	Good	–
			Well-graded gravel – sand mixtures, little or no fines	GW	Mechanical anlysis	None to very slight	Almost none	Excellent	> 2.00 $e < 0.35$
		Soils with an appreciable fraction between 75 mm and 2 mm	Well-graded gravel – sands with small clay content	GC	Mechanical analysis, liquid and plastic limits on binder	Medium	Very slight	Practically impervious	> 2.08 $e < 0.30$
	Gravel and gravelly soils		Uniform gravel with little or no fines	GU	Mechanical analysis	None	Almost none	Excellent	> 1.75 $e < 0.50$
			Poorly-graded gravel – sand mixtures, little or no fines	GP	Mechanical analysis	None to very slight	Almost none	Excellent	> 1.85 $e < 0.45$
			Gravel – sand mixtures, with excess of fines	GF	Mechanical analysis, liquid and plastic limits on binder if applicable	Slight to medium	Almost none to slight	Fair to practically impervious	> 1.92 $e < 0.40$

COARSE-GRAINED SOILS	Sands and sandy soils	Soils with an appreciable fraction between 2 mm and 63 μm	Well-graded sands and gravelly sands, little or no fines	SW	Mechanical analysis	None to very slight	Almost none	Excellent	> 1.92 $e < 0.40$
			Well-graded sand with small clay content	SC	Mechanical analysis, liquid and plastic limits on binder	Medium	Very slight	Practically impervious	> 2.00 $e < 0.35$
			Uniform sands, with little or no fines	SU	Mechanical analysis	None to very slight	Almost none	Excellent	> 1.60 $e < 0.70$
			Poorly-graded sands, little or no fines	SP	Mechanical analysis	None to very slight	Almost none	Excellent	> 1.60 $e < 0.70$
			Sands with excess of fines	SF	Mechanical analysis, liquid and plastic limits on binder if applicable	Slight to high	Almost none to medium	Fair to practically impervious	> 1.68 $e < 0.60$
FINE-GRAINED SOILS Containing little or no coarse-grained material	Fine grained soils having low plasticity (silts)	Soils with liquid limits less than 35 per cent and generally with less than 20 per cent of clay	Silts (inorganic), rock flour, silty fine sands with slight plasticity	ML	Mechanical analysis, liquid and plastic limits if applicable	Medium to very high	Slight to medium	Fair to poor	> 1.60 $e < 0.70$
			Clayey silts (inorganic)	CL	Liquid and plastic limits	Medium to high	Medium	Practically impervious	> 1.60 $e < 0.70$
			Organic silts of low plasticity	OL	Liquid and plastic limits from natural conditions and after oven drying	Medium to high	Medium to high	Poor	> 1.44 $e < 0.90$
	Fine grained soils having medium plasticity	Soils with liquid limits between 35 and 50 per cent and generally containing between 20 and 40 per cent clay	Silty clays (inorganic) and sandy clays	MI	Mechanical analysis, liquid and plastic limits if applicable	Medium	Medium to high	Fair to poor	> 1.60 $e < 0.70$
			Clays (inorganic) of medium plasticity	CI	Liquid and plastic limits	Slight	High	Fair to practically impervious	> 1.52 $e < 0.80$
			Organic clays of medium plasticity	OI	Liquid and plastic limits from natural conditions and after oven drying	Slight	High	Fair to practically impervious	> 1.52 $e < 0.80$

TABLE 22.1 (*cont.*)

FINE-GRAINED SOILS
Containing little or no coarse-grained material

Major divisions	Description and identification	Sub-groups	Casagrande group symbol	Applicable classification tests (carried out on disturbed samples)	Potential frost action	Shrinkage or swelling properties	Drainage characteristics	Bulk dry density at optimum compaction, t m^{-3} and void ratio e
Fine grained soils having high plasticity	Soils with liquid limits greater than 50 per cent and generally with a clay content greater than 40 per cent	Highly compressible micaceous or diatomaceous soils	MH	Mechanical analysis liquid and plastic limits if applicable	Medium to high	High	Poor	> 1.60 $e < 0.70$
		Clays (inorganic) of high plasticity	CH	Liquid and plastic limits	Very slight	High	Practically impervious	> 1.44 $e < 0.90$
		Organic clays of high plasticity	OH	Liquid and plastic limits from natural conditions and after oven drying	Very slight	High	Practically impervious	> 1.60 $e < 0.70$
Fibrous organic soils with very high compressibility	Usually brown or black in colour. Very compressible. Easily identifiable visually	Peat and other highly organic swamp soils	Pt	Moisture content	Slight	Very high	Fair to poor	

*Based on table 3, CP 2001

more subjective than those of the laboratory tests, some practice and experience are necessary to obtain reliable results – preferably gained by making both the laboratory and the simple tests.

Particle Size Distribution

The gravels and sandy soils comprise particles which are distinguishable by eye so that visual examination should enable an assessment of the classification to be made. Particles of 63 μm size are at the limit of visibility unaided and materials around this size are judged by 'feel', on rubbing the moist material between the fingers. At 63 μm a fine sand feels harsh but not gritty. The presence of coarser sizes makes the material feel more gritty, and finer sizes make it less harsh. The finer silt-sized material feels smooth and only slightly sticky. Clay sizes give a greasy texture to the soil and the material is noticeably sticky.

Plasticity

The coarser soils may contain just sufficient fine material that when moist a cohesive mass of the soil can be formed. With a greater content of fines or more clay sizes among them the soil may also show plasticity. The fine fraction of a coarse soil can thus usually be distinguished as either nonplastic or plastic.

The clays and silty soils are assessed on the basis of their toughness, dry strength and dilatancy. To make these tests a sample of the soil is first mixed to a water content just greater than the plastic limit. In the simple test for *toughness* a portion of the sample is rolled to a thread of diameter just over 3mm as in the plastic limit test, then it is kneaded to a ball and rolled again to a thread. This is continued until the water content of the soil is reduced to its plastic limit and the thread crumbles at just over 3 mm diameter. At this water content the highly plastic soils have fairly stiff threads and tough lumps whereas those of low plasticity have weak threads and are more crumbly.

For *dry strength,* a portion of the sample is formed into a pat about 25 - 30 mm in diameter and 5 - 8 mm thick and this is allowed to dry completely. The strength of the dried soil is judged by breaking the pat and crushing the fragments with the fingers. Nonplastic fine soils have little or no dry strength and dry strength increases with the liquid limit of the soil.

The *dilatancy* test distinguishes the silts and fine sands from the clays. A portion of the sample is taken and more water mixed in until the consistency is soft but not sticky. A pat of about the same dimensions as in the previous test is formed in the open palm of the hand. With the hand held horizontal it is struck against the other hand several times. If the soil is a dilatant material water will flow to the surface of the pat giving it a shiny 'livery' appearance. Further, on squeezing the pat with the fingers the water recedes into the pat and the surface dulls, and the pat stiffens and crumbles. The property of dilatancy belongs to the granular materials, and the silts and very fine sands show marked reactions to the test. If the material is substantially clay, then the surface of the pat will not alter in appearance in this test.

Organic Soils and Peat

The presence of organic materials is indicated by a distinctive odour and to a certain extent colour – usually dark greys and browns. Plant remains may be visible in the soil. In peats, plant remains predominate so that these are readily recognisable. One group symbol is allocated to peats although many forms of the material occur.

22.3 Extended Description of Soil

A full description of a soil sample will extend beyond its group classification to include where possible its water content and liquidity index, its colour and shape, the texture and mineral composition of its particles, and its appearance and strength as it exists in place in the site.

Colour

Generally the brighter colours are associated with the inorganic soils and the darker duller colours with the organic soils. For accurate colour descriptions reference may be made to standardised colour charts. Usually a simple description of the style 'light greenish grey' will be sufficient.

Activity

The affinity of a clay for water will depend in part on the mineralogical composition of the clay and on the fineness of particle size, composition and fineness being related. Fineness of a particulate material is often quantified as its specific surface, that is, the total surface area of all its particles per unit of mass or absolute volume. Montmorillonite clays have a high specific surface relative to the kaolinite clays. A simple ratio giving a measure, for engineering purposes, of the mineralogical composition of a clay was proposed by Skempton (1953). This ratio is called the 'activity' A of a clay and

$$A = \frac{I_p}{(\text{per cent clay})}$$

where I_p is the plasticity index of the clay, and 'per cent clay' is the mass of clay fraction (finer than 2 μm in size) expressed as a percentage of the dry mass of the soil. For a clay of particular physicochemical characteristics A has been found to be a constant and a measure of the colloidal activity of the clay. Skempton gave the following classification of activity for clays

inactive	$A < 0.75$
normal	$0.75 < A < 1.25$
active	$A > 1.25$

There appears to be a relationship between the components of shear strength of clays and their activity, and between activity and difficulties in obtaining deep samples of some clays. The largest group of clays has activity in the normal range and for these $A = 1$ approximately. Wilun (1975) quotes an approximate classification from Polish standards, based on clays for which $A = 1$ and this is given in table 22.2.

TABLE 22.2
Description of cohesive soils

Description of cohesiveness of soil	Clay fraction (per cent)	I_p (per cent)
Cohesionless	0 - 2	< 1
Slightly cohesive	2 - 10	1 - 10
Medium cohesive	10 - 20	10 - 20
Cohesive	20 - 30	20 - 30
Very cohesive	30 - 100	> 30

In the case of a discrepancy in the description of a soil by the plasticity index and the clay fraction, the colloidal activity should be quoted after the description. For example, a soil containing 25 per cent clay and having $I_p = 38$ per cent would be described in terms: heavy cohesive silty clay (25 per cent clay fraction, $A = 1.5$).

Site Condition

A description of the site conditions of soils is usefully based on the terms given in table 22.3. Taken together with the particle size distribution of a coarse-grained soil, a value of relative density D_r adds much to the classification of the material in place. Table 22.4 gives a descriptive scale of D_r values.

22.4 Chemical Composition

In chemical composition soils are complex and variable minerals. For engineering purposes, however, tests are usually restricted to the determination of organic matter content, sulphate content of soil and groundwater, and determination of the pH value. Details of test procedures are given in BS 1377. The carbonate content may also be of interest since chalk soils are known to be particularly frost-susceptible.

TABLE 22.4
Classification of soil density

Relative density (per cent)	Description
0 - 15	Very loose
15 - 35	Loose
35 - 65	Medium
65 - 85	Dense
85 - 100	Very dense

see pg 263

TABLE 22.3

Description of site condition of soils

Soil types	Strength		Structure	
	Term	Definition	Term	Definition
Coarse soils	Loose	Can be excavated with spade, 50 mm wooden peg easily driven	Weathered	Particles are weakened, and may show concentric layering
	Compact	Requires pick for excavation, 50 mm peg hard to drive more than a few inches		
	Weakly cemented	Pick removes soil in lumps that can be abraded with thumb and broken with hands		
	Strongly cemented	Cannot be abraded with thumb or broken with hands		
	Indurated	Broken only with sharp pick blow, even when soaked. Makes hammer ring		
Coarse soils and fine soils			Homogeneous	Material essentially of one type
			Stratified	Alternating layers of various types
			Laminated	Stratified with thin layers (less than 6 mm)
Fine soils	Very soft	Exudes between fingers when squeezed	Aggregated	Strength decreases on working
	Soft	Easily moulded with fingers	Weathered	Usually exhibits crumb or columnar structure
	Firm	Moulded only by strong pressure of fingers	Fissured	Breaks into polyhedral fragments along fissure planes
	Stiff	Cannot be moulded with fingers	Intact	Not fissured
	Hard	Brittle or very tough		
Peat	Firm	Fibres compressed together	Fibrous	Plant-remains easily recognisable, retains structure and some of original strength
	Spongy	Very compressible and open structure	Pseudo-fibrous	Plant-remains recognisable, but plastic when moist
	Plastic	Can be moulded in hands and smeared between fingers	Amorphous	Recognisable plant-remains absent

The presence of appreciable organic matter indicates a compressible soil and even a small percentage of decomposed organic matter is sufficient to react unfavourably with Portland cement when the latter is used as an admixture in soil stabilisation. Taken together with the pH value, the sulphate content gives a measure of the aggressiveness of the soil chemicals towards Portland cement concretes and other materials of construction. Sulphates occur in a number of clay soils and may also be troublesome in fills of colliery shale, brick rubble, ash and some industrial wastes. Sulphate can only continue to reach concrete by movement of groundwater so that concrete which is wholly and permanently above the water table is unlikely to be seriously attacked.

The reaction of most natural soils in Britain to pH tests varies between about pH 4.0 for very acid fen, heath and peat soils to about pH 8.0 for chalky alkaline soils. A knowledge of the acidity or alkalinity is of interest to the engineer mainly because of its influence on corrosion of concretes or metalwork laid in contact with the ground.

23

Water in Soils

In outlining the processes of formation of soils, the products were visualised in their complexity as composed largely of solid particulate matter. While this is usually so, the presence of water in the pores between the solid particles may have a great influence on the behaviour of the composite material. Much of the thinking of the geotechnical engineer is directed towards the understanding of this behaviour. The porous nature of soils permits the storage and flow of groundwater in the void spaces and these voids may be wholly or partly water-filled. In the coarser-grained soils this water may be a basic resource for industry, agriculture and human consumption to be drawn off from wells and recharged naturally or with man's intervention. Many situations arise in engineering design and practice where the flow, or even simply the presence, of water must be taken into account in analysis. The coarser soils generally do not experience much volume change as a result of movements of groundwater but in the finer-grained soils swelling or shrinkage will occur when some change is made in the environment of the soil. The presence of water has important effects on the states of stress in soils and on their compressibility and strength. In discussing water in soils, therefore, a distinction is usefully made in behaviour between the coarser-grained soils (silts and larger sizes) and the finer-grained soils (clays).

23.1 Transfer of Stress through Soil

The engineer is concerned with the influence of *stress* on his materials: stress changes may produce volume changes, changes in strength and may lead to yield or failure of materials. Although concretes, ceramics and timbers are porous materials it is usual in practice to deal with them as if they are *isotropic* single-phase materials such that definition of the *total normal stress* σ on a plane through a point in the material is all that is required to describe that normal stress. This is not sufficient in soils which, as we have seen, are compressible particulate assemblies comprising the three phases of solid particles, water and air or gas. The stresses in the pore spaces in the water and the air are measurable as u and u_a respectively where, owing to surface tension, u is always less than u_a. The total stress σ is also determinable. It is one of the problems in soil mechanics to determine what part of σ is supported by the soil particle assembly or skeleton.

For saturated soils it was shown by Terzaghi in 1923 that

$$\sigma' = \sigma - u$$

where σ' is termed the *effective stress.* It is changes in σ' which produce the changes in the volume and strength of such soils. This expression has been studied theoretically and experimentally and for practical engineering usage has stood the test of time. The effective stress σ' is closely related to the stress transmitted through the soil skeleton and is often called the intergranular stress. The two terms are usually used interchangeably. The unit of stress in soils is usually kN m^{-2} or kPa (1 kN m^{-2} = 1 kilopascal).

23.2 Coarser-grained Soils – Static Groundwater

It will usually be possible in the coarser-grained soils, by inserting a borehole casing tube or standpipe down into the soil, to determine a water surface level. This water level at the standpipe location is variously described as the *groundwater surface* in the soil (in static conditions defining the level of the groundwater table), the *piezometric surface* or the *phreatic surface.* The position of the water level in the standpipe can be readily determined using, say, an electrical 'dipper' with its connecting cable marked in length units. At this groundwater level the absolute porewater pressure is equal to atmospheric pressure. The *porewater pressure u* at any place in a soil is defined as the excess pressure in the porewater, above atmospheric pressure, thus, $u = 0$ at groundwater level. Where there is no flow of groundwater taking place through the pores of the soil, u increases linearly with depth z below the groundwater level as illustrated in figure 23.1. That is

$$u = +\gamma_w z$$

Capillary Water

Above the groundwater level the soil may be saturated for some distance owing to *capillary rise* of water pushed up from the water table or retention of water percolating down to the water table. This capillary rise is held by surface tension at the air - water interface from where the water is continuous to the groundwater level. Surface tension at an air - water surface generally tends to keep the surface flat, but when the water is in contact with soil particles, to which it acts as a wetting liquid, the water surface meets the particle surfaces tangentially or at a very small contact angle. This gives an air - water interface of a most complicated form in the soil. A simpler picture is obtained by considering a single vertical open-ended fine-bore glass capillary tube (*capilla,* Latin for hair) whose lower end is dipped in water. The tiny meniscus takes up a hemispherical concave form as shown in figure 23.2 and a pressure difference is set up across the meniscus, which is pushed up the tube until an equilibrium position is reached where the weight of the water column is just supported by the surface tension force. It is a simple matter to show that $z_s \propto 1/D$ where z_s is the height of capillary rise in a tube of

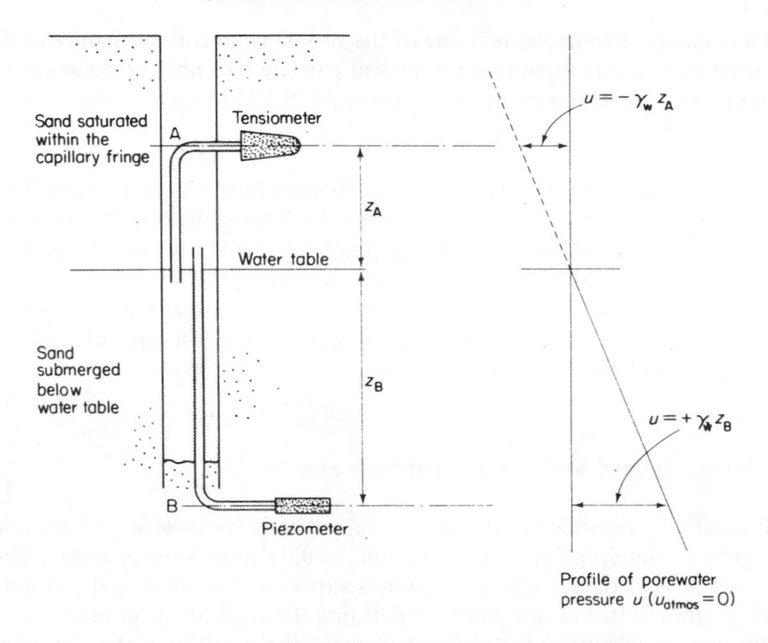

Figure 23.1 *Vertical distribution of porewater pressure – static groundwater in equilibrium*

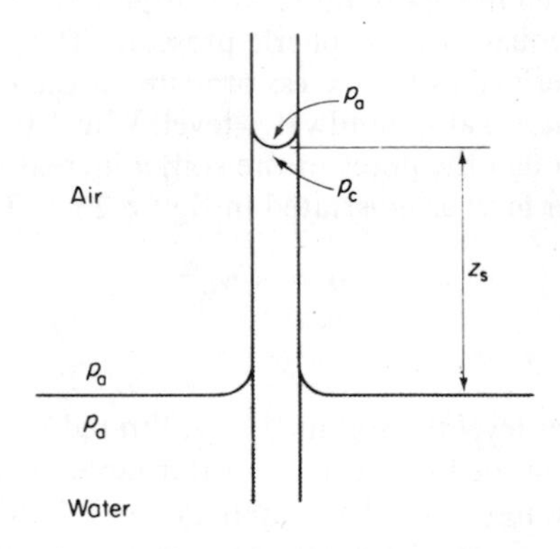

Figure 23.2 *Capillary rise of water in small-bore tube.* p_a = *atmospheric pressure;* p_c = *pressure in capillary water;* $p_c < p_a$ *and as* $z_s \to 0$, $p_c \to p_a$

diameter D. Within a real soil the meniscus or interface is of a complicated form with saddle-shaped masses of water forming at the points of contact of the interface and the soil particles. The greatest suction (pressure below atmospheric) which can be maintained at such an interface corresponds to the sharpest curvature the interface can form in the complex pore spaces. The sharpness of this curvature depends on the smallness of size of the pore spaces which in turn is related in some manner to the sizes of particles present in the soil and their closeness of packing. Thus, in a general way, the force raising the water in the soil capillaries above the groundwater level is inversely proportional to the particle size in the soil. The weight of water hung up at the interface must tend to pull the supporting soil particles closer together, increasing the effective stress σ' and compressing the soil itself. In the coarser-grained soils the particles are already predominantly in physical contact with each other and little volume change occurs on capillary rise or retention, beyond that due to a slight rearrangement of the particles.

Capillary water that is continuous through the pores of the soil to the water table, will again show a linear variation of u with height z above the groundwater level, but with negative porewater pressures, that is, below atmospheric pressure. Thus

$$u = -\gamma_w z$$

These values of u are also illustrated in figure 23.1. The maximum height z_s of capillary rise or retention in the coarser-grained soils depends on a number of factors, but principally on the 'diameter' of the irregular pore spaces, the nature of the soil grain surfaces and the history of wetting of the soil (whether by capillary rise alone from the groundwater level or by percolation). Capillary rise and capillary retention are illustrated in the simple model in figure 23.3.

In a natural soil comprising a range of particle sizes a zone of full saturation may exist just above the groundwater level and this zone may merge into an upper zone of partial saturation. In such soils z_s is the combined height of these zones, sometimes called the *capillary fringe.* This height can only be measured approximately or estimated.

Since water to a height z_s is held in the soil pores, it will not drain out under gravity unless the groundwater level itself is lowered. An interesting phenomenon attributed to capillarity is described by Terzaghi and Peck (1948). Water pushed up by capillary action on one side of an impermeable barrier within soil may connect over the top of the barrier with pore spaces on the other side which are subject to gravity drainage, so forming a self-priming capillary siphon. Leakage over the impermeable core of the dyke on the Berlin – Stettin canal was apparently due to this phenomenon. When the core was 0.3 m above the free water surface in the canal an estimated loss of water of over 3500 m^3 per day occurred along a canal length of some 19 km. When the height of the core was increased by 0.4 m the loss was reduced to less than 800 m^3 per day.

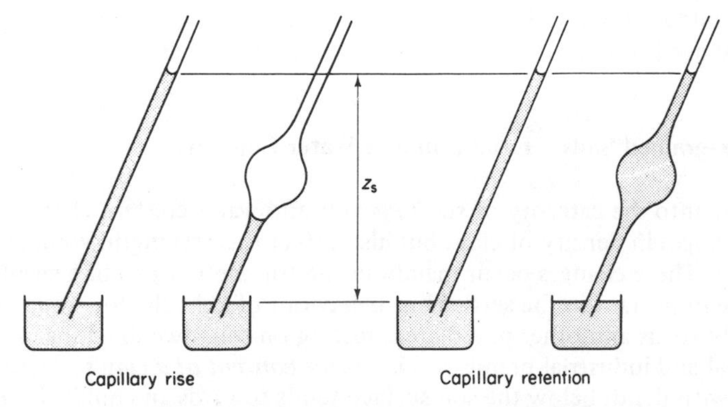

Figure 23.3 *Diagrammatic representation of capillary rise and capillary retention*

Contact Moisture

Above the capillary zone the soil may contain moisture of a discontinuous nature in the form of traces of water at the points of contact of grains. This *contact moisture* produces surface tension forces which pull the soil grains together at points where they are bridged by a trace or droplet of water. This leads to the soil acquiring a strength which has a similar effect to that of cohesion in clays and this strength is sometimes described as *apparent cohesion.* Because of its origin, this apparent cohesion disappears completely on drying the soil or immersing it in water.

The phenomenon of 'bulking' of moist sands is well known in concrete technology and leads to gross inaccuracies in volume batching of moist sand aggregates. This is due to apparent cohesion from contact moisture enabling sands to remain stable at lower densities than would be practicable if they were dry or immersed.

Height of Capillary Rise

For most purposes it will be sufficient to estimate rather than attempt to measure the approximate height z_s of the capillary zones.

Taylor (1948) gives a relationship between z_s and the coefficient of permeability k (see section 23.5) of soils, and k in turn can be related approximately and empirically to D_{10}.

$$k = D_{10}{}^2 \times 10^4 \ \mu\text{m s}^{-1}$$

where D_{10} is in millimetres.

$$\frac{700}{\sqrt{k}} < z_s < \frac{2400}{\sqrt{k}}$$

where z_s is in millimetres. A coarse silty sand material with $D_{10} = 0.02$ mm would have a k value in the neighbourhood of 4 μm s^{-1} so that $0.35 < z_s < 1.2$ m.

It is only in the siltier types of coarse-grained soils that capillarity will have significant effects.

These soils are sometimes described as *nonswelling soils,* in that no significant volume change is observed as a result of water content changes.

23.3 Finer-grained Soils – Equilibrium of Water Contents

Clays come into the category of *swelling soils* and water content changes lead not only to changes in density of clays but also affect their strength, volume and compressibility. These changes occur mainly in the top metre or so of a swelling clay and can be important in the settlement behaviour of light shallow foundations – particularly roads and other paved areas resting on soil, dwelling houses and light commercial and industrial premises. *The water content of a clay soil* and its distribution with depth below the soil surface tends towards an equilibrium which *is essentially the result of competition for the water:* the force of gravity on the water induces drainage of the soil, the force of attraction or soil suction, if any, draws moisture into the soil, and the force of gravity on self-weight or surface loadings on the soil tends to expel water from the soil. On reaching this equilibrium the soil has experienced water content and volume changes. A further change in the environment produces another unsteady system tending with time towards a new equilibrium. For example, paving over a clay soil alters the environment by reducing or eliminating ground-surface evaporation and if the soil is susceptible to volume changes it will swell or shrink, depending on the weight of the paving and the original seasonal condition of the soil, until a new equilibrium is reached. Other environmental changes producing similar water movements and volume changes can be visualised. The effects of these ground movements in expansive clay soils are often sufficient to cause damage to buildings and other engineering works of construction.

This competition for the porewater in a soil element may be expressed as an equation

$$s = \alpha\sigma - u$$

where s is the *soil suction*, u the porewater pressure with respect to atmospheric pressure, σ the vertical total pressure due to overburden surface loading and α the fraction of this pressure transmitted to the porewater.

Soil Suction

This is a measure of the capacity of a given clay at a given water content to retain its porewater when the soil is free from external stress. A range of equipment has been developed for measuring soil suction (Croney and Coleman, 1961). α can be measured by a simple loading test on a laboratory specimen but it is mainly dependent on the plasticity characteristics of the soil.

$$I_p < 5 \qquad \alpha = 0$$
$$I_p > 40 \qquad \alpha = 1$$
$$5 \leqslant I_p \leqslant 40 \qquad \alpha = 0.027 I_p - 0.12$$

Soil suction is expressed as a pressure, namely the pressure s (below atmospheric pressure) which has to be applied to the porewater to overcome the capacity of the soil to retain the water. The pressure is often referred to a pF scale where the common logarithm of the suction expressed in multiples of 10 mm head of water is described as its pF value. Thus a suction s of

1 × 10 mm of water equals pF 0
10 × 10 mm equals pF 1
100 × 10 mm equals pF 2

and so on. Saturated soils will show a pF value corresponding to zero porewater tension. There is no zero on a logarithmic scale but this condition would be described as pF 0. When oven dry the soil will have a pF in the region of 7 (that is, some 10 000 atmospheres of suction). One atmosphere of suction corresponds to nearly pF 3. An example of the relation between pF and w is given in figure 23.4.

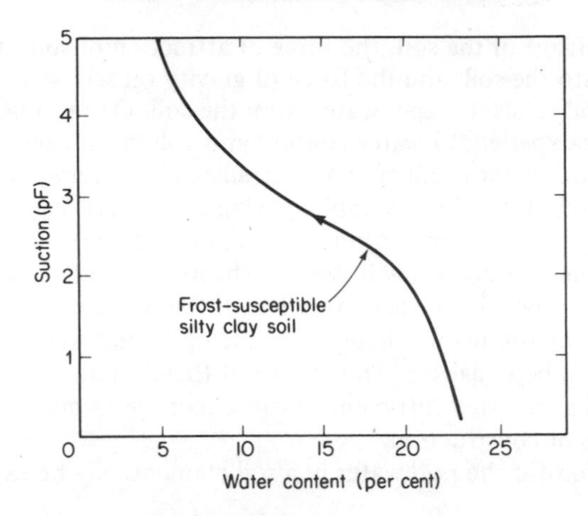

Figure 23.4 *Typical suction curve for a silty soil*

Strictly the relation is not a single curve and values measured differ on wetting and drying cycles. Except for salty materials, soils feel wet when below pF 2.5, moist in the range pF 2.5 to pF 4.5 and dry above pF 4.5, when they also appear lighter in colour.

Since a relationship of the form of figure 23.4 can be established between s (or pF) and w for a clay soil and in turn w can be related to soil strength and volume, it is possible by applying the above equation to explore the consequences of changes in the environment of a clay – changes in loading, in surface evaporation and transpiration, in position of groundwater table and so on. For example, in road pavement design the evaluation of the improvement in soil strength produced by lowering the groundwater table by drainage requires the application of soil suction characteristics. Prediction of shrinkage and swelling of clays which result from changes arising from construction works can be made with some success. Because of the impervious character of clays it may take some time to reach an equilibrium water content profile, indeed at shallow depths fluctuations in the climatic environment may cause continuous fluctuations in the water contents.

Plants, trees and general vegetation take part strongly in the competition for the porewater in soil, and roots exert a considerable suction. Unless the water drawn out by the vegetation is replaced, the clay will reduce in water content, with consequent shrinkage, until the wilting point of the vegetation is reached at about pF 4.18 (nearly 15 atmospheres of suction head). It follows that the removal of trees which draw off substantial quantities of porewater will produce swelling in such soils.

23.4 Frost Susceptibility of Soils

Another aspect of soil suction is shown when soils or other porous materials are subjected to freezing conditions. This is common in roads and paved areas resting on soil and in shallow foundations. When ice and water are present in a soil the soil suction experienced in the unfrozen water becomes dependent on the tem-

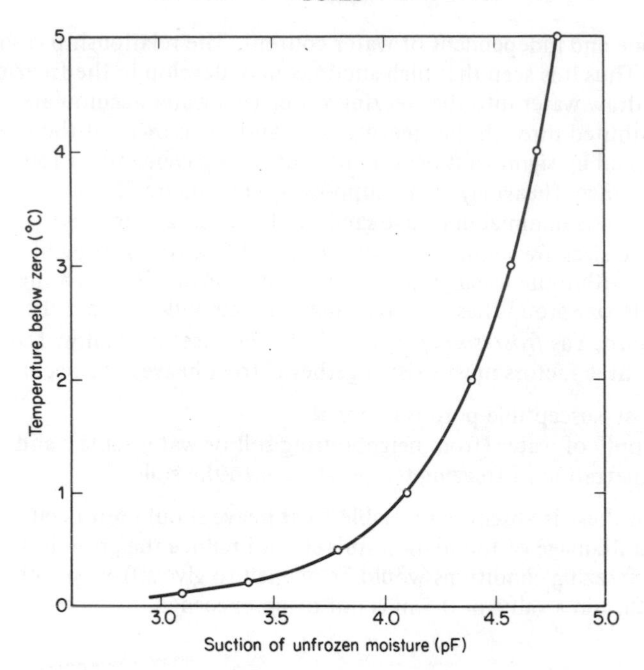

Figure 23.5 *Relationship between temperature below freezing point and suction*

Figure 23.6 *Frost heave in a silty soil*

perature alone and independent of water content. The relationship is shown in figure 23.5. Thus it is seen that high suctions may develop in the freezing zone in soils and so draw water into the freezing zone. This water accumulates in ice 'lenses' distributed through the freezing zone and the growth of these lenses may be accompanied by significant heaving pressures on pavements or foundations and by volume changes (heaving) of the supporting soil (figure 23.6).

Such suction is minimal in coarse sands and gravels, and in clays ($I_p > 15$ per cent) the pore sizes are so minute that the rate of flow is insufficient to build up ice lenses within the usual climatic period of freezing. Between these extreme particle size groups lies the silty range of soils and these are the materials broadly described as *frost-susceptible soils.* In the cases of shallow foundations mentioned, three factors must exist together if frost heave is to occur

(1) frost-susceptible porous material
(2) supply of water (from neighbouring soil or water table); and
(3) penetration of freezing temperature into the soil.

If any one of these is absent appreciable frost heave should not occur.

Improved drainage of foundation soils should reduce the growth of ice lenses. Very severe freezing conditions would be needed to give a frost penetration of 1 m and in Britain a value of 0.5 m is not often exceeded.

Heaving Pressures

The direction of *frost heaving* of soil is usually vertical, corresponding to the direction of frost penetration and heat flow, and ice lenses form with their long axes practically horizontal. The heaving pressures are of two main types: those associated with the 'normal' expansion of water on freezing (ice having a volume about 9 per cent greater than the water from which it is formed), and those associated with the increase of water content (as ice) during freezing. Only the latter is regarded as the 'heaving' pressure in soils engineering. The process of heaving of freezing soil against a foundation or in lifting foundation piles continues to exert heaving pressures up to the maximum value for as long as the freezing conditions are maintained. These maximum heaving pressures can be considerable and substantial heaving displacements may therefore take place during a prolonged cold spell. Because particle size distributions and compositions are so variable precise values of pressure cannot be determined. Table 23.1 gives some values for illustration only. This pressure is also the pressure required just to prevent frost heaving.

Thawing

The intake of water and growth of ice lenses produces problems when the soil thaws. A silty clay soil is broken up by the ice lenses to form a flaky structure and immediately after the thawing of the ice lenses the soil has a very high void ratio and water content and its strength is substantially reduced. Thawing will usually

Soil water, permeability & flow.

TABLE 23.1

Heaving pressures exerted by freezing soils

Soil type	Heaving pressure p_i for $u = 0$ (kN m^{-2})
Coarse sands or coarser material only	0
Medium and fine sands or coarse silty sands	0 – 7.5
Medium silts or mixed soils with small amounts < 0.006 mm	7.5 – 15
Largely fine silts or silts with some clays	15 – 50
Silty clays	50 – 200
Clays	> 200

From Williams (1967)

occur from the top downwards so that initially the water is prevented from draining away. The overburden in time may reconsolidate the soil but in its initially weakened condition it may easily suffer damage, for example, by traffic movement over the soil.

23.5 Flow of Groundwater

In practice the effects of capillary water are principally related to the vertical profile: water content and strength variations with depth below ground surface, vertical shrinkage and swelling of the top 2 – 3 m in clay soil.

Consider now the porewater below the groundwater surface in a continuously porous soil mass – the pore spaces are likely to be saturated or nearly so. If the porewater at a point in the soil mass possesses greater potential energy, or *total head H* (metres or other length units) above a datum level, than exists at an adjacent point in the mass, a flow will be induced towards the latter point. The flow of water in soils is accompanied by a loss of hydraulic head as energy is taken up in overcoming resistance to flow through the pores.

The total head H of water at a point in a soil is simply

$$H = \frac{u}{\gamma_w} + z$$

where u is the porewater pressure at the point, u/γ_w is the *pressure head* and z is now the height of the point above the reference datum level, and is the *elevation head* as in figure 23.7. This is the familiar Bernoulli equation from fluid mechanics in which the velocity head term $v^2/2g$ also appears, v being the velocity of porewater flow and g the acceleration due to gravity. In practical terms in groundwater flow, $v^2/2g$ is triflingly small numerically and is ignored.

In geotechnology the prediction of flow patterns in soils, and the resulting distribution of seepage forces and porewater pressures, forms an important area of

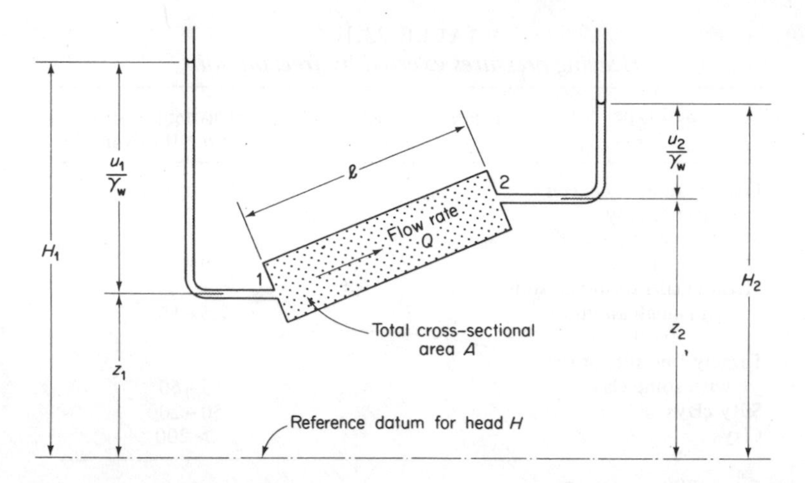

Figure 23.7 *Diagrammatic representation of flow of water through a porous material defining total head H, pressure head u/γ_w and elevation head z at two points 1 and 2. The flow rate is Q through a total area A normal to the direction of flow and the points 1 and 2 are a distance l apart in the direction of flow*

expertise. This lends itself to mathematical analysis and the study has long exercised mathematicians and physicists so that the flow patterns associated with many practical engineering problems now have a broad theoretical background. The successful matching of the theoretical to the practical continues to depend on the validity of the assumptions made in the analysis – and foremost among these is the description in quantitative terms of the drainage characteristics of the soils.

Permeability of Soils

The first scientific investigation into the movement of water in sands is attributed to the famous railway engineer Robert Stephenson. During construction of the Kilsby tunnel, near Rugby, completed in 1838, extreme difficulties were encountered in waterlogged coarse sands. Stephenson had shafts sunk along the line of the tunnel and pumping from these took place. He took careful measurements of water levels and, among other observations, noted that the resistance which the water encountered in its passage through the sand was related to the angle or inclination of the groundwater surface towards the pumped shafts. In 1856, Darcy reported on experimental studies of flow in saturated sands and noted that the flow rate ($m^3\ s^{-1}$) per unit of total cross-sectional area of soil, at right angles to the direction of flow, was proportional to the rate of loss of total head H with distance l along the average flow path of the water. This is known as *Darcy's law* and may be expressed

$$\frac{Q}{A} = k\,\frac{(H_1 - H_2)}{l} \quad (\mathrm{m\ s^{-1}})$$

where Q is the rate of flow of water through a total area A, the conditions being illustrated in figure 23.7. Any elevation may be taken for the reference datum since the actual value of z, the elevation head, has little meaning; it is the difference in elevation head that is of interest. The quantity Q/A has units of velocity (m s^{-1}) and is sometimes referred to as the *superficial velocity* or specific discharge v. The ratio $(H_1 - H_2)/l$ is dimensionless and gives the rate of energy loss in flow through the soil and is called the *hydraulic gradient* i. Thus the above equation is commonly written

$$v = ki$$

Since flow only occurs through the voids spaces in the soil, v will not be a measure of the actual velocity of flow of water particles. The average water particle velocity can be estimated from $v_s = Q/nA = v/n$ where n is the porosity of the soil. It is usual, however, to work in terms of v. The coefficient k, termed the *coefficient of permeability*, is considered to be a soil property. Strictly it is influenced by variations in the porewater or pore fluid since $k = K\gamma_w/\mu$ where K is the permeability (unit, m^2 or equivalent) and is the soil property; μ is the coefficient of viscosity of water. For fresh groundwater at ordinary temperature ranges γ_w and μ are practically constant and the use of K as the soil property is acceptable and usual in practice.

Of much more significance and importance is the appreciation that the value of k appropriate to the analysis of groundwater flow problems is that representative of the soil mass as it exists in place. A review has already been given of the complex nature of soils, their stratified, fissured and otherwise anisotropic character, and the influence of their state of stress on their engineering behaviour. This has pointed to the importance of careful and perceptive sampling and of testing soils in place where practicable. This property of permeability is strongly influenced by the effects of anisotropy and the state of stress.

The validity of Darcy's law $v = ki$

When water flows through a pipe or in an open channel the relationship between the average flow velocity and the hydraulic gradient is often expressed as $v \propto i^n$. Students of fluid mechanics will know that such flows are well described by putting $n = \frac{1}{2}$ and are classed as *turbulent* flows. When water flows through the pore spaces in sands, however, the flow is usually classed as *laminar* for which $n = 1$. Further, in pipe and channel flows the velocities at the walls or boundaries are measurably less than at the centre of the flow, whereas in an ideal porous material of sandy texture which is subject to uniform hydraulic gradients the average velocities of flow at the boundaries of the material are no different from those at the centre. It is seen then that there are two kinds of flow, turbulent flow and laminar flow. Turbulent flow is usual in pipe and open channel situations in civil engineering practice; laminar flow is usual in groundwater situations. In groundwater flow the flow conditions in the coarser soils may occupy a transition stage between laminar and turbulent flow when Darcy's law will not apply. A clear separation between laminar and turbulent conditions cannot be defined and a conservative limit for laminar flow is usually adopted. Taylor

(1948) proposed that, under a hydraulic gradient of 100 per cent, uniform sands with an effective size D_{10} of 0.5 mm or less always have laminar flow. This means that in soils coarser than coarse sands or fine gravels the validity of Darcy's law should be considered questionable. It would be exceptional and in any case undesirable to place a water-retaining hydraulic structure on such pervious soil.

Many investigations following those of Darcy have given results substantially in accord with the equation $v = ki$ for sands. In soils containing fines there is some evidence that particle migration within the soil may cause departures from Darcy's law. Departures have also been observed when sands have been tested at very high hydraulic gradients. However, most problems of drainage of sands, flow of water beneath foundations or floors of hydraulic structures, where the hydraulic gradient is usually less than 100 per cent, fall well within the range of validity of Darcy's law.

Factors Influencing the Coefficient of Permeability k

Assuming laminar flow, the relation $v = ki$ applies to groundwater flow in a given direction in a soil. We have already noted that k is a parameter which is a function of the pore fluid as well as of the soil. When the pore fluid is water and its temperature range not large, k is considered to describe fully the permeability of the soil in a given direction of flow.

Other factors influence k, principally

(1) particle size, shape and grading;
(2) arrangement of particles, and stratification and fissures;
(3) density of packing;
(4) composition;
(5) consolidation pressure;
(6) migration of particles; and
(7) presence of air or gas.

Particle size, shape and grading

These characteristics of the soil and, to a lesser extent, the density of packing of the particles, largely determine the size, shape and continuity of the pore spaces in the soil. Since k is a measure of the ease with which water can flow through a soil, these characteristics largely determine the value of k. It is reasonable that in a graded soil the coefficient of permeability should be mainly influenced by the finer sizes present. From experimental work on filter sands of medium permeability, Hazen showed that k could be roughly estimated from

$$k = D_{10}^2 \times 10^4 \ \mu\text{m s}^{-1}$$

where D_{10} is the effective size of the sand in millimetres. This expression contains no parameter describing either density of packing or particle size distribution, but for estimation purposes gives good results.

The coefficient of permeability is a parameter set apart from most other soil

TABLE 23.2
Classification of degrees of permeability

Degree	k (μm s^{-1})
High	Over 1000
Medium	10 - 1000
Low	0.1 - 10
Very low	0.001 - 0.1
Practically impermeable	below 0.001

TABLE 23.3
Typical values of coefficient of permeability

Soil type	Particle size range $D_{max.}$ (mm)	$D_{min.}$	Effective size D_{10} (mm)	Measured coefficient of permeability k (μm s^{-1})
Uniform, coarse sand	2	0.5	0.6	0.4×10^4
Uniform, medium sand	0.5	0.25	0.3	0.1×10^4
Clean, well-graded sand and gravel	10	0.05	0.1	100
Uniform, fine sand	0.25	0.05	0.06	40
Well-graded, silty sand and gravel	5	0.01	0.02	4
Silty sand	2	0.005	0.01	1
Uniform silt	0.05	0.005	0.006	0.5
Sandy clay	1.0	0.001	0.002	0.05
Silty clay	0.05	0.001	0.0015	0.01
Clay (30 to 50 per cent of clay sizes)	0.05	0.0005	0.0008	0.001
Colloidal clay (over 50 per cent finer than 2 μm)	0.01	10 Å*	40 Å	10^{-5}

*1 Å = 10^{-10} metre

parameters in engineering usage in that it has a great range of numerical values. Table 23.2 lists the terms describing degrees of permeability and tables 23.3 and 25.1 give values of k typical of certain soils. Hazen's relation of k and D_{10} appears to fit quite well.

Arrangement of soil particles

The processes of soil formation have been commented on several times in this section, the resulting soils often showing complex and erratic variations. Alluvial soils are likely to show less resistance to horizontal flow of water along the strata than to vertical flow, particularly where the particles are flake-shaped and tend to lie with their largest area nearly horizontal. Particle arrangements may even be such that coefficients measured in two horizontal directions at right angles to each other may differ, particularly in sloping strata.

A particle arrangement of importance to the engineer is that of *open-work gravel.* This term is used for the more uniform gravels, and layers of this type of highly pervious material may occur among deposits of less pervious sand - gravel mixtures. The presence of seams of such gravels, even of small thickness, enormously increases seepage flows and if undetected may lead to serious and unexpected water losses and water pressures in hydraulic construction works. Detection is difficult and trial pits, borings, pumped wells and observations during construction may all contribute information.

Problems of a similar kind may occur in *fissured clays* and *jointed rocks*, the fissures and joints often rendering the material considerably more pervious than the intact material between the fissures. It is obvious therefore that tests for permeability on the intact material would be misleading and that the importance of representative sampling and testing in place should again be recognised.

Density of packing of soil

As the pore volume in a given soil decreases as a result of densification or compaction the coefficient k also decreases. A number of empirical or semi-empirical expressions have been proposed for expressing the relation between permeability and porosity, or void ratio. For example Terzaghi and Peck (1948) quote a formula of Casagrande

$$k = 1.4k_{0.85}\, e^2$$

where k is the coefficient of permeability at void ratio e in a clean sand whose permeability is $k_{0.85}$ at a void ratio of 0.85. From this it would appear that $k \propto e^2$. There are considerable experimental data however to show that a plot of e against log k for a soil is a straight line for nearly all soils.

Composition of soil

Groundwater flow studies are mainly applied to the size range silts to small gravels and in this range the mineral composition of the soil has little significance as regards permeability. In clays, however, something has been seen of the importance of mineral composition on a range of properties – swelling, consolidation and strength – and permeability is also dependent on this composition. Montmorillonite clays exist at high void ratios yet their permeability is very low. A sodium montmorillonite has a permeability less than 10^{-3} μm s^{-1} at a voids ratio of about 15. Kaolinite clays show k-values of about 10^{-1} to 10^{-2} μm s^{-1} at lower void ratios of less than 2. Much depends too on microstructure and it is observed that for a given void ratio the higher permeabilities are associated with the more flocculated clays having the larger flow passages and the lower values with the more dispersed clays.

When clayey soils are used as fill materials the water content at the time of mechanical compaction appears to influence the microstructure formed and hence the permeability – material compacted on the wet side of optimum (see chapter 26) being significantly less pervious than material compacted on the dry side.

Consolidation pressure

It was stated earlier that it is desirable for the testing of soils to be carried out on specimens subjected to the same state of stress that exists in the soil in place. This requirement aimed to prevent disturbance of the soil. Stress changes may also occur *in situ* during or as a consequence of construction works. Increases in effective stress lead to volume decreases in soil and hence to void ratio reductions and reductions in k. These effects are not usually of much significance in coarser-grained soils but in plastic clays, particularly if organic or if fissured, the influence of consolidation on the coefficient of permeability is quite marked owing to a reduction in void ratio and a closing or tightening of fissures.

Migration of fine particles

The drag or seepage force exerted on the soil particles by the flowing water may be sufficient to transport some of the fines through the void spaces in the coarser fraction. In most soils the particle size distribution will be such that this does not happen but in coarse uniform sands containing fines such migration is possible. Once the fines have been removed the remaining material is much more permeable. The phenomenon is known as internal erosion and may occur locally within a soil causing a 'pipe' of high permeability to develop. If this is allowed to grow in or beneath a hydraulic structure, the high flows may lead to undermining of the structure and eventual failure. This migration can be controlled to a large extent by preventing the loss of fines at the outlet boundary of the soil by means of filters (see section 23.6). Foreign matter in the flowing water, from solution, suspended solids and organic matter may also be present to clog the pore spaces and so reduce the permeability.

Air or gas in the soil

If the sand is not fully saturated, substantial reductions occur in the coefficient of permeability, a drop from 100 per cent to 85 per cent in the degree of saturation being accompanied by a drop of around 50 per cent in the value of k. The permeability will also change with changes in the pore pressure.

Determination of the Coefficient of Permeability k

In relatively uniform coarse-grained soil formations satisfactory estimates of k may be made based either on the particle size, shape and size distribution of the soil or on the results of laboratory permeability tests. Because of the variable and anisotropic nature of soils and because of sampling disturbance in obtaining core samples it is preferable to determine average values for k using tests carried out on the material *in situ.*

Indirect methods

The dependence of k on the particle size of a soil has been discussed and the simple and approximate expression

$$k = D^2{}_{10} \times 10^4 \ \mu\text{m s}^{-1}$$

presented, in which D_{10} is the effective size of the soil in millimetres. The expression was originally limited in application to clean fairly uniform sands having uniformity coefficients C_u less than 5, but recent work indicates that it is useful over a much wider particle size range and for nonuniform, well-graded sands.

The Kozeny - Carman expression (Loudon, 1952)

$$k = \frac{g}{k'S^2\eta} \ \frac{e^3}{(1+e)}$$

is a development which takes account of the particle size distribution and the density of packing of the soil particles, although not the orientation and arrangement of the particles, and gives values of k to better accuracy than the previous equation. k' is a factor which accounts for particle shape and hence pore shape and varies somewhat with porosity. S is the specific surface (particle surface area per unit *volume*) of the sand, η the viscosity of water, e the void ratio of soil and g the acceleration due to gravity.

Indirect estimates of k in the finer-grained soils are also deduced from information obtained in the oedometer or consolidation test in which the porewater is induced to flow out of the parent soil by squeezing or loading (chapter 24).

Direct methods in the laboratory

The more pervious soils can be directly tested for permeability using either the constant head or the falling head methods.

In the *constant head permeability test* (figure 23.8*a*) a prepared specimen of the soil to be tested is placed in the cell or mould of the permeameter. After saturating the specimen with de-aired water (the permeant) a difference H in total head is applied between the ends of the specimen or between two points l apart in the direction of flow and a steady state of flow is given time to develop.

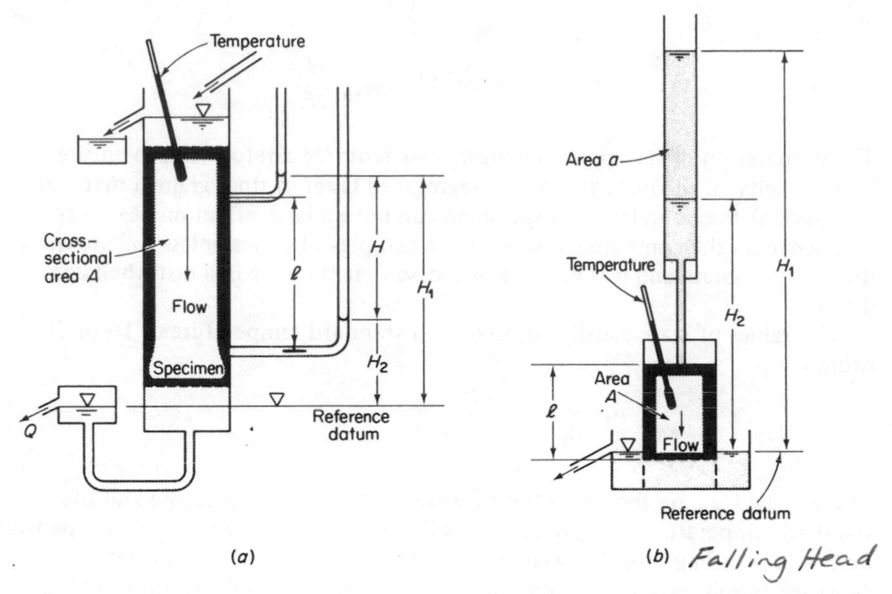

Figure 23.8 *Permeameters (a) constant head and (b) falling head*

The volume Q of water then passing through the specimen of cross-sectional area A in time t is measured.

Hence from Darcy's law a value of k can be calculated since

$$v = ki$$

$$\frac{Q}{At} = k \times \frac{H}{l}$$

or

$$k = \frac{Ql}{HAt}$$

The *falling head permeability test* arrangement is shown in figure 23.8*b*. Again the soil specimen is placed in the permeameter and saturated. A standpipe is connected to the specimen and filled with water. The water is allowed to drain from the standpipe through the soil specimen and the rate of fall of water level in the standpipe is measured. Usually several diameters of standpipe are available from which one is selected as giving a suitable rate of fall for measurement purposes with the particular soil under test. From continuity of flow in the standpipe and in the specimen

$$\mathrm{d}Q = -a\,\mathrm{d}H = k\frac{H}{l}A\,\mathrm{d}t$$

with symbols as defined on figure 23.8, from which

$$k = \frac{al}{At}\log_e\frac{H_1}{H_2}$$

or

$$k = \frac{2.3\,al}{At} \log_{10} \frac{H_1}{H_2}$$

The preparation of specimens for these tests requires careful work to ensure homogeneity of the material. A thin segregated layer of finer grained material within or at the boundary of a specimen can have a large effect on head loss and hence on the computed k-value. Core samples of cohesionless soil are quite difficult to obtain and it is likely in any case that they are in a disturbed condition.

The values of k are usually adjusted to a standard temperature of 10 or 20°C, using

$$k = \frac{\mu_t}{\mu_{st}} k_t$$

where μ_t and μ_{st} are the viscosities of water at the test temperature and the standard temperature respectively, the adjusted values being used for comparison of k-values among soils. If specimens are tested for permeability at several densities or void ratios, the value of k may also be adjusted to correspond to the density of the soil in the field.

Direct methods in the field

Because of the shortcomings of the sampling methods and methods of specimen preparation neither of the above tests can give results reliably representative of the *in situ* soil. The average flow in the permeameter is limited to one direction only through a specimen where in the field situation flow may occur in many directions. The hydraulic gradient in the laboratory test is usually much greater than that typically experienced in actual hydraulic structures and hence untypical internal erosion may occur in the test specimen.

It is possible to determine an average coefficient of permeability of a soil from field measurements in observation wells during pumping of water into or out of the soil.

Coefficient of Permeability of a Layered Soil

In an aquifer of total thickness H composed of n distinct horizontal layers of thickness $H_1, H_2, \ldots, H_n$ having isotropic coefficients of permeability $k_1, k_2, \ldots, k_n$ the equivalent coefficient of horizontal permeability of the aquifer H is

$$k_h = \frac{1}{H} \int_0^H k \, dH$$

$$= \frac{1}{H} (k_1H_1 + k_2H_2 + \ldots + k_nH_n)$$

since for horizontal flow the hydraulic gradient of the flow is the same along each layer.

The coefficient of vertical permeability of the aquifer is

$$k_v = \frac{H}{\int_0^H \mathrm{d}H/k}$$

$$= H \text{recip} (H_1/k_1 + H_2/k_2 + \ldots + H_n/k_n)$$

since, for a steady state with no volume change in the soil, vertical flow continuity gives the same flow rate across each layer. The permeability ratio k_h/k_v exceeds unity and may reach high values of a hundred and more.

23.6 Soils as Drainage Filters

When groundwater emerges from a soil through an exposed soil face or passes from a soil into a drainage system the groundwater is likely to carry with it some of the fines from the soil. This is called *internal erosion* or *piping* and if it is allowed to continue over a period of time it may lead to instability in the soil, because the loss of fines produces zones or 'pipes' of increased permeability and flow rate extending in upstream directions. The hydraulic gradient is increased in the approaches to these permeable zones and so the tendency to erosion is aggravated. Flow from an open outlet face of soil, for example, at a side slope of an earth dam, may cause *surface erosion*, also leading to instability by sloughing of the surface of the slope. These situations are commonplace in engineering and for safe and durable earthworks and for their efficient drainage the erosion must be largely prevented.

By placing a layer or layers of selected natural soil at the outflow surface, protection from erosion may be obtained. These selected soils are called *filters* and their use is important in the control and proper functioning of drainage works, wells and indeed in any case where seepage water emerges from soil. Some examples of their use are given in figure 23.9. A suitable filter material will have pore spaces too small to admit the larger particles of the soil it is protecting and these particles will then partially block the pores of the filter, so reducing the size of the passages into the filter. This prevents the entry of successively finer particles from the soil and soon a zone is developed at the soil - filter boundary which largely prevents the erosion of the soil. The size fractions most likely to migrate through and from soils are the silts and fine sands. These are coarse enough to allow fairly high seepage velocities and yet fine enough to be readily transported. Coarser soils are usually too large to be transported by flow through void spaces and the fine silts and clays do not permit sufficiently high seepage velocities to cause appreciable internal erosion.

The *selection of a filter* is therefore based on ensuring that the pore spaces in the filter are small enough to prevent the migration of the coarser sizes of the soil being protected. For efficient conveyance of the water emerging from the soil the filter should nevertheless be as permeable as possible. These two requirements are

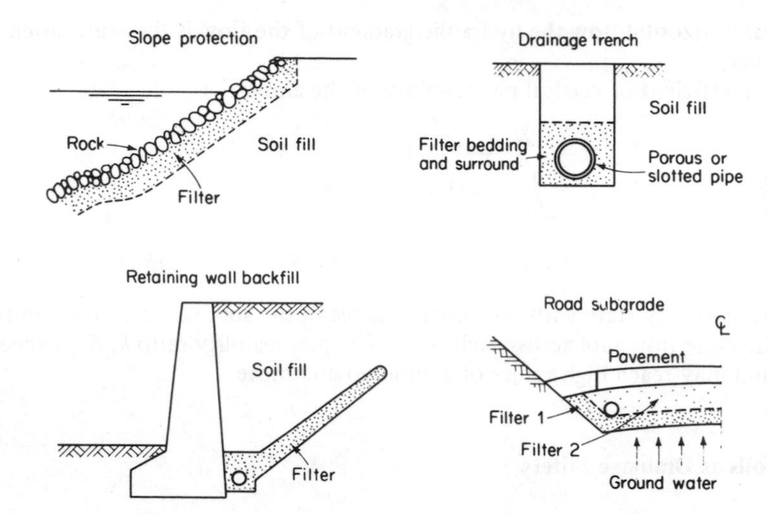

Figure 23.9 *Use of filters in the control of drainage*

in conflict and a compromise is necessary. A number of criteria exist to guide in the selection of filter soil gradings. The U.S. Waterways Experiment Station criteria are much used. These require that a filter should satisfy the following

(1) Piping requirement

$$\frac{D_{15}\text{ filter}}{D_{15}\text{ soil}} \leqslant 20$$

$$\frac{D_{15}\text{ filter}}{D_{85}\text{ soil}} \leqslant 5$$

$$\frac{D_{50}\text{ filter}}{D_{50}\text{ soil}} \leqslant 25$$

except that for uniform soils ($C_u \leqslant 1.5$)

$$\frac{D_{15}\text{ filter}}{D_{85}\text{ soil}} \leqslant 6$$

and for well-graded soils ($C_u > 4$)

$$\frac{D_{15}\text{ filter}}{D_{15}\text{ soil}} \leqslant 40$$

(2) Permeability requirement

$$\frac{D_{15}\text{ filter}}{D_{15}\text{ soil}} \geqslant 5$$

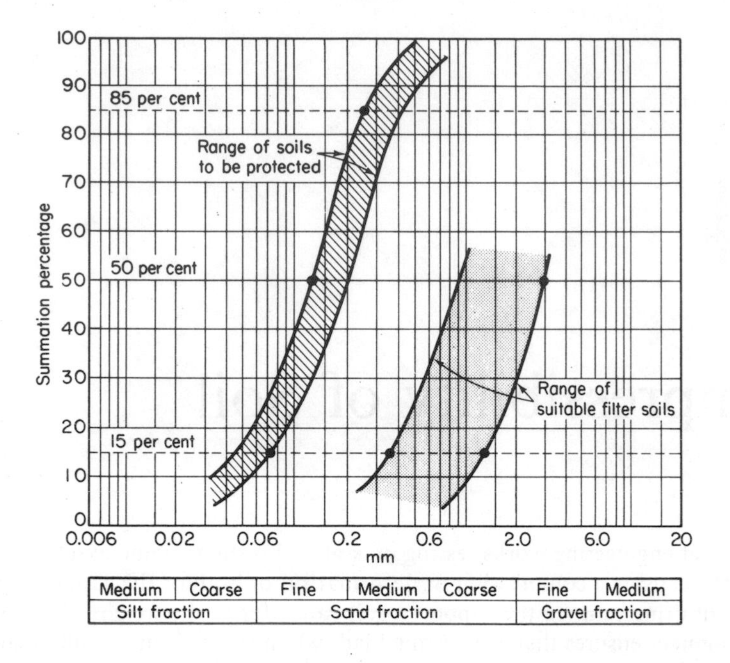

Figure 23.10 *Selection of suitable filter soils*

The filter material should not be gap graded. It is sometimes recommended that the particle size distribution curves of filter and soil should be approximately 'parallel'. An example of the range of filter soils for the protection of a base soil is given in figure 23.10.

Often a single filter is sufficient but two or more layers may be used to form a *graded filter* with greatly increased effectiveness in drainage. Each filter layer should meet the above criteria with respect to the next layer upstream. It has also been seen that seepage exerts a force on soil. If that force is upwards, countering the weight of the soil, instability may occur. Filters are sometimes used to improve safety in such cases by acting as a surcharge on soil. Since a filter is by design relatively highly permeable the seepage forces in it will be low and its weight will be effective as a surcharge. Such filters are called *loaded filters*.

Where filter flow is collected in perforated or porous pipes for conveyance, the openings in the pipes must also be selected to prevent inflow of fines and clogging. The following criterion has been found satisfactory

$$\text{circular hole diameter} \leqslant D_{85} \text{ filter}$$

$$\text{slot width} \leqslant 0.83\, D_{85} \text{ filter}$$

24

Compressibility of Soil

Buildings and engineering works resting on soil will settle to some extent owing to their weight causing a compression or deformation or both of the supporting soil. There is a limiting value to the support a soil can offer to the weight of a building and the engineer ensures that at working load, when the building is fully loaded, an adequate factor of safety exists. That is

$$\text{working load} = \frac{1}{F} \times \text{failure of limiting load}$$

where F will usually have a value of about 2.5 for ultimate failure.

Settlements will occur at the working load and lesser loads and the engineer makes predictions of the amounts and rates of these settlements based on measurements or estimates of the compressible quality of the soil. To do this he requires a stress – strain relationship for soil. In principle this is extremely complex and has not yet been fully developed for soil. For practical usage therefore some simplifications are made to allow predictions of vertical settlement. One simplification is to make use of the theory of elasticity for isotropic materials. In the application of the theory of elasticity it is not necessary that a material should show elastic rebound characteristics but simply that over a range of a stress increase (or decrease) the stress is proportional to strain.

The apparatus generally used in the laboratory for the study of the stress – strain behaviour of soils is the *triaxial compression apparatus*. The features of this are shown in figure 24.1 and these will receive further reference in chapter 25. The test specimen of soil is cylindrical in form and of diameter usually 100 mm or more. The length is commonly twice the diameter, but if lubricated end platens are used, as is preferable (Rowe and Barden, 1964), the length may equal the diameter. The specimen is enclosed in a rubber membrane and the whole placed in a cell which is filled with a liquid (water, usually) and the latter put under pressure σ_3. In this condition the specimen is subjected to a *hydrostatic state of stress*, σ_3 all round. An additional vertical stress $(\sigma_1 - \sigma_3)$, termed the *deviator stress*, is applied axially on to the end faces of the cylindrical specimen. The stress system is then one of bi-axial symmetry, as in figure 24.2.

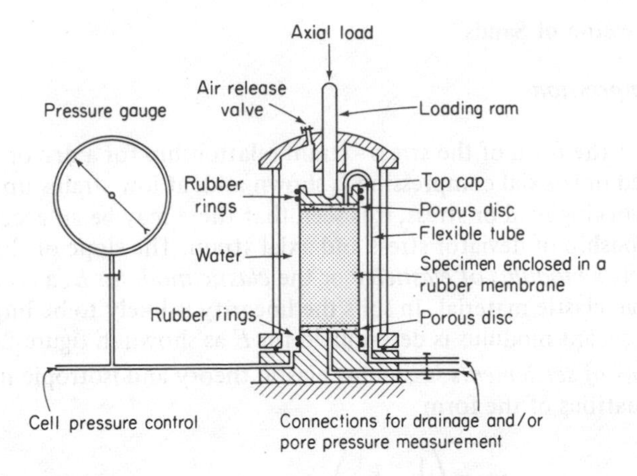

Figure 24.1 *Arrangement of the triaxial compression cell, showing the environment of the test specimen*

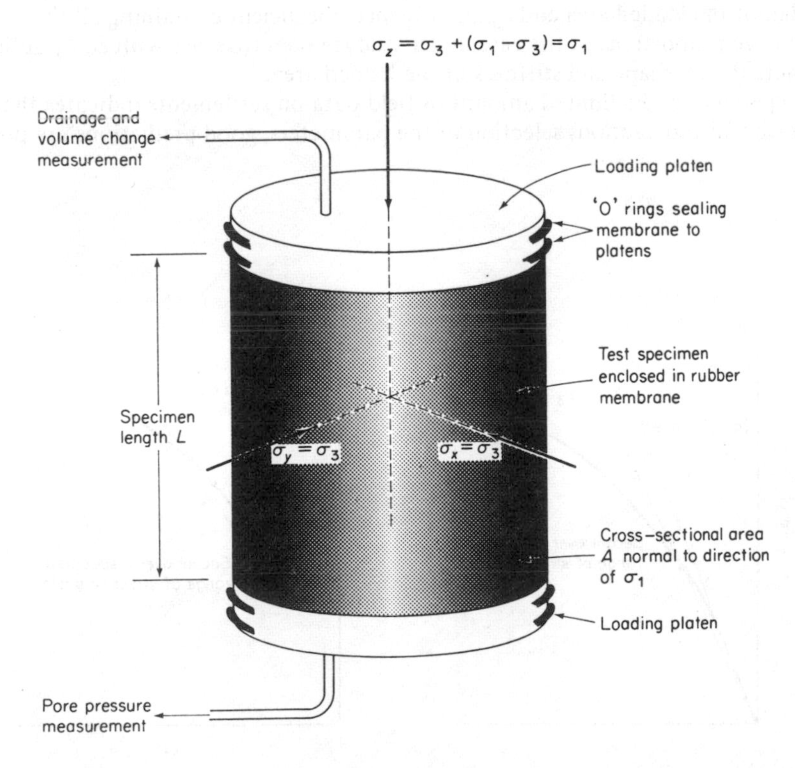

Figure 24.2 *Stress system on specimen in the triaxial compression test ('hidden' reaction vectors σ_x, σ_y, and σ_z omitted)*

24.1 Compression of Sands

Triaxial Compression

In figure 24.3 the form of the stress - strain relationship for a dry or freely draining soil tested in triaxial compression is shown and, at low strains up to the maximum working load or stress, it is seen that there may be an acceptable linearity to the relationship of deviator stress and axial strain. The slope of this initial part defines *Young's modulus of elasticity* or the *elastic modulus* E, a constant for an ideal isotropic elastic material. In soils the linearity is likely to be imperfect and a tangent or a secant modulus is determined for E as shown in figure 24.3.

Predictions of settlements, assuming elastic theory and isotropic material, are based on equations of the form

$$\rho = \Delta q\, B \left(\frac{1 - \nu^2}{E}\right) I_\rho$$

where ρ is now the vertical settlement of a loaded area on the horizontal surface of the material, Δq the increment of vertical stress causing the settlement, ν Poisson's ratio, the other elastic constant of an elastic material, B the least dimension in plan of the loaded area and I_ρ an influence coefficient containing all the geometric proportions of the case under study; taken together with B, I_ρ defines the actual size, shape and stiffness of the loaded area.

Experience and a limited amount of field data on settlements indicates that with careful and cautious selection of the parameters, good predictions are possible.

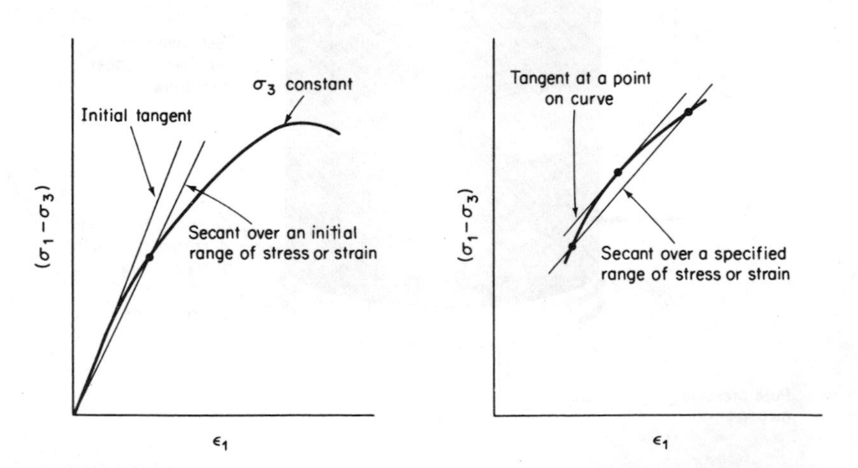

Figure 24.3 *Estimation of elastic modulus E from the stress - strain relationship of a soil*

Confined or Constrained Compression

If the cylindrical test specimen is constrained such that $\epsilon_x = \epsilon_y = 0$ the ratio of axial stress to axial strain is now called the *constrained modulus* E^*. These conditions are met in the oedometer test described in section 24.3.

The use of a Young's modulus E or a constrained modulus E^* in predicting settlements in an isotropic material would be quite straightforward. The value of the modulus, however, depends on sampling disturbance and on many other factors, not all clearly understood. The soil itself is unlikely to be isotropic and, even if homogeneous in appearance, will have a history of stressing which exerts an influence on its compressibility. In sands, provided crushing of the grains does not occur, the modulus E or E^* tends to increase with increasing relative density D_r; with increasing applied and confining pressures and hence with increasing depth; and with successive cycles of loading (the increase per cycle being large on the first cycle and diminishing on successive cycles until a fixed stress - strain curve is reached). Grain shape, grading and composition all have an influence on the modulus.

The stress - strain data from confined compression tests on soils are also commonly presented graphically as the void ratio e plotted against the corresponding effective vertical stress σ_z' (= σ_z when $u = 0$) or against log σ_z'. These plots for a sand and a clay are shown in figures 24.4*a* and *b*. In dry or freely draining sands the void ratio change which accompanies the application of a stress change occurs almost without delay so that for these soils the data in figure 24.4 are not time dependent. The slope of the $e - \sigma_z'$ curve in figure 24.4*a* at any stress σ_z' or stress range $\Delta\sigma_z'$ is described as the *coefficient of compressibility* a_v.

$$a_v = -\frac{de}{d\sigma_z'}$$

or

$$a_v = -\frac{\Delta e}{\Delta\sigma_z'}$$

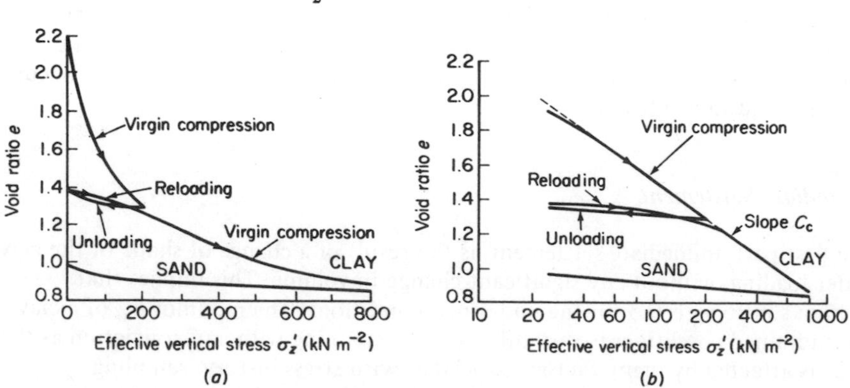

Figure 24.4 *Confined compression test data plotted as (a) void ratio e against effective vertical stress* σ_z' *and (b) void ratio e against log* σ'_z

and for that stress or stress range a_v is a parameter defining a soil property (compressibility). In figure 24.4 it should be noted that the unloading curve does not coincide with the preceding limb of the loading curve and the reloading curve does not coincide with either of these until at higher stresses it follows the extension of the preceding loading curve.

Another parameter in common use is the *coefficient of volume compressibility* m_v where

$$m_v = \frac{a_v}{1 + e_0} = \frac{1}{E^*}$$

where e_0 is the initial void ratio and E^* the constrained modulus of elasticity. The plot in the form of figure 24.4*b* enables the soil compressibility to be displayed over a wide range of stresses. Usually this plot becomes linear at the higher stresses and the slope of the curve, usually its linear part, is described as the *compression index* for the soil.

$$C_c = \frac{de}{d(\log \sigma'_z)}$$

or

$$C_c = -\frac{\Delta e}{\Delta(\log \sigma'_z)} = -\Delta e_c$$

where Δe_c is the change of void ratio on the linear portion of the plot, in one logarithmic cycle of stress. The semi-logarithmic plot of figure 24.4*b* is particularly useful in presenting the compressibility of clay soils. The slope of the swelling limb of the plot of figure 24.4*b* is a *swelling index* C_s.

24.2 Compression of Clays – Short Term

In clays the problem of estimating settlement is especially difficult and the amount of ultimate settlement is usually considered as comprising two principal parts – an *immediate* or *undrained settlement* occurring as the loading is applied to the saturated clay, and a long term settlement called *consolidation settlement* occurring under sustained loading.

Immediate Settlement

The first part, immediate settlement, is the result of a change of shape of the clay under loading, without any significant change in volume. This implies that Poisson's ratio ν_u is 0.5 for the undrained conditions. The modulus E_u of a clay for undrained conditions is very difficult to determine with any conviction as the value is affected by many factors associated with stress history, sampling disturbance and test procedure.

If a natural stratum of saturated clay is relatively thin and is subjected to a stress increment over a large area in plan, that stratum is fully constrained and practically incompressible on loading. In this case the constrained modulus E^* is infinite and immediate settlements negligible. Where this constraint does not exist, that is, in a thicker stratum or one stressed over a small area, estimates of immediate settlement based on E_u and ν_u may be required. For a saturated clay ν_u is taken as 0.5 and E_u may be estimated from the results of laboratory triaxial compression tests on specimens from carefully extracted block samples or large diameter piston samples. Until a generally agreed test procedure is available it would seem reasonable first to restore the *in situ* stress system on each test specimen as closely as possible and to allow the specimen to stabilise (consolidate). The specimen is then subjected to loading up to the working stress and then unloading. This is repeated for as many as five cycles of loading and unloading and what should be a reasonable estimate of E_u is then estimated from σ_z/ϵ_z on the last loading. Use of a value of E_u from the first loading appears to give excessive values of immediate settlement and the effect of applying one or two cycles of loading may be to offset sampling disturbance.

In situ Tests for Immediate and Short-term Compression

The difficulties of obtaining reliable elastic parameters in the laboratory to define soil behaviour together with the nonlinear and anisotropic behaviour of real soils, have led to the use of *in situ* tests to determine these elastic parameters or to classify soils sufficiently closely to allow predictions of immediate settlement to be made. These *in situ* tests include plate loading penetration tests and pressure meter tests.

24.3 Compression of Clays – Long Term

The compression producing *consolidation settlement* in a clay is due to a gradual reduction in the volume of the voids accompanied by a compression of the assembly of solid particles forming the clay. The water and the solid matter from which the particles are formed are themselves relatively incompressible whereas the assembly of solid particles forming the clay is relatively compressible.

For example, Costet and Sanglerat (1969) list the changes in volume under a pressure of 100 kN m^{-2} as follows

water	1 : 22 000
solids comprising the particles	1 : 100 000
assembly of clay particles	1 : 100

Any air or gas in the voids spaces would be highly compressible.

When a sustained increment of loading, a stress increment, is applied to a saturated clay this increment is initially supported by the porewater – which is relatively incompressible, as just discussed. The porewater in the stressed volume of the clay is thus at a greater pressure than the porewater in the surrounding clay

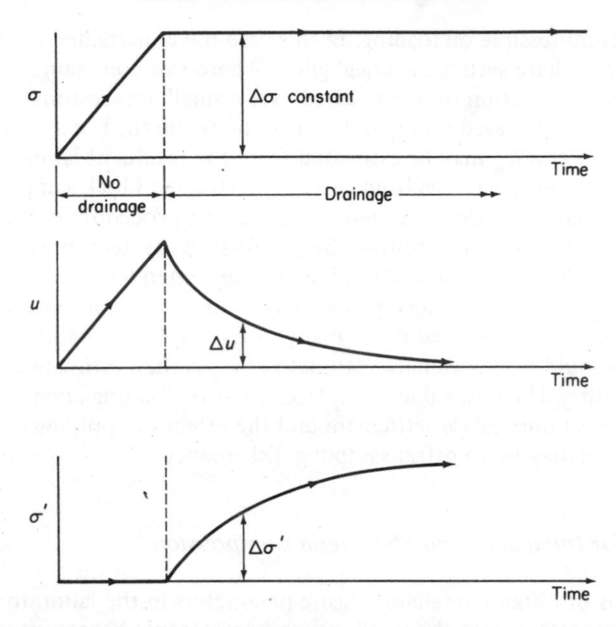

Figure 24.5 *Pore pressure dissipation during drainage, producing increase in effective stress (and also a decrease in volume)*

and a hydraulic gradient now exists. If drainage is free to take place, porewater flows from the stressed zone with a reduction in volume occurring in the voids of the clay in that zone, and the stress increment is gradually transferred to the particle assembly as an increment of effective stress. Since clays are relatively impervious this flow and stress transfer take some time. The process is illustrated in its simplest form in figure 24.5 where at any time t, $\Delta\sigma = \Delta\sigma' + \Delta u$. At $t = 0$ $\Delta\sigma = \Delta u$ and as $t \to \infty$, $\Delta\sigma \to \Delta\sigma'$. Predictions of the rate of settlement require a theoretical model of the process which, through appropriate parameters, applies to any given clay.

One-dimensional Consolidation

The laboratory apparatus used for studying consolidation behaviour in constrained compression is the *oedometer* and two forms of this are shown diagrammatically in figure 24.6. The detailed procedure for using the form of oedometer in figure 24.6*a* is given in BS 1377. A specimen of the saturated clay to be tested is accurately trimmed to cylindrical form and confined in a metal ring or cell, as illustrated, to prevent strains taking place laterally when a vertical external stress is applied. This stress is applied to and sustained on the specimen through rigid porous discs which permit vertical drainage of porewater from the clay. The vertical compression of the clay is measured at suitable intervals of time until the compression is completed. This process is described as *one-dimensional consolidation* and is often explained by reference to the *spring analogy* (figure 24.7). The

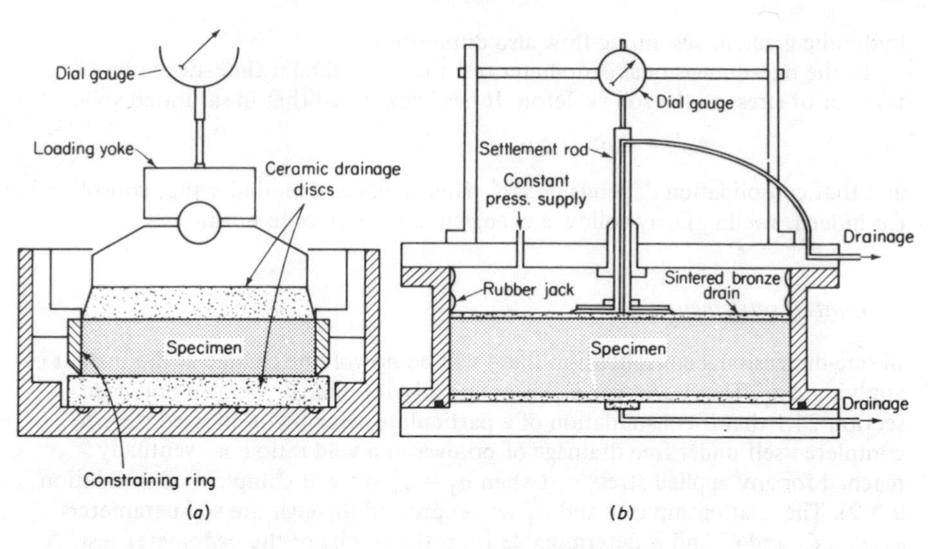

Figure 24.6 *Oedometers (a) conventional fixed ring and (b) Rowe types*

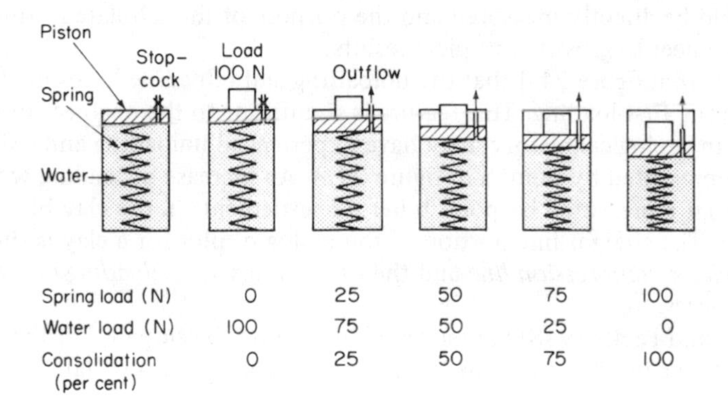

Figure 24.7 *Piston-and-spring analogy*

spring represents the compressible assembly of particles forming the soil, sometimes called the soil skeleton; the water in the cylinder represents the porewater in the soil; the stopcock on the piston represents the permeability of the soil. Thus if no drainage is permitted as a stress increment $\Delta\sigma_1$ is applied, this stress increment is carried by the pore pressure, as in the first stage of figure 24.5. Because the water is incompressible none of the stress $\Delta\sigma_1$ is carried by the spring at this stage and an equal stress increment Δu is experienced in the porewater, that is, $\Delta u/\Delta\sigma_1 = 1$. Opening the stopcock permits drainage to occur and as the pore pressure dissipates the external stress is transferred to the spring as in the second stage of figure 24.5, the rate depending on the stopcock opening. The rate diminishes asymptotically with time since with dissipation of pore pressure the hydraulic gradient sustaining flow also diminishes.

In the one-dimensional oedometer test there is a similar time-dependent transfer of stress to the soil skeleton. It has been noted that in saturated soils

$$\Delta\sigma' = \Delta\sigma - \Delta u$$

and that consolidation depends on $\Delta\sigma'$. This serves as a reminder that consolidation (or indeed, swelling) may follow a change in σ or in u, or in both.

Amount of settlement

In one-dimensional consolidation there will be no volume change at the instant of applying $\Delta\sigma_1$. That is, E^* (or E_u) is practically infinite. It was also noted in section 24.1 that if consolidation of a particulate assembly is allowed time to complete itself under free drainage of porewater a void ratio e is eventually reached for any applied stress σ_1 (when $\sigma_1 = \sigma_1'$ since at complete consolidation $u = 0$). The relationship of e and σ_1' was expressed through the soil parameters a_v, m_v, C_c and C_s and is determinable from the results of the oedometer test. A knowledge of these parameters enables the engineer to answer questions about how much settlement is likely to occur.

Table 24.1 lists approximate values of m_v and C_c for a range of soils. The values should be directly measured and the purpose of the tabulated values is to alert the engineer to grossly untypical results.

It is seen from figure 24.4 that the unloading and reloading limbs of the curve differ from the first loading. This feature is significant to the engineer in that a given clay, in geological history, may have experienced unloading and exists today in a state represented by point a in figure 24.8. An increase of loading would need to exceed that represented by point b before settlements in the clay become appreciable. The straight-line portion of the e - log σ' plot for a clay is therefore called the *virgin compression line* and the other limbs are *unloading (swelling)* and *reloading curves*.

Terzaghi and Peck (1948) have assembled data on the slope C_c of the virgin compression line in relation to the liquid limit of clays to show that

TABLE 24.1
Compressibility of clays

Soil description	Compressibility	m_v (m^2 MN^{-1}) or (MPa^{-1})	C_c
Heavily overconsolidated clays	Very low	< 0.05	< 0.10
Very stiff to hard clays	Low	0.05 - 0.10	0.10 - 0.25
Medium clays	Medium	0.10 - 0.30	0.25 - 0.80
Normally consolidated clays	High	0.30 - 1.5	0.80 - 2.50
Very organic clays and peats	Very high	> 1.5	> 2.50

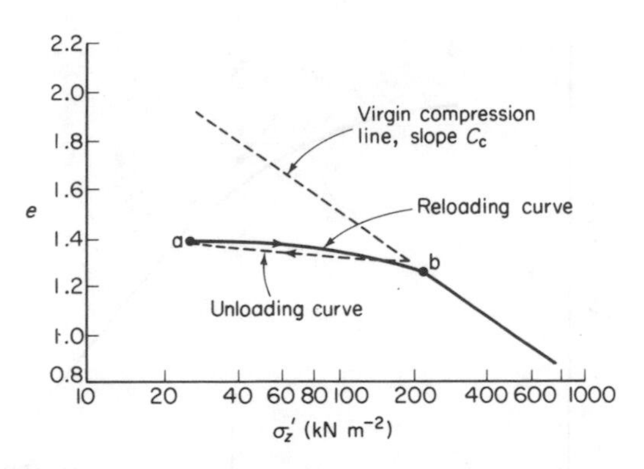

Figure 24.8 *Consolidation settlement on reloading a clay*

$$C_c = 0.009\,(w_l - 10) \text{ for undisturbed soil}$$

or Cc = (L.L. − 10) %

and

$$C_c = 0.007\,(w_l - 10) \text{ for remoulded soil}$$

where w_l is the liquid limit in per cent. This usefully supplements values of C_c measured in the oedometer test. Because of its importance in the prediction of settlements, the virgin compression line has received considerable study. Two principal items arise: first, that the slope C_c is appreciably affected by sampling disturbance and, to be representative of field behaviour of the clay, some adjustment of the laboratory data is required; second, since the settlements on reloading are much less than those on virgin compression it is necessary to classify the present state of the actual clay which may be

(1) *Normally consolidated*: that is, the clay is consolidated under its present state of stress in place and this has never been exceeded. Its present void ratio is therefore represented by a point on the virgin compression line.

(2) *Overconsolidated*: that is, the clay has been consolidated in the past under loading greater than the existing loading today. Its present void ratio is represented by a point on a reloading curve.

(3) *Incompletely consolidated*: this means that excess pore pressure exists in the clay and has not yet dissipated. Future settlements will therefore comprise the remaining settlement associated with the existing stress conditions together with settlement arising from new loading.

The *maximum past consolidation pressure* σ_c' of a clay is the maximum vertical effective stress experienced by the clay in its geological history. If σ_0' denotes the current vertical effective *overburden stress*, then for a normally consolidated clay $\sigma_c'/\sigma_0' = 1$ and for an overconsolidated clay $\sigma_c'/\sigma_0' > 1$. For an incompletely consolidated clay $\sigma_c'/\sigma_0' < 1$. The ratio σ_c'/σ_0' is known as the *overconsolidation ratio* (OCR) of the clay. The maximum past pressure is commonly estimated using a method proposed by Casagrande. Referring to figure 24.9 the method consists of

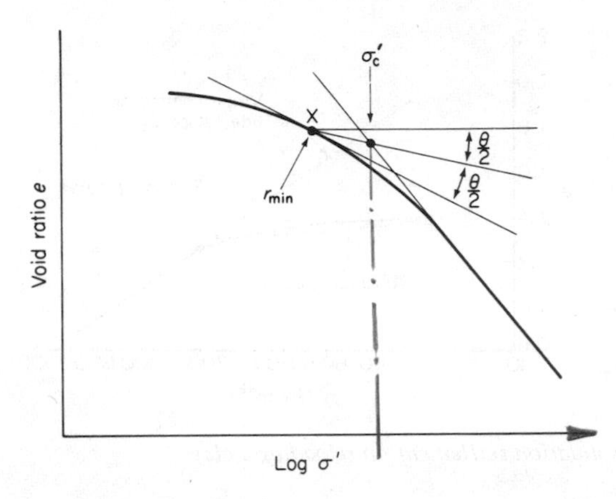

Figure 24.9 *Estimation of the maximum past consolidation pressure σ'_c experienced by a clay*

determining the point X of minimum radius of curvature on the laboratory e - log σ' curve. A line is then drawn through X parallel to the log σ' axis and through X a tangent is also drawn to the curve. The bisector from X of the angle between these lines is drawn and the point where the bisector cuts the extension of the virgin compression line is considered to represent the state of the clay at the maximum past pressure σ'_c. There are other methods but the above is simple, if approximate.

For field conditions which satisfy the assumptions of one-dimensional consolidation, the *total vertical settlement* ρ of a clay layer of thickness H is given by

$$\rho = \sum_{0}^{H} m_v \, \Delta\sigma'_v \, \Delta z$$

where m_v, $\Delta\sigma'_v$ may vary from element to element Δz of H. If conditions of stress and soil property at the middle of the layer are representative of the whole layer, this may be written

$$\rho = m_v \Delta\sigma' H$$

where m_v and $\Delta\sigma'$ are now the values at the middle of the layer. Alternatively, $\rho = H(e_0 - e)/(1 + e_0)$ where e_0 is the void ratio at the initial state of stress σ' and e is the void ratio after complete consolidation under the increment $\Delta\sigma'$. This can be written

$$\rho = \frac{H}{1 + e_0} C_c \log \frac{\sigma' + \Delta\sigma'}{\sigma'}$$

The soil parameters have been determined on the basis of one-dimensional compression and drainage.

Skempton and Bjerrum (1957) have proposed a method of using data from oedometer tests together with pore pressure parameters to predict settlements in three-dimensional conditions.

Rate of settlement

Questions relating to the rate and the duration of the settlement process often have to be answered by the engineer but this aspect of settlement is beyond the scope of this book.

25

Shear Strength of Soil

A soil may be considered to have failed to support a built structure if the soil compresses or settles (or swells) to an extent which causes damage to the structure. This aspect was the subject of the previous chapter. When reference is made to *failure of a soil,* however, its failure in shear is usually meant, that is, the state of stress in the soil is such that the shearing resistance of the soil is overcome and a relative and significant displacement occurs between two parts of the soil mass. If this shearing resistance, or *shear strength*, of the soil is measured or predicted the geotechnical engineer is then able to analyse problems of stability of soil masses and to estimate factors of safety against the occurrence of failure by shear within such masses.

From the emphasis already placed on the complex nature of soil formations it will be understood that the measurement and representation of shear strength present difficult problems to the engineer. The strength is likely to differ from one major stratum to another in a natural soil formation and will differ also at any one location with the direction of measurement of the strength. The fabric of laminations and fissures in some clay soils makes it necessary to test quite large elements of the mass in order to obtain a representative value of strength. In a fissured material it is once more the defects which decide its behaviour, not the quality of the intact material between fissures.

A further complication lies in predicting the future behaviour of the soil. It has been noted that the clay soils are the principal group affected by time-dependent changes. If a clay is subjected to an increased sustained loading it will in time consolidate and become stronger. This behaviour is acknowledged in methods of improving (stabilising) clays by preloading or controlled loading – that is, by applying a loading for the purpose of consolidating the soil in advance of building construction, or by controlling the rate of increase of loading, from storage tanks say, to allow the increasing strength of the clay to keep pace with the increasing loading upon it. If the clay is subjected to a sustained relief of loading the soil will in time swell and become softer. In a relatively uniform clay this may take a number of years but in a clay formed with a fabric of fissures and drainage channels the strength may be affected quite quickly, even within the construction period. A relief of loading would occur in cases such as cuttings for side slopes in highways, trench and retaining-wall excavations.

A close study of shear strength of soils and its relationship to the consolidation process is beyond the scope of this book and only some of the main characteristics will be explored here.

25.1 Graphical Presentation of Stress

A full description of the state of *stress at a point in a material*, soil in the present case, is given by the components of the stress tensor (see, for example, Schofield and Wroth, 1968) and from this the actual stress on any plane through the point may be deduced. The point may be any point within the mass of soil or it may simply be representative of conditions in the laboratory test specimen. The three mutually perpendicular planes at the point on which the shearing stresses are zero are defined as principal planes and the (normal) stresses on them are principal stresses. In soil mechanics literature these are conventionally described as *positive if compressive*. The largest of these stresses is called the major principal stress σ_1, the smallest is the minor principal stress σ_3 and the remaining principal stress is the intermediate principal stress σ_2.

The influence of the intermediate principal stress σ_2 is obscure and in this chapter the study is limited to the two-dimensional stress state, this being wholly defined by σ_1 and σ_3.

The *Mohr diagram* is much used to display one given state of stress at a point in a material. Figure 25.1 shows the components of two-dimensional stress at a point whose stress tensor is (σ_1, σ_3). The Mohr circle is the locus of the values of (σ, τ) on all possible planes at the point.

Sometimes it is desired to follow the state of stress through a series of controlled changes of stress and, to avoid obscuring the pattern of change in a large number of Mohr circles, a *p-q diagram* may be used. In this diagram only the points $p = \frac{1}{2}(\sigma_1 + \sigma_3)$, $q = \frac{1}{2}(\sigma_1 - \sigma_3)$ are plotted throughout the change of stress. The locus of points p, q for a stress change is called a *stress path* and at any point on this locus the coordinates p, q enable the Mohr diagram to be drawn for that particular state of stress at the point. For illustration, the stress path in a simple triaxial compression test (figures 24.1 and 24.2) is shown in figure 25.2, the

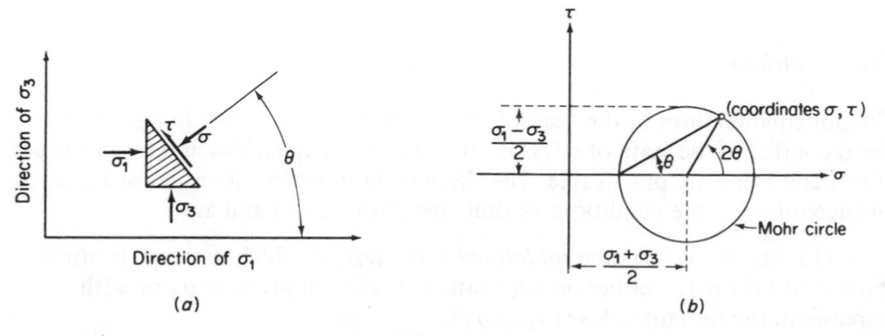

Figure 25.1 *Representation of two-dimensional stress at a point : (a) directions of stresses at the point and (b) Mohr diagram for stress at the point*

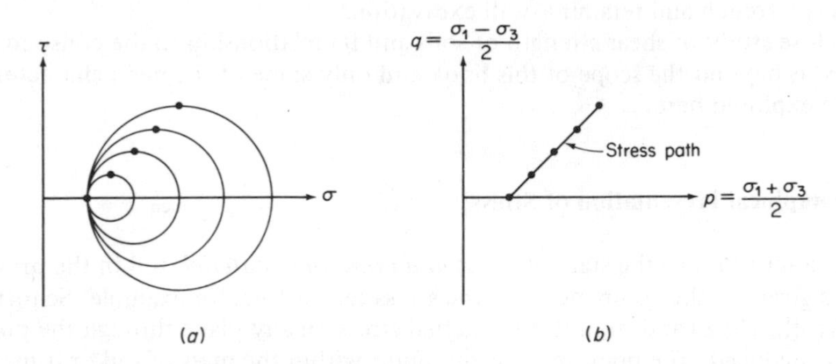

Figure 25.2 *Representation of a changing state of stress : (a) Mohr circles for successive states and (b) p – q diagram. The corresponding points • on the two diagrams represent the same stress conditions*

initial application of cell pressure σ_3 being followed by an increasing deviator stress $(\sigma_1 - \sigma_3)$ until a maximum (failure) condition is reached. It is the failure state of soils that is the main concern in this chapter on shear strength.

25.2 Shear-strength Testing in the Laboratory

The triaxial compression apparatus, including unconfined compression, is the most common laboratory apparatus for determining the stress – strain and shear-strength properties of soils. The simpler direct shear apparatus retains a place in the laboratory for testing the coarser sands, gravels, crushed rock and like materials. Because they are usually limited to the testing of rather small and sometimes rather disturbed specimens these tests may not be capable of giving results of value to the engineer. The obvious procedure to avoid disturbance is to test soils *in situ*. Such tests still only test rather a small volume of soil and the advancing of pits or bores to the point of test subjects the soil to some disturbance. The control of tests and boundary conditions are often uncertain so that problems remain in the interpretation of results. An outline of some *in situ* tests is given in section 25.3.

Test conditions

The principal features of the *triaxial test apparatus* were shown in figure 24.1. Good control of the state of stress is attained in this apparatus and several types of test condition are practicable. The three main types in common use are classified according to the conditions of drainage during a test and are

(1) *the unconsolidated undrained* (UU) *test*, in which drainage is not allowed to take place, either on application of the cell pressure σ_3 or with increases in the deviator stress $(\sigma_1 - \sigma_3)$;

(2) *the consolidated – undrained* (CU) *test*, where the specimen is first allowed to consolidate (drain) on application of σ_3 and when excess pore pressures

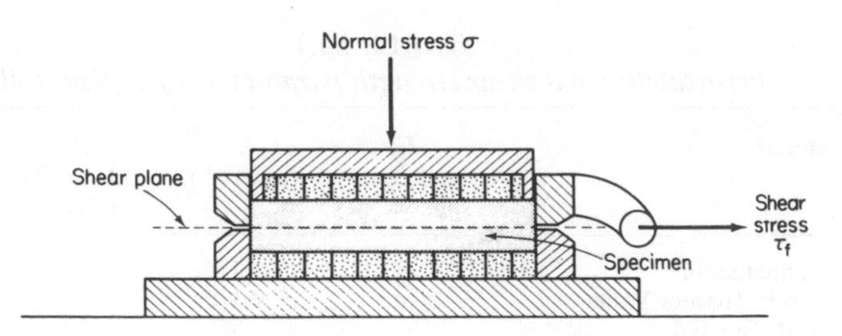

Figure 25.3 *Direct shear box, vertical section showing stressing arrangement. The top half of the box and the normal stress platen are free to move up or down as volume changes occur during shearing of the soil specimen*

are fully dissipated, that is, consolidation is complete, the deviator stress is applied under conditions of no drainage;

(3) *the drained or consolidated - drained* (CD) *test*, in which drainage is allowed throughout the test and no excess pore pressure is allowed to build up at any time.

In the triaxial compression test the specimen is cylindrical and usually 38 mm or 100 mm in diameter, as described in chapter 24. Sizes up to 250 mm in diameter are sometimes required to obtain sufficiently representative specimens.

The limited control of drainage available in the *direct shear test* means that this method is usually used for drained tests on sands, gravels and similar coarse-grained soils. The direct shear apparatus is simple in concept. The soil to be tested is placed in a square box-shaped container which is split horizontally across at mid-height, as shown diagrammatically in figure 25.3. A vertical stress σ is applied and the maximum horizontal shearing stress τ_f is measured.

The soil particle size determines the appropriate size of the shear box. For tests on fine to medium sands a box 60 mm square by about 20 mm thick is often used, but for testing gravels, broken rock and like materials a much larger box is required.

Coarser-grained Soils

When their strengths are expressed in terms of *effective stresses* there is a great similarity between the behaviour of sands and gravels on the one hand and clays on the other. There is also, however, a great difference in the permeabilities of these two extremes of particle size, as shown in table 25.1, and this leads to widely differing rates of response to the application of stress. It has been noted that a change $\Delta\sigma$ in total stress applied to a freely draining sand produces an immediate response in a corresponding effective stress change $\Delta\sigma'$ and also a volume change. In a saturated relatively impervious clay, however, the response to a total stress change is not complete until sufficient time has elapsed to allow the clay to drain or absorb water and reach a new equilibrium under the changed stress.

It is convenient therefore to look first at the stress - strain and strength characteristics of the coarser-grained soils straining under drained conditions in which $\sigma = \sigma'$ and $u = 0$.

TABLE 25.1

Permeability and shear strength parameters of typical soils

Material	I_p (per cent)	k (μm s^{-1})	c' (kN m^{-2}) or (kPa)	ϕ' (degrees)
Rockfill: tunnel spoil	–	5×10^4	0	45
Alluvial gravel: Thames Valley	–	5×10^2	0	43
Medium sand: Brasted	–	–	0	33
Fine sand	–	1	0	20 – 35
Silt : Braehead	–	3×10^{-1}	0	32
N.C. clay of low plasticity (undisturbed samples)	20	1.5×10^{-4}	0	32
N.C. clay of high plasticity (undisturbed samples)	87	1×10^{-4}	0	23
O.C. clay of low plasticity (undisturbed boulder clay samples)	13	1×10^{-4}	8	$32\frac{1}{2}$
O.C. clay of high plasticity (undisturbed London clay samples)	50	5×10^{-5}	12	20
Quick clay (undisturbed samples)	5	1×10^{-4}	0	10 – 20

From Bishop and Bjerrum (1960).

Internal friction and cohesion

The shear resistance of an assembly of particles is essentially frictional in character. The oversimple analogy to a sliding block is commonly made (figure 25.4). When a normal compressive stress σ is applied through the block to the supporting surface, a relative displacement in shear between the block and the surface will not occur until the shear stress τ reaches a value τ_f at which the available shearing resistance between the material surfaces is fully mobilised and the obliquity of the

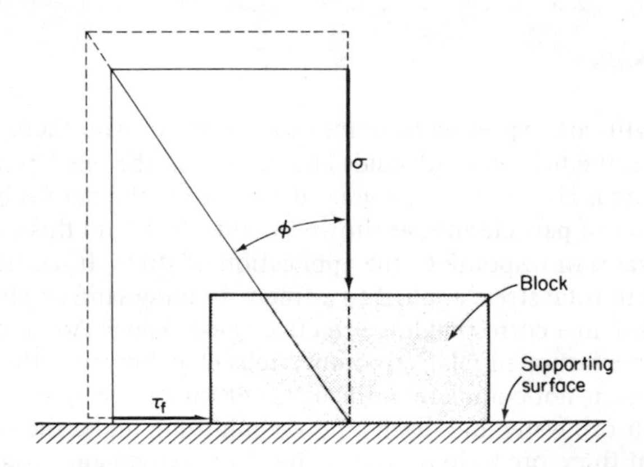

Figure 25.4 *Sliding block analogy*

resultant stress is then ϕ, as shown. A change in the value of σ would lead to a change in τ_f but in every case the obliquity of stress at the failure condition would remain at ϕ. The angle ϕ is thus the parameter which determines the shearing resistance and τ_f/σ = constant = $\tan \phi$ for all values of σ. Tan ϕ is the coefficient of friction familiar to students of engineering mechanics. When a shear failure occurs within a soil a zone or surface of shearing exists between the parts experiencing relative displacement and in this zone the shearing stress τ_f is proportional to the normal stress σ as in the sliding block analogy. If there is still a shearing resistance c in a soil when this normal stress is reduced to zero that resistance is described as *cohesion*. The shearing resistance is then generally expressed by the Mohr - Coulomb failure criterion as

$$\tau_f = c + \sigma \tan \phi$$

At this stage a coarse-grained soil is being considered in which $u = 0$ where the value of c is denoted by c' and is insignificant so that

$$\tau_f = \sigma' \tan \phi'$$

in which ϕ' expresses the frictional characteristic of the soil and is known as the *angle of shearing resistance* of the soil in terms of effective stresses σ'. It may also be noted that in the failure state the principal stress ratio

$$\sigma'_1/\sigma'_3 = \frac{1 + \sin \phi'}{1 - \sin \phi'} = \tan^2 [45^\circ + (\phi'/2)]$$

where $[45^\circ + (\phi'/2)]$ is the angle between the major principal plane and the theoretical plane of shear at the point.

It should be understood throughout this introduction to shear strength that the parameters c and ϕ for a soil are not unique to the soil alone but are functions of both the soil and the conditions of testing.

The angle of shearing resistance ϕ' of a coarse-grained soil is more than simply a measure of the friction between two sliding surfaces; it embraces effects of interlocking of particles and their rolling and sliding actions when subjected to shear. It is apparent then that while the value of ϕ' is dependent on the soil minerals, it is also a function of the particle shape, size and grading, and the initial density of packing of the soil.

Maximum stress and volume change on shearing

Stress - strain curves for loose and dense sand specimens tested in triaxial compression have the general form shown in figure 25.5. The dense specimen shows a peak resistance in terms of deviator stress, thereafter continuing to shear at a lower stress. The loose specimen shows an increasing resistance until a maximum deviator stress is reached and thereafter the deviator stress remains more or less constant.

When the volume changes are examined it is found that the dense specimen expands on increasing the axial compressive strain while the loose specimen decreases in volume at first, followed by a partial recovery in volume. These

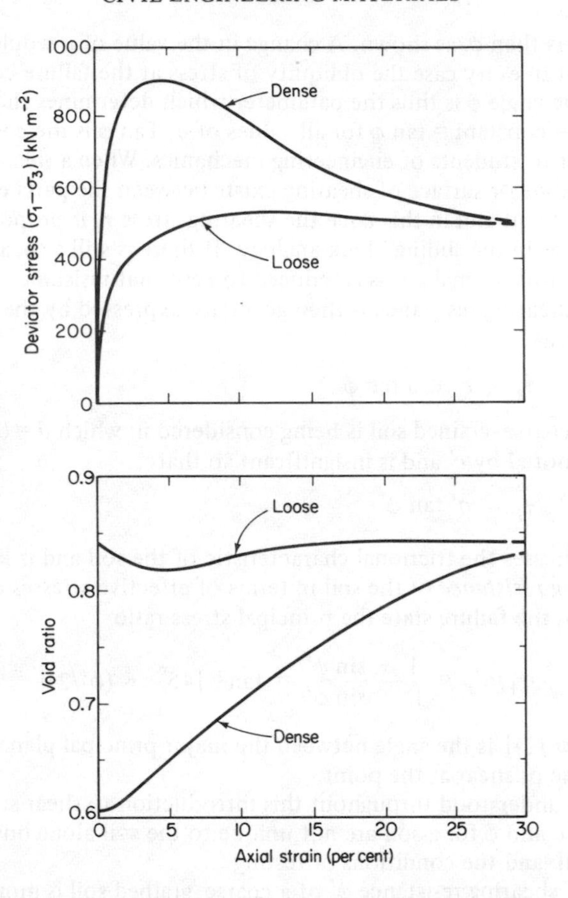

Figure 25.5 *Stress - strain curves and volume change curves for loose and dense specimens of a medium fine sand tested in triaxial compression at* $\sigma_3 = 200$ *kN* m^{-2}

volume changes, particularly in the loose sands, have implications in practice in their effect on the behaviour of saturated sands under shock loading. There is apparently a density, or void ratio, at which a given sand will neither expand nor decrease in volume on shearing. This void ratio is known as the *critical void ratio* and is a function of the sand itself and also of the effective confining pressure on the sand.

If the peak values of deviator stress for tests run at several values of cell pressure on dense specimens, at a given initial density, are plotted on a Mohr diagram as in figure 25.6*a* the line or curve tangent to the circles is called the *Mohr failure envelope.* This may be nearly a straight line passing through the origin of the diagram and having the equation

$$\tau_f = \sigma' \tan \phi'$$

where ϕ' is the slope of the tangent and is the measured angle of shearing resistance of the dense sand. If the envelope is curved a straight-line approximation is

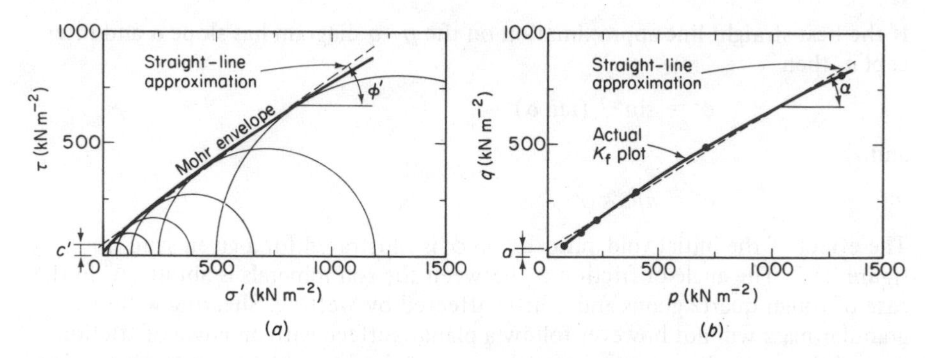

Figure 25.6 *Shear strength data for a dense coarse-grained soil plotted on (a) a Mohr diagram and (b) a p – q diagram*

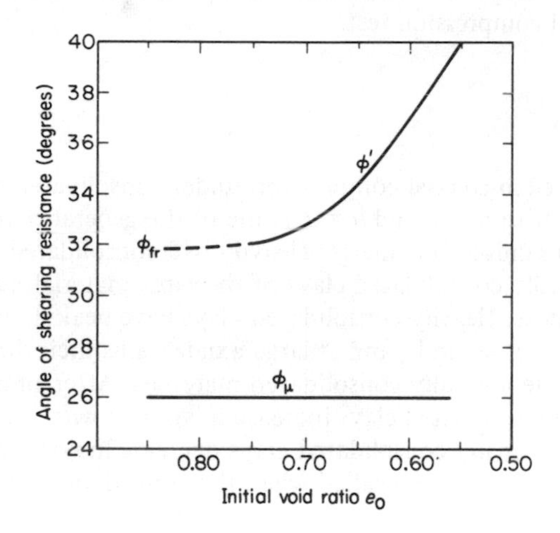

Figure 25.7 *Effect of initial void ratio of a sand on the measured angle of shearing resistance*

usually required for practical usage, covering a specified range of direct stress associated with a particular design problem. For this approximation

$$\tau_f = c' + \sigma' \tan \phi'$$

and usually c' is small. A similar plot using the ultimate value of deviator stress gives the value of ϕ' for the loosened material, equalling the value from tests on initially loose material.

The data for the failure conditions may also be plotted on a $p - q$ diagram as in figure 25.6*b*, again giving a line or curve through the points. This line is called the K_f line and is sometimes found to be easier to plot than the Mohr envelope.

If the best straight line approximation on the $p - q$ diagram has slope α and intercept a, then

$$\phi' = \sin^{-1} (\tan \alpha)$$

and

$$c' = a/\cos \phi'$$

The effect of the initial void ratio e_0 on ϕ' is illustrated for a given sand in figure 25.7. The angle of friction ϕ_μ between the soil minerals is about 26° in the case of rough quartz grains and is little affected by wetting. Shearing within a granular mass will not however follow a planar surface with an angle of friction ϕ_μ and the many directions of particle contacts lead to a higher angle of shearing resistance ϕ_{fr} at the critical void ratio. The remaining part of ϕ' derives mainly from particle interlocking, which depends on grain shape and grading and on density of packing of grains.

Data from the direct shear test are also easily plotted to give estimated ϕ' values and volume change behaviour. The same pattern emerges as has just been seen for the triaxial compression test.

Finer-grained Soils

When clays are tested in triaxial compression under consolidated drained (CD) conditions, that is, with $\sigma = \sigma'$ and $u = 0$, some of the general characteristics just seen in the tests on sands again emerge. Heavily overconsolidated clays are stronger than normally consolidated clays of the same material, just as dense sands are stronger than loose. Heavily consolidated clays have peak drained shear strengths just as for dense sands, and at large axial strains their drained strengths approach those of the normally consolidated materials. After an initial decrease in volume, the overconsolidated clays increase in volume with increasing axial strain whereas the normally consolidated clays decrease in volume. These volume changes during shearing will be recalled when the conditions during the undrained tests are examined.

The Mohr failure envelope for the results of a set of drained tests on a clay passes near the origin, and in the case of an overconsolidated clay may be slightly curved. It is not usually difficult to fit a straight line to represent the envelope giving $\tau_f = c' + \sigma' \tan \phi'$ as before. The value of ϕ' for the clay is then established and also c' is estimated. The intercept c' is usually rather small. Table 25.1 shows permeability k and shear strength parameters c' and ϕ' of typical soils. The range of k-values is seen to be large whereas the range of ϕ' is not.

Drained or CD tests in clays must be run at very slow rates in order to keep pore pressures negligibly small as the test proceeds. For this reason they are sometimes called slow tests.

Residual or ultimate strength

When the shearing deformation in a clay soil is continued well beyond the relative displacement at which the peak resistance is mobilised, a lower ultimate resistance remains and this is termed the *residual strength.* It has been noted that the peak resistance is usually pronounced in the more heavily overconsolidated clays, less so in the normally consolidated and remoulded soils. The loss in strength is partly attributable to a change in water content but also to a reorientation of the minute clay platelets in the zone of shearing to a parallel, dispersed arrangement aligned in the direction of the deformation and of notable smoothness. A clay containing a proportion of silt sizes and coarser, which inhibits this reorientation, shows an ultimate strength in terms of ϕ' which is closer to that for non-platy particles, say about 30°. If, however, a clay is almost wholly composed of material less than 2 μm in size the reorientation dominates the strength and the reduced value of ϕ' for the residual state (now termed ϕ'_r) may be as low as 15° or so, and the cohesion intercept c' virtually disappears. The large deformations required to attain this loss of strength *in situ* may be part of a long term process currently taking place unnoticed; or they may have occurred in the geological past in the course of ground movements which have since stabilised. Ground which has experienced past movements may be in a delicate state of stability today and any slickensided shear surfaces observed during site investigation or subsequent works should be noted as indicators that the residual strength may already be the maximum available resistance.

Undrained tests

While CD tests are sometimes made on fine-grained soils it is much more common practice to carry out triaxial compression tests in which drainage is prevented. These undrained tests may be made with or without the measurement of porewater pressures and may in fact be applied to any soil type. They are usually associated however with the finer grained soils of relatively low permeability.

In the UU test, specimens as sampled are brought to failure under undrained conditions (Bishop and Henkel, 1962; BS 1377). If a set of specimens from a point in a stratum of saturated clay is taken each specimen has been consolidated *in situ* to the same pressure. Removal of the clay from the stratum reduces to zero the external total compressive stresses on each specimen and this stress change results in an equal reduction in the porewater pressure since undrained (constant volume) conditions prevail. Thus there is no change in the effective stress in the specimen. Application of a cell pressure to a specimen in the triaxial apparatus now results in an equal increase in the pore pressure, but again no change occurs in the effective stress if drainage is prevented.

When the deviator stress is applied and increased until failure takes place changes occur in the porewater pressure which depend both on the initial effective stress in the clay and on the consolidation history of the clay. The maximum deviator stress and hence the *undrained shear strength* is a function of these two factors so that even if each specimen of the set is tested at a different cell pressure, the maximum deviator stress should be constant. This is found to be so in the test

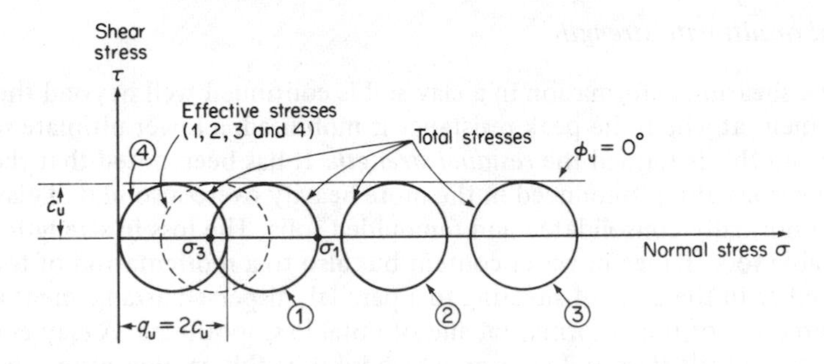

Figure 25.8 *Undrained test data, for specimens of a saturated clay, plotted on a Mohr diagram*

conditions described and the Mohr circle plot would have the form of figure 25.8. This is the $\phi = 0°$ condition where $\phi = \phi_u$ and $\tau_f = c_u = \frac{1}{2}(\sigma_1 - \sigma_3)_f$ for the undrained tests. It is interesting to note that specimens tested at zero cell pressure (atmospheric pressure) should also yield the same result $\tau_f = c_u$, as in circle 4. This simple version of the triaxial test is called the *unconfined compression test* and for saturated clays provides a rapid way of measuring c_u. The unconfined test may be carried out in the laboratory in the triaxial compression apparatus but for use in either laboratory or field a simple and inexpensive form of portable apparatus is available, which is suitable for testing saturated nonfissured clays. One such apparatus is described in BS 1377. Specimens may be obtained from a borehole, trial pit or open excavation and tested on the spot. The *unconfined compressive strength* q_u of a soil is sometimes given as its measure of strength, where $q_u = 2c_u$ as shown.

If pore pressures are measured during the tests at the various cell pressures the total stress data at failure may be expressed in terms of effective stress and it is found, as expected from the above argument, that only one effective stress Mohr circle is obtained from the entire set of specimens. The total stress data cannot therefore be interpreted to give values of c' and ϕ'.

If a soil is only partially saturated the strength measured in undrained testing is not independent of the total stress conditions and, although the test results may be expressed by a curved Mohr failure envelope, and values of c_u and ϕ_u may be selected, the interpretation is rather difficult.

Table 25.2 lists a classification of consistency and undrained shear strength of clays from CP 2004.

Sensitivity

Disturbance of saturated clay soils usually leads to a rapid loss of undrained shear strength and at large amounts of disturbance a remoulded strength is reached. In clays with a flocculated structure this is often much smaller than the peak undisturbed strength. Any disturbance leading to remoulding, for example, by kneading or rolling, will have the effect of changing the whole structure of the clay to a more oriented, dispersed particle arrangement. An initially dispersed clay would show little change in strength on remoulding.

TABLE 25.2
Undrained shear strength of clays. From CP 2004

Consistency			Undrained shear strength c_u (kN m^{-2}) or (kPa)
In accordance with CP 2001	Widely used	Field indications	
Very stiff	Very stiff or hard	Brittle or very tough	greater than 150
Stiff	Stiff	Cannot be moulded in fingers	100 to 150
	Firm to stiff		75 to 100
Firm	Firm	Can be moulded in the fingers by strong pressure	50 to 75
	Soft to firm		40 to 50
Soft	Soft	Easily moulded in the fingers	20 to 40
Very soft	Very soft	Exudes between the fingers when squeezed in the fist	less than 20

The ratio of the undisturbed (peak) undrained strength of a soil to its fully disturbed and remoulded strength at the same water content is known as its *sensitivity* S_t. A classification of degrees of sensitivity is given in table 25.3. The low sensitivities would be associated with the more highly consolidated clays.

Recent marine clays deposited with flocculated structures would be expected to be quite highly sensitive (see section 19.2). Clays of extreme sensitivity are termed quick clays. Some of the leached marine clays of Norway and the Leda clays of Canada are striking examples of clays which are altered by disturbance

TABLE 25.3
Degrees of sensitivity in soils

S_t	Classification	S_t	Classification
1	Insensitive	8 - 16	Slightly quick
1 - 2	Slightly sensitive	16 - 32	Medium quick
2 - 4	Medium sensitive	32 - 64	Very quick
4 - 8	Very sensitive	over 64	Extra quick

from soft cohesive soils to viscous liquids. These clays commonly have sensitivities exceeding 100 and pose very difficult geotechnical problems in dealing with landslides and also during site investigation and the construction of works.

Some partial recovery of strength with time after completion of remoulding has been observed in clays. This characteristic is termed *thixotropy* and reasons for the behaviour are speculative meantime.

Consolidated undrained test

The consolidated undrained (CU) test extends the information from the undrained test for wider application to stability problems, particularly when pore pressures are also measured. Several sets of specimens are prepared and each set is allowed to consolidate under a different cell pressure σ_c' before testing under undrained conditions to obtain a value c_u for that set. When the test results are plotted in the form of figure 25.9*a* it would appear that the relationship has the form of a strength c_u - depth profile (with σ_c' representing depth). Laboratory measurements of the slope c_u/σ_c', however, give higher values than those estimated in the field from *in situ* measurements. This is attributable to a number of factors including sampling disturbance, differing consolidation conditions in laboratory and field. Figures 25.9*a* and *b* are alternative ways used to convey the results of the CU test in terms of c_u and σ_c'.

If pore pressures are measured in the CU tests, the effective stress Mohr circles may be plotted as shown in figure 25.10. The failure envelope enclosing the circles is usually closely linear like that for the coarse-grained soil in figure 25.6*a*. The intercept and slope give values for c' and ϕ', the strength parameters in terms of effective stress. This envelope expresses the same information as the values of c_d and ϕ_d that are obtained in slow, drained (CD) tests on the same soil. For practical purposes the values of c' and ϕ' from the two types of test are the same.

It is worth reiterating that the values of c and ϕ are not unique for a given soil but depend on the method of testing, which must be stated when describing the strength of a soil.

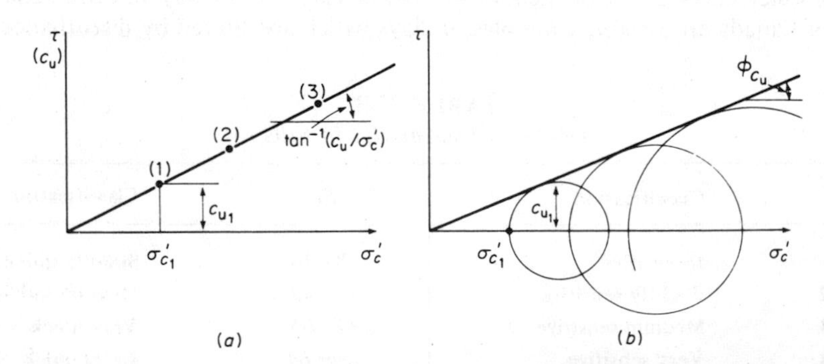

Figure 25.9 *Alternative ways of presenting results of CU tests in terms of* c_u *and* σ_c'

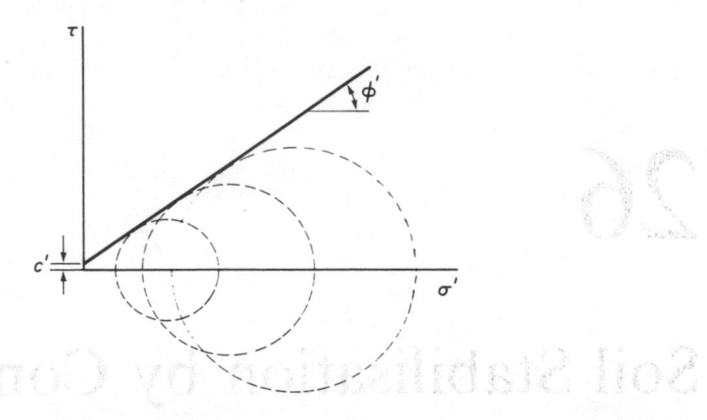

Figure 25.10 *Mohr diagram in terms of effective stress. Data from CU tests, on specimens of a saturated clay, with pore pressure measurement*

25.3 Shear-strength Testing in the Field

The difficulties of sampling soils without seriously disturbing their fabric and structure have led to the development of a number of field or *in situ* methods of assessing their quality.

Among these tests are the standard penetration test and the California bearing ratio test, both as detailed in BS 1377, cone penetration tests, plate loading tests, the vane shear test and the pressure meter test.

26

Soil Stabilisation by Compaction

In the wider sense, soil stabilisation is a term used for the improvement of soils either as they exist *in situ* or when laid and densified as fill. The purpose of stabilisation is to make a soil less pervious, less compressible or stronger, or all of these. This is often achieved by injecting a fluid into the pore spaces in soil, the fluid gelling or hardening in the pores. In suitable soils stabilisation may be achieved by introducing admixtures and then applying mechanical work or vibration to densify the material. For further information see Ingles and Metcalf (1973).

26.1 Mechanical Compaction

In this chapter the study of stabilisation is limited to processes of densification or *compaction by mechanical equipment* such as vibrators or rollers. The treatment of the subject is necessarily brief but the subject itself is important. Extensive developments have occurred and are continuing in excavating, transporting, spreading and compacting machinery enabling large quantities of material to be economically handled in earthworks. Use of such plant is seen in the construction of highway embankments, earth dams, and general infilling. The fill, when placed in loose layers and compacted under controlled conditions, can be made up into a relatively uniform strong, incompressible and impervious material with predictable properties. As such it will be preferred as an engineering material to randomly dumped fill and even to some natural soils if these are erratic or poor in quality and distribution. Some natural soils and existing fills may be improved by compaction in place – although usually to very shallow depths of under 1 m – the process often being used to improve road foundations. Greater depths have been successfully compacted using very high tamping energies, the final densities being measured using *in situ* tests.

Consolidation and compaction have similar meanings but in geotechnical engineering the term consolidation is usually reserved for the time-dependent reduction in voids volume accompanying an expulsion of porewater due to the application of a loading. The term *compaction* is used for the rapid reduction in voids volume in a soil arising from the application of mechanical work – the reduction being due to an expulsion of air from the voids. The state of compaction of a soil is described quantitatively in terms of the *dry density* of the soil. That is, although the soil itself is moist, a greater or lesser compaction means that a greater or lesser mass of solids is packed into a unit total volume, with a consequent influence on the engineering quality of the soil. The compacting energy is applied to the soil in the field by one of three principal methods: static (smooth-wheeled roller, pneumatic-tyred roller, grid roller and sheeps foot roller), dynamic (falling weight, rammers) and vibrating (vibrating smooth-wheeled or pneumatic-tyred roller, vibrating plate). Choice of field compacting plant depends on the soil to be compacted – kneading methods work best in the more clayey soils, vibration in the sandy soils.

26.2 Compaction Properties of Soil

Laboratory testing for the compaction properties of soil is detailed in BS 1377. This testing involves the application of standard compaction energies to the soil, which is contained in a standard mould. Batches of soil are prepared to cover a range of water contents. Three tests are described, differing mainly in their compaction method, two in which dynamic compaction is applied through blows from standard rammers falling on to the soil and one, for cohesionless soils, in which a

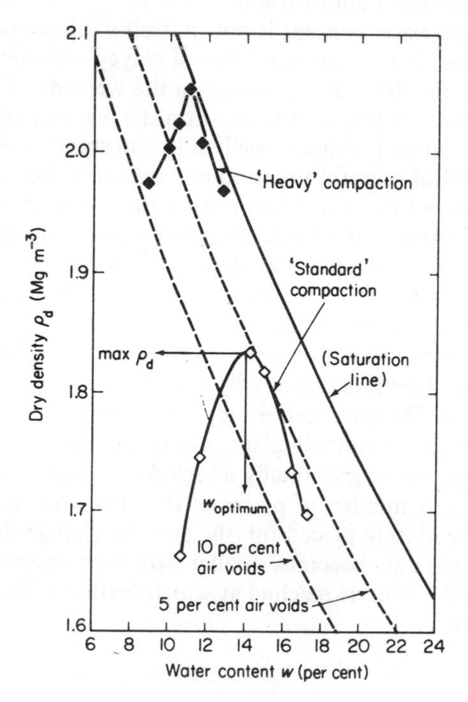

Figure 26.1 *Relation of dry density to water content for a sandy clay compacted at two different compaction energies*

vibrating rammer is used. For a given compacting effort, the dry densities ρ_d corresponding to several water contents w in the soil are determined. The relationship of these is shown typically for a cohesive fill in figure 26.1. The water content at which the maximum value of dry density is obtained is called the *optimum water content* for the compacting energy used. With a given soil, an increase in the compacting energy leads to a shift of the entire curve, upwards and to the left – that is, giving higher maximum density and lower optimum water content.

For a soil of specific gravity G the relation of ρ_d to w per cent for zero air voids may be calculated from

$$\rho_d = \frac{100G}{100 + wG} \rho_w$$

Lines corresponding to air voids contents of 0, 5 and 10 per cent are shown in the figure. Lines of equal degree of saturation S_r are sometimes given. It is seen that the maximum dry density occurs at an air voids content of 5 per cent or less. Investigations reported by Sherwood (1970) suggest that with skilled operators following the BS 1377 procedures, the tests for maximum dry density and optimum water content may be expected to give reproducible results just within acceptable requirements for design purposes. For control of compaction, however, the variation in results among test operators was found to be high. This means that for effective control on site a considerable number of tests must be undertaken even for the rare case of a uniform soil.

Although the optimum water content is not in itself a unique parameter there are some broad differences in the characteristics of clayey soils between those which are compacted on the dry side and those on the wet side of optimum. In British climatic conditions soil fills tend to exist on the wet side of optimum. If just on the wet side they should compact well and form stable soils.

The laboratory test fulfils a useful function in *comparing* the compaction behaviour of fills. It does not necessarily reproduce the compaction characteristics of any given mechanical plant in the field. There is no unique optimum water content and density – as we have seen they depend on the compacting energy applied. This means that tests should be run in the field, if possible at several water contents, using the selected mechanical plant to give a portion of the density - water-content curve of the type shown in figure 26.1. Because of the quantities of material involved it is not usually practicable to manipulate the water content in the field. The question usually is to select the most stable fill material available and the most suitable plant, sometimes operating with more than one stage of compaction to give finally a high percentage of the maximum soil density in an acceptable number of passes of the plant. As the plant makes the initial passes over the loosely placed fill, the greatest changes in density occur. The increase of density per pass becomes smaller with each successive pass, as shown in figure 26.2, and a stage is reached at which further rolling produces little further compaction.

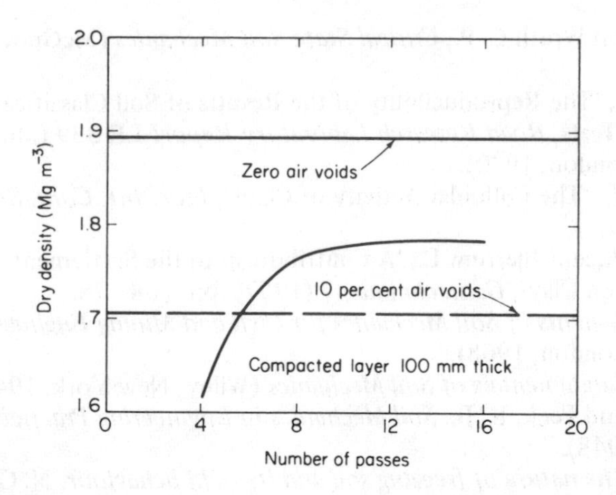

Figure 26.2 *Effect of the number of passes of a roller equipment on the dry density achieved in a sandy clay (w = 16 per cent) under field compaction*

References

Bishop, A. W., and Bjerrum, L., 'The Relevance of the Triaxial Test to the Solution of Stability Problems', *Proceedings of the Research Conference on Shear Strength of Cohesive Soils,* (Am. Soc. civ. Engrs, Boulder, Colorado, 1960).

Bishop, A. W., and Henkel, D. J., *The Measurement of Soil Properties in the Triaxial Test* (Arnold, London, 1962).

BS 812: Part 1: 1975 Sampling, size, shape and classification

BS 1377: 1975 Methods of tests for soils for civil engineering purposes

CP 2004: 1972 Foundations

CP 2001: 1957 Site investigations

Costet, J., and Sanglerat, G., *Cours Pratique de Mechanique des Sols* (Dunod, Paris, 1969).

Croney, D., and Coleman, J. D., 'Pore pressure and suction in soil', *Proceedings of the Conference on Pore Pressure and Suction in Soils* (London, 1961) pp. 31 - 7.

Dumbleton, M. J., 'The Classification and Description of Soils for Engineering Purposes: a suggested revision of the British System, *Road Research Laboratory Report* LR 182 (Ministry of Transport, London, 1968).

Ingles, O. G., and Metcalf, J. B., *Soil Stabilisation, Principles and Practices* (Butterworth, London, 1973).

Kolbuszewski, J. J., 'An Experimental Study of the Maximum and Minimum Porosities of Sands', *Proc. Int. Conf. Soil Mech.,* 1 (Rotterdam, 1948), p. 158.

Lambe, T. W., and Whitman, R. V., *Soil Mechanics* (Wiley, New York, 1969).

Loudon A. G., 'The Computation of Permeability from Simple Soil Tests, *Géotechnique,* 3 (1952) pp. 165 - 83.

Rowe, P. W., and Barden, L., 'The Importance of Free Ends in Triaxial Testing, *Proc. Am. Soc. civ. Engrs,* 90 (1964) pp. 1 - 27.

Schofield, A., and Wroth C. P., *Critical State Soil Mechanics* (McGraw-Hill, New York, 1968).

Sherwood, P. T., 'The Reproducibility of the Results of Soil Classification and Compaction Tests, *Road Research Laboratory Report* LR 339 (Ministry of Transport, London, 1970).

Skempton, A. W., 'The Colloidal Activity of Clays', *Proc. Int. Conf. Soil Mech.*, 1(1953) p. 57.

Skempton, A. W., and Bjerrum L., 'A Contribution to the Settlement Analysis of Foundations on Clay', *Géotechnique*, 7 (1957) pp. 168-78.

Smith, G. N., *Elements of Soil Mechanics for Civil and Mining Engineers* (Crosby Lockwood, London, 1968).

Taylor, D. W., *Fundamentals of Soil Mechanics* (Wiley, New York, 1948).

Terzaghi, Karl, and Peck, R. B., *Soil Mechanics in Engineering Practice* (Wiley, New York, 1948).

Williams, P. J., *The nature of freezing soil and its field behaviour*, N. G. I. Publication No. 72 (1967) pp. 90-119.

Wilun, Z., and Starzewski, K., *Soil Mechanics in Foundation Engineering*, vols 1 and 2 (Intertext, London, 1975).

Index